R for Programmers

R for Programmers
Quantitative Investment Applications

Dan Zhang

CRC Press
Taylor & Francis Group
Boca Raton London New York

CRC Press is an imprint of the
Taylor & Francis Group, an **informa** business

Published in 2018 by CRC Press
Taylor & Francis Group
6000 Broken Sound Parkway NW, Suite 300
Boca Raton, FL 33487-2742

© 2018 by Taylor & Francis Group, LLC
CRC Press is an imprint of Taylor & Francis Group

No claim to original U.S. Government works

Printed in the United States of America on acid-free paper
10 9 8 7 6 5 4 3 2 1

International Standard Book Number-13: 978-1-4987-3689-3 (Paperback)

Library of Congress Cataloging-in-Publication Data

Names: Zhang, Dan, 1983- author.
Title: R for programmers. Quantitative investment applications/Dan Zhang.
Description: Boca Raton, FL : CRC Press/Taylor & Francis Group, 2018. | "A CRC
title, part of the Taylor & Francis imprint, a member of the Taylor &
Francis Group, the academic division of T&F Informa plc." | Includes
bibliographical references and index.
Identifiers: LCCN 2018000799 (print) | LCCN 2018006702 (ebook) |
ISBN 9781315382197 (e) | ISBN 9781498736893 (pbk. : acid-free paper) |
ISBN 1498736890 (pbk. : acid-free paper)
Subjects: LCSH: Investment analysis--Data processing. | R (Computer
program language)
Classification: LCC HG4529 (ebook) | LCC HG4529 .Z48 2018 (print) | DDC
332.60285/5133--dc23
LC record available at https://lccn.loc.gov/2018000799

Visit the Taylor & Francis Web site at
http://www.taylorandfrancis.com

and the CRC Press Web site at
http://www.crcpress.com

Contents

SECTION III FINANCIAL STRATEGY PRACTICE

Foreword

It's my pleasure to have this opportunity to recommend Mr. Dan Zhang's new book in quantitative trading using R.

I knew Dan Zhang from his blog and his books about R. Since then, we started discussing trading ideas via emails and phone calls. His R books have been very useful to me.

He is a very sincere person and is willing to share what he has learned with other people. As far as I know, he is well known and very active in the R community in China.

He began his career as an IT engineer and now focuses on bank IT infrastructures and trading.

His new R book covers several topics, including financial markets in China, data management and high performance computing, financial theory and quantitative investment strategies.

Here are some topics I find very useful:

- A chapter focuses on Fast Data Handling and Processing with C ++ procedures.
- A chapter discusses the Docker architecture, which provides guidelines on the automated deployment of applications.
- An excellent example of using R on a convertible bond system, which forms the basis for designing a more advanced bond trading model.
- A chapter on the fundamental principles of quantitative trading and how to apply them to the real world.
- A mutual fund IT system which is very useful for the operation and management of funds.
- A case study on a China Robo Advisor company which gives us insights of wealth management business in China.

To be honest, it's rare to see a book which covers so many useful topics. I have learned a lot from Dan's new book and I highly recommend that everyone should keep one copy on the bookshelf.

Max Chen, Ph.D.
University of Washington
Founder, Magic Flute Capital Management

Preface

Why Did I Write This Book?

This is the third book of the series *R for Programmers*. It combines R language with quantitative investment of finance, mainly to achieve the combination of the technology of R language and the practical cases of financial quantification and to help readers experience how to turn knowledge into real productivity.

Traditional traders watch the quotes and market situations every day by using their transaction experience of years. A good trader can simultaneously observe dozens of trading instruments in several financial markets. As the development of financial products, there are already over 3000 A-shares, more than 7000 bonds, almost 4000 publicly offered funds and various financial derivatives in China's market. There are too many for manual labor to digest and analyze.

However, with a computer, we can scan the whole market to discover the irrational pricing and the arbitrage opportunities, which will tremendously improve the work efficiency of traders. An ideal design to deal and make money with programs will free us to do whatever we like. The realization of technology will liberate our life.

The original thoughts and methods stated in this book are what I learned from the application of theory research. Actually, I have been, for a long time, looking for such a book, in which theoretical models and practical cases are combined. Unfortunately, I didn't find any; or rather, I didn't find any that contain practical cases suitable for China's market. Therefore, I had to write one myself. The contents of this book are, kind of, notes I kept. I review them a lot to help clear my mind.

Features of This Book

A main thought of writing this book was, from the perspective of IT practitioners, to introduce technology to the financial market and make quantitative investments with it. Then with the spirit of IT practitioners that concentrate on learning and are willing to share and with the rapid spread of the Internet, we would break the traditional financial barriers. Exploit the creativity of geeks, turn knowledge into productivity and enable more ambitious IT practitioners to have the opportunities to enter the financial field, thus to propel the revolution and innovation of financial industry.

However, this is not a book that can be easily understood, for quantitative investment is a cross-discipline field. You need to prepare yourselves with knowledge of multiple disciplines to be competent in quantitative investment jobs. To comprehend the contents of the book, knowledge of many books may be needed.

To understand this book, not only the use experience of R language but also the understanding of the financial market is required. This book mainly includes three parts, i.e. financial market, statistical knowledge and IT technology. The financial market section contains the elementary knowledge of the financial industry, such as the introduction to the secondary market environment, the usage of trading tools, the trading rules of financial products, the research and investment thinking of institutions in China, strategy and backtesting, fund accounting etc. The section on statistical knowledge includes the algorithms of statistics and metrology models such as time series, simple linear regression, multiple linear regression, autoregression etc. The IT technology section refers to the programming technology related to R language, the usage of financial quantification packages, the processing of financial data, the modeling of financial data, the implementation of quantitative strategy, the coding of R language etc.

One of the main features of this book is that it employs many real cases. With the real financial market of China as a background, you will feel the volatility caused by the market, the influence that the state macro policies have on the market, the difference between individual and professional investors and the different understandings of the market from the perspective of quantification and that of subjectivity.

This book is a summary of my investment researches on financial theoretical models, market characteristic tests, mathematical formulas, modeling of R language, historical data backtesting, asset accounting and firm bargain. With R language, it can be simple to make our investment ideas come true. Everyone has some investment ideas, but our IT guys use our technical advantages to actually combine the ideas with real practice.

The financial products involved in this book, such as stocks, futures, bonds, funds, cash management etc., cover wide ranges of financial subject matters in multiple financial markets. As for the trading models and strategies mentioned, some are based on the price-volume strategy of market technical indicators, some on the arbitrage strategy of statistical theories, some on the event strategy of financial product rules, some use the stock picking strategy by scanning the whole market and some use the market timing strategy of high frequency trading. I believe that this book can help you feel the charm of the financial market and get to know the value that our technology advantage can bring us.

To profoundly understand every chapter of the book, you may need to (as I did) acquire some knowledge of technology, make some real trades in financial market, communicate frequently with people in financial industry and continuously learn and think.

Applicable Readers of This Book

This book is suitable for all R language practitioners, including but not limited to:

- Users of R language
- Software engineers with computing background
- Data scientists with data analysis background
- Practitioners of financial industry, researchers of securities companies, analysts, fund managers, quants
- Students majored in finance, statistics, data science and computing

How to Read This Book

There are six chapters in this book, categorized into three parts: Financial Market and Financial Theory, Data Processing and High Performance Computing of R, and Financial Strategy Practice. Every chapter is a holistic knowledge system.

Section One is Financial Market and Financial Theory (including Chapters 1 and 2), which starts with an understanding of finance to establish a basic idea of financial quantification. Chapter 1, Financial Market Overview, is the opening chapter of this book, which mainly introduces the ideas and methods of how to use R language to make quantitative investments. Chapter 2, Financial Theory, mainly introduces the classic theoretical models of finance and the R implementation methods.

In Section Two, Data Processing and High Performance Computing of R (including Chapters 3 and 4), essential tools of R language for data processing and their usage are introduced in detail. Chapter 3, Data Processing of R, cored with the data processing technology of R, introduces the methods of processing different types of data with R language. In Chapter 4, High Performance Computing of R, three external technologies are introduced to help the performance of R language meet the production environment requirements.

Section Three, Financial Strategy Practice (including Chapters 5 and 6), combines the R language technology and the financial market rules to solve the practical problems in financial quantification field. In Chapter 5, Bonds and Repurchase, readers can learn the market and the methods of low-risk investment. In Chapter 6, Quantitative Investment Strategy Cases, the investment research methods from theory to practice are introduced in whole.

Since knowledge of different areas is comprehensively applied in this book, it is suggested that you read all the chapters in order. Some of the technical implementations mentioned in this book use the information from the other two books of the series, *R for Programmers: Mastering the Tools* and *R2 for Programmers: Advanced Techniques*, so it is recommended that you read those two as well.

Corrigendum and Resources

Due to my limited capacity and the constraint of writing time, there could be something wrong or incorrect in this book. Please point it out and correct it if you find any. For this purpose, I have established an online official website of the book to conveniently communicate with readers (http://fens.me/book). All the source codes in the book can be downloaded from this website. If you encounter any problem during the reading, you can leave a message on it. I will try my best to give you a satisfactory explanation. Any advice is highly appreciated. Please feel free to email me via bsspirit@gmail.com. I look forward to your advice and suggestions.

Acknowledgments

Thanks to Zhang Song, general manager of Beijing QianZhuangZhiJin Technologies Co. Ltd; Liu Yafei, CEO of Quantum Financial Service; and Xu Bin, my colleague in China, Minsheng Bank, for their help and support that got me through difficulties during the most frustrating times. Thanks to all the readers of R language and every friend in the community. We started to know each other through R language and then we spread the knowledge together. Thank you Liang Yong, CEO of TianShan community, for his sponsorship and promotion of this book. Also, thanks to Dr. Max Chen of University of Washington and Professor Huang Da of Fudan University, for their recommendation prefaces for this book. Thank you Yang Fuchuan, the chief editor of China Machine Press HZ Book, and Li Yi, the editor, for their help in reviewing all the chapters and sections of this book and also the publishing work. Thank you Wang Shihua, the translator of this book, for his elaborate translation that enables the book to have English readers.

Special thanks to my wife for her indefatigable encouragement which finally carried me out of the shadow of my frustration. Thanks to my mom and dad for their support for my work and for taking care of my life.

I dedicate this book to my beloved families and many R language fans.

Dan Zhang
Beijing, China

Translator

When I first read the book, I was worried, for as a person who majored in language and literature, I doubted my ability to understand this professional book on computer science and I did not know if I could successfully accomplish its translation. However, once I tried to translate some part of it, I found myself attracted by its content and amazed by the author's wide span of knowledge and his persistence in learning. This book starts with the methodology of quantitative investment, then introduces the basic theory of statistics, the theory of finance, the IT technology of R language and finally implements the theories in financial market. The style of writing in this book is simple and with humor, the chapters and sections are properly cohesive and it adopts the progressive illustration technique, all of which makes it hard to stop reading. You can even take it as a reference book and read it at any time.

I believe that China will gradually take the lead in science, economy and culture and will be recognized around the world. More and more excellent works are emerging in China. It can be a new opportunity in the worldwide Internet industry development to translate those excellent works into other languages, to spread them beyond China, to show the peers or scholars abroad the charm of Chinese technology and the current situation of Chinese Internet or finance.

It was already late at night and I had read the last page. My mind was at ease. I closed the book and I made up my mind to translate the book for the author. It became the wish of both the author and myself to complete the English version of this book.

I am an associate senior translator and interpreter, with a securities practice qualification; I used to work as an English interpreter in NDRC, then as the secretary to the CEO of CNPC. I am good at Chinese-English translation and interpretation in IT technology, finance, animation, economics, etc.

Wang Shihua

Author

Dan Zhang is a senior expert in R language technology and a preacher of it; he is at the forefront of the R language technology community in China. He has worked in software development for 12 years, is experienced in the Internet application architecture and has rich knowledge accumulation in Java, NodeJS, big data, statistics, data mining etc.

Dan is also an expert in financial big data, working for China Minsheng Bank now and familiar with the secondary financial market, trading rules and investment research system in China.

He has obtained 10 technology certificates from Sun and IBM and was recognized as an MVP by Microsoft in 2017. He has been a keynote speaker for many technical conferences on the Internet and data analysis and participated in the "Chinese Academy of Sciences Sugon Big Data–One Hundred Universities" project sponsored by the Chinese Ministry of Education, to design the curriculum of the big data major for undergraduates.

Dan is passionate about sharing. He has authored *R for Programmers: Mastering the Tools* and *R for Programmers: Advanced Techniques*, and he is the co-author of *Beautiful Data Practices*.

Personal blog: http://fens.me (Alexa: 70k).

FINANCIAL MARKET AND FINANCIAL THEORY

Chapter 1

Financial Market Overview

This is the opening chapter of this book, which mainly introduces the ideas and methods of how to use R language to make quantitative investments. Quantitative investment is one research direction of interdisciplinary knowledge combination that includes the R language knowledge of technology level, the application of basic science and the financial market conditions. The R language community provides abundant financial toolkits, which enable us to quickly build the architecture of quantitative investments. In this chapter, I will, based on my working experience, observe China's financial market from data aspect and find out the investment opportunities and the booming sectors.

1.1 R for Quantification

Question
Why do we need to use R language for quantitative investments?

R For Quantification
http://blog.fens.me/r-finance/

Introduction
Those who analyze data must have known R language. R language is a statistical language, which shows remarkable advantages in the area of data analysis. Finance is an industry that plays with

data, and data analysis is the greatest advantage of R, so it is really natural to match finance with R language and to make quantitative investment strategy with R. If you've already started using it, you know that it's easy and convenient.

This section, starting with "R for Quantification," mainly stresses that R language, an important quantitative investment tool, has a broad application prospect in the field of quantitative investment.

1.1.1 Why R Language?

R language is a data-oriented programming language. In its early days, it was only used by statisticians in the field of statistics. Recently, as big data develops and the ecology of R language rapidly grows, R language is not only seen in statistical area, but also in main industries, such as Internet, data science, artificial intelligence, machine learning, biomedical, games, e-commerce, global geography, data visualization, etc.

R language not only does an excellent job in data analysis, but also makes us think by connecting us with our daily life data. For instance, we get receipts recording what we buy when we go shopping in a supermarket. If we collect all the receipts, we can analyze the data by frequent item set algorithm of Apriori associated rules and thus know our shopping habits.

Every day, we receive nuisance calls for house selling, bank loans, stock recommendations, false invoice issuance and some even for guessing who I am, which are typical in China. We can gather our phone records, analyze the phone numbers, registration locations and talk durations and find out what calls are useful and what are a waste of time. Then how much time is wasted can be figured out by dichotomous classification of effective and ineffective calls with Bayes algorithm. Finally, crank calls can be prevented.

If we collect our phone records as well as our relatives' and friends', we can build a data base. Crank calls can be marked and put on a blacklist. It can go public, so that more people can prevent crank calls.

Food, clothing, housing and transportation all produce data and where there is data, there is data analysis. Data analysis will improve the efficiency of our lives and help us save our precious time.

In human society, besides the basic necessities of life – food, clothing, housing and transportation – money is in much greater demand now. We deal with money every day. We have consumption needs and we have income at the end of every month. When our income surpasses our consumption and the capital accumulation reaches a certain stage, there will be demand for investment. Chinese usually invest their money in houses, gold, stocks, insurance and financial products. Most of these investors are irrational. They will buy something on a sudden impulse without any careful study in the market and may get held-up soon after the investment. Media hype like "Chinese Dama Rush to Gold," "Chinese Dama is the Bailout Main Force," etc. are good examples of this.

Opportunities are what the investment market never lacks. Whatever you invest in, the outcome can be different if you give it a second thought. With the knowledge of data analysis and the understanding of financial market rules, we can analyze the data with R language, seize the chances and then keep the value of or even add value to what we've earned through hard work.

1.1.2 Cross-Disciplinary Combination

The major difference between R and other programming languages is that R language is data-oriented, which is very important and may influence the thinking pattern. I have been programming in Java for more than 10 years, which makes the thinking pattern of a programmer ingrained in my mind. I have gotten used to building models in an object-oriented thinking pattern, in which objects, actions and connections of the world should be expressed in an object-oriented way. I have also become accustomed to the top-down approach from architecture design to development in IT projects and the completely demand-driven agile development where the development route changes with business needs. As a programmer, I have a dream. I think technology will change the world. It is the influence that the programming thinking pattern has on me. If you are not familiar with the concept of object-oriented thinking, please refer to Chapter 4 of *R for Programmers: Advanced Techniques*.

Then I got to know R language, which opened a new window for me to understand the world. I found a different world. In a world without programming, you can express your manners by words, you can understand the world perceptually and you can let data tell the meaning of the world. You can watch Chinese CCTV news every day to follow state affairs, or you can update your MicroBlog or WeChat Moments to get information around. However, those are the popular ideas of the public, not the programmers. In this society, programmers are the minority.

Besides technology, the way to change the world can be through policies, capital, markets, laws, human resources, etc., all of which seem to be more important than technology in the sense of company operation. This may not be the concern of programmers. However, when confronted with the market, even the most powerful technology can be constrained by other factors.

I once started my own business, from which I learned a profound lesson that I needed to break my mindset and view the world and myself from a multi-dimensional perspective. R language has helped me open the window to viewing the world through data. I evaluate my capability improvement no longer by how beautiful my program is or how to solve $O(N^2)$ with $O(\log(N))$ in time complexity, but by how to collect data, how to improve the data quality and how to apply it in practical business problems. Of course, I still optimize my 50-line code into 40-line, even if it forces me to stay late.

The problem has been converted and it needs specialists in the area. However, as ambitious geeks, our goal is to have a comprehensive grasp of cross-disciplinary knowledge. When I started my business, I was always asked by the investors this question: how hard was it to get the quantitative investment project done? If a few people with a finance background and a few with an IT background worked together on the project, could it be finished soon? Actually, no. It is difficult for people with different backgrounds to communicate with each other, especially those with finance and IT backgrounds, for one is unwilling to share and the other is open without any reservation; one will speak it out first and the other will do it first; one wants to be served by everyone and the other focuses only on his interests. Therefore, we cannot do the quantitative investment projects well without someone who specializes in both disciplines.

We are determined to be a cross-disciplinary person, which is the irreplaceable core point. As for IT technology, you need solid programming ability and architecture thinking, scientific project management methods as well as strict product design logic; you also need knowledge of basic subjects like statistics, mathematics, etc. and data processing experience in data mining and machine learning; most importantly, you need to understand the market rules of your business, in quantitative investment or in other areas (Figure 1.1).

It sounds difficult and it is. It is difficult for you and for all. However, if you are persistent enough, your fate will be certainly changed by this knowledge.

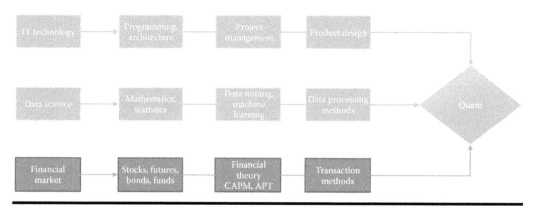

Figure 1.1   Cross-disciplinary combination.

1.1.3 Quantification Packages of R

What is the advantage of R language in quantitative investment? R can help us understand quantitative investment from multi-dimensional aspects. R language itself is an IT technique that we need to master. It is a technique that needs programming and yet significantly lowers the difficulty of programming. A code of 20–30 lines in R language is able to accomplish a lot. R language is data-oriented with convenient data processing operations; it contains abundant functions to support basic subjects like mathematics, statistics, etc. and it provides various algorithm libraries of data mining and machine learning for direct use.

Moreover, R language is strong in this quantification field. There are packages for every aspect of quantitative investment, such as investment research, backtesting analysis, financial products pricing, portfolio optimum and risk management, some of which, when applied in China, may need modification and optimization according to the characteristics of China's market.

R language provides abundant packages, which can be classified as follows. No other language can provide such support.

- Data management: Includes the capture, storage, read, time series, data processing, etc. of datasets. The R packages concerned are zoo (time series object), xts (time series processing), timeSeries (Rmetrics-time series object), timeDate (Rmetrics-time series processing), data. table (data procession), quantmod (data downloading and graph visualization), RQuantLib (QuantLib data interface), WindR (Wind data interface), RJDBC (database access interface), rhadoop (Hadoop access interface), rhive (Hive access interface), rredis (Redis access interface), rmongodb (MongoDB access interface), SparkR (Spark access interface), fImport (Rmetrics-data access interface), etc.
- Index calculation: Includes all kinds of technical index calculation methods in the financial market. The R packages concerned are TTR (technical index), TSA (time series calculation), urca (unit root test), fArma (Rmetrics-ARMA calculation), fAsianOptions (Rmetrics-Asian options pricing), fBasics (Rmetrics-calculation tool), fCopulae (Rmetrics-financial analysis), fExoticOptions (Rmetrics-options calculation), fGarch (Rmetrics-Garch model), fNonlinear (Rmetrics-non linear model), fOptions (Rmetrics-options pricing), fRegression (Rmetrics-regression analysis), fUnitRoots (Rmetrics-unit root test), etc.
- Backtesting trading: Includes financial data modeling and model reliability verification with historical data. The R packages concerned are FinancialInstrument (financial products),

quantstrat (strategy model and backtesting), blotter (account management), fTrading (Rmetrics-transaction analysis), etc.

■ Portfolio: Used for multi-strategy or multi-model management and optimization. The R packages concerned are PortfolioAnalytics (portfolio analysis and optimization), stockPortfolio (stock portfolio management), fAssets (Rmetrics-portfolio management), etc.

■ Risk management: Used for positions' risk indicator calculation and risk warning. The R packages concerned are PerformanceAnalytics (risk analysis), fPortfolio (Rmetrics-portfolio optimization), fExtremes (Rmetrics-data processing), etc.

The introductions to the packages above can be found in the three books from the series *R for Programmers*. You can refer to the packages according to their names.

1.1.4 Practical Applications

With the convenience of R language, we can easily build a trading model by using the packages mentioned above. A simple investment strategy can be built without many codes.

In the following, I will illustrate with an example how to comprehensively use the packages R provides. Please follow the steps to make a simple trading strategy or an investment plan which is based on the moving average MACD (moving average convergence divergence) and aims at the global indexes. If you are still a beginner in finance, knowing little about MACD strategy, please refer to Section 2.3 of *R for Programmers: Advanced Technique*s.

The system environment used here:

■ Win10 64bit
■ R: 3.2.3 x86_64-w64-mingw32/x64 b4bit

Below is the simplified research process of quantitative strategy. There are six steps:

1. Download the data with quantmod package.
2. Standardize the data format with zoo package and xts package.
3. Calculate the model with TTR package.
4. Analyze the risk indicator with PerformanceAnalytics package.
5. Give the visualized output with ggplot2 package.
6. Analyze the result.

First, we need to obtain data. Individuals can download free data from the Internet, while institutions usually buy specialized database. We can download data from Yahoo Finance with quantmod package. Here I've selected five global market indexes for comparison. Their codes and the names are listed in Chart 1.1.

Then we continue to build the model of this trading strategy with codes in R language. The strategy details will be left out here in order to intuitively illustrate the whole process.

```
# Load the library
> library(quantmod)
> library(TTR)
> library(PerformanceAnalytics)
> library(ggplot2)
> library(scales)
```

Chart 1.1 Index Names and Codes

Index Name	Index Code (Yahoo Finance Code)	Abbr.
S&P 500 Index	^GSPC	GSPC
Nikkei 225 Index	^N225	N225
Hang Seng Index	^HSI	HSI
STI Index	^STI	STI
Shanghai Composite Index	000001.SS	SSE

```
# Download data of global indexes from Yahoo Finance
> options(stringsAsFactors = FALSE)
> symbols<-c("^GSPC","^N225","^HSI","^STI","000001.SS")
> suppressWarnings(getSymbols(symbols,src = "yahoo",from="2012-01-01"))
[1] "GSPC"      "N225"      "HSI"       "STI"       "000001.SS"

# Take the data of index prices after adjustment and merge the dataset
> df<-merge(GSPC$GSPC.Adjusted,HSI$HSI.Adjusted,N225$N225.
Adjusted,STI$STI.Adjusted,`000001.SS`$`000001.SS.Adjusted`)

# Rename the columns
> names(df)<-c("GSPC","HSI","N225","STI","SSE")
Now let's check the data and draw the global index plot.
# View data of the first six lines
> head(df)
              GSPC      HSI     N225     STI      SSE
2012-01-03 1277.06 18877.41       NA 2688.36       NA
2012-01-04 1277.30 18727.31 8560.11 2711.02 2169.39
2012-01-05 1281.06 18813.41 8488.71 2713.02 2148.45
2012-01-06 1277.81 18593.06 8390.35 2715.59 2163.40
2012-01-09 1280.70 18865.72       NA 2691.28 2225.89
2012-01-10 1292.08 19004.28 8422.26 2719.83 2285.74

# View data of the last six lines
> tail(df)
              GSPC      HSI      N225     STI      SSE
2017-02-24 2367.34 23965.70 19283.54 3117.03 3253.43
2017-02-27 2369.73 23925.05 19107.47 3108.62 3228.66
2017-02-28 2363.64 23740.73 19118.99 3096.61 3241.73
2017-03-01 2395.96 23776.49 19393.54 3122.77 3246.93
2017-03-02 2381.92 23728.07 19564.80 3136.48 3230.03
2017-03-03 2383.12 23552.72 19469.17 3122.34 3218.31

# View data type and it is xts
> class(df)
[1] "xts" "zoo"
```

The whole dataset is made of data from January 3, 2012 to March 3, 2017. The data type is xts, which is the special time series type of R. For more details about xts, please refer to Section 2.2 of *R for Programmers: Mastering the Tools.*

```
# Draw the global indexes plot
> g<-ggplot(aes(x=Index,y=Value, colour=Series),data=fortify(df,melt=TRUE))
> g<-g+geom_line(size=1)
> g<-g+scale_y_continuous(breaks = seq(1000,30000,4000))
> g<-g+ggtitle("Gloabel Index")
> g
```

Due to the difference of index establishing time in countries and the diversity of their constituent stocks, some of the index values are very large and some are comparatively small. We cannot judge if it's good or not according to the values. Usually, the values will be converted to the return rates for measurement and comparison, so that all the subjects are under a unified dimension (Figure 1.3).

```
# Cumulative Daily Returns for Global Indexes
> ret_df<-Return.calculate(df, method="discrete")
> chart.CumReturns(ret_df,legend.loc="topleft", main="Cumulative Daily
  Returns for Gloabel Index")
```

The higher the returns, the better the index performed during the period, which means it was worth investing in. As Figure 1.3 shows, Nikkei 225 Index (N225) magnificently surpassed other indexes, Shanghai Composite Index (SSE) fluctuated greatly with a dramatic rise and fall and S&P 500 Index (GSPC) was steady.

Figure 1.2   Global indexes.

Figure 1.3   Cumulative daily returns for global indexes.

Let's calculate the average annualized returns of these indexes. If we continue investing money in these indexes, what average yearly returns will we get?

```
> Return.annualized(ret_df)
                       GSPC       HSI      N225       STI        SSE
Annualized Return 0.1133813 0.0619811 0.1927681 0.03696703 0.04817027
```

We can see that Nikkei 225 Index (N225) has the highest average annualized return of 19.28%, which is consistent with its accumulative return in Figure 1.3. Shanghai Composite Index (SSE) makes an annualized return of 4.82%, which is not high and with great fluctuations. Overall, S&P 500 Index (GSPC) is the best choice of investment, with an annualized return of 11.34% and a low volatility, i.e. it yields well with a low risk.

Next, let's build a simple MACD model and then with it, we will make the trading strategy of the five indexes above.

```
> # MACD strategy model
> MACD<-function(dt,n=30){
+    names(dt)<-c('close')
+
+    # The moving average of MACD
+    dat<-na.locf(dt)
+    dat$ma<-SMA(dat$close,n)
+
+    # Trading signals
+    sig_buy<-which(dat$ma-dat$close>0)
+    sig_Sell<-which(dat$ma-dat$close<0)
+    sig_buy<-sig_buy[which(diff(sig_buy)>1)]
+    sig_Sell<-sig_Sell[which(diff(sig_Sell)>1)]
+    if(first(sig_Sell)<first(sig_buy)) sig_Sell<-sig_Sell[-1]
+    if(last(sig_Sell)<last(sig_buy)) sig_buy<-sig_buy[-length(sig_buy)]
+
+    # Trading list
+    trade_dat<-do.call(rbind.data.frame, apply(cbind(sig_buy,sig_
    Sell),1,function(row){
+      dt[row[1]:row[2],]
+    }))
+
+    # Calculate the returns
+    ret_trade<-Return.calculate(trade_dat, method="discrete")
+    return(ret_trade)
+ }

# Daily returns of MACD strategy
> macd_ret<-lapply(df, function(col) MACD(col,30))

# Annualized returns of MACD strategy
> t(do.call(rbind.data.frame, lapply(macd_ret,Return.annualized)))
           GSPC       HSI      N225        STI       SSE
close 0.2137435 0.2406476 0.2261996 0.01869112 0.2817241
```

We write a function of MACD strategy, which can be taken as the modeling process. The output of the function is the strategy returns. Then we pass the index data to MACD() function and after calculation, we get the strategy returns. At last, we compare the strategy returns with the pure index returns, as shown in Chart 1.2.

It is obvious in Chart 1.2 that the average yearly returns by MACD strategy, which uses a moving average, are much higher than those by pure index. We will earn more if we trade using MACD strategy. This is actually the thinking pattern of quantitative investment, i.e. to discover rules in data. Maybe it is confusing to you. How can it earn money with the process and the codes above? Take it easy. This is just the opening of this book. You will get it after you carefully read the whole book.

A code of 40 lines in all has accomplished a lot. The whole process can be subdivided and listed as follows: data acquisition, data cleaning, data standardization, index calculation, data modeling, backtesting, portfolio construction, portfolio optimization, evaluation of calculation result, data visualization, etc. See Figure 1.4.

To build a quantitative trading system is mainly to complete the steps in Figure 1.4. For individuals, to simply follow the pattern of the above codes in R language is enough to implement the steps. However, if you want to do better and be more than just an individual and you wish to establish a company and build a powerful system to serve more people in quantification, you will need a whole set of platform architecture schemes to work for your financial services.

Chart 1.2   Returns Comparison

Strategy	GSPC	HSI	N225	STI	SSE
Index returns	0.1133813	0.0619811	0.1927681	0.03696703	0.04817027
Strategy returns	0.2137435	0.2406476	0.2261996	0.01869112	0.2817241

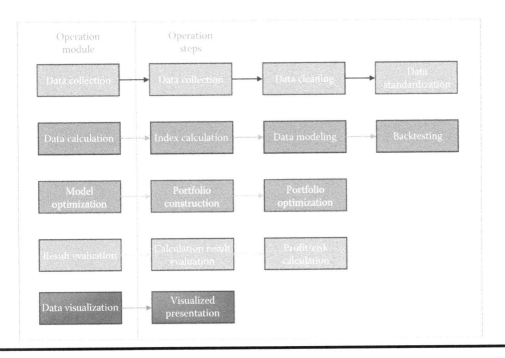

Figure 1.4   Operation steps of quantitative process.

1.1.5 System Architecture of Quantitative Trading Platform

Though R language is powerful, it is not omnipotent. To establish an integrated quantitative investment platform, we need not only R language to build the model, but also IT technology to set up the whole system architecture. First, we need to know what is a quantitative investment platform and what platform modules it needs. See Figure 1.5.

The operation modules shown in Figure 1.5 are defined as follows:

■ Data acquisition system: Mainly used to collect trading data of stocks, futures, bitcoin, etc., financial data of stock products, macro data of indexes, Internet data and so on.

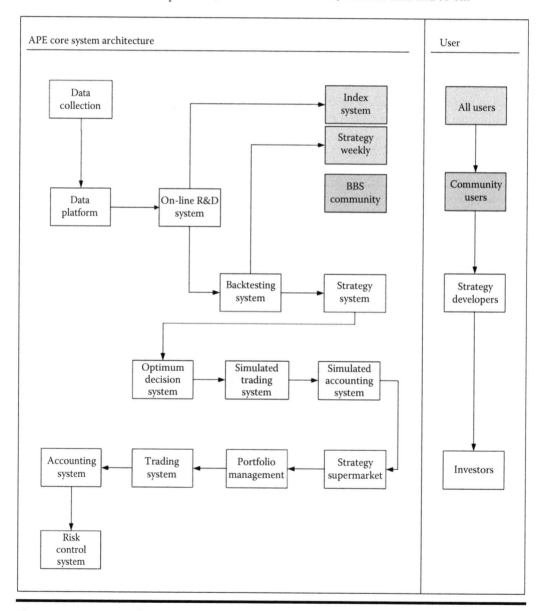

Figure 1.5   System architecture of quantitative trading platform.

- Data platform system: Used for data storage and data access, so as to unify the data access interface.
- Online R&D system: Enables the developers to develop strategies online with API (application programming interface). Developers can directly access the data of the platform and backtesting can be auto-run after the strategy submission.
- Backtesting system: Runs backtesting with historical data for the strategies that users submit, and then grades and evaluates the strategies.
- Strategy system: Used to visually present the signals of strategies verified by the backtesting and to update data in real time during the trading period, so as to help developers understand strategy signals more intuitively.
- Optimum decision system: Selects some strategies that match certain conditions from those verified by backtesting and then uses them in mock tradings.
- Mock trading system: Runs the strategies, outputs strategy signals and makes mock tradings.
- Simulated accounting system: Used for daily accounting calculation of mock tradings.
- Strategy supermarket: Selects some strategies that match certain conditions in simulated accounting and then puts them in the strategy supermarket for investors.
- Portfolio: Investors choose multiple strategies from the strategy supermarket and makes them a portfolio.
- Firm bargain system: For auto-trading in a firm offer according to the portfolio strategy signals.
- Firm accounting system: For daily accounting calculation of firm bargains.
- Risk control system: For strategy risk monitoring in a firm offer. If these strategies encounter any "black swan" event, the trading should be cut off immediately and a notice would be sent to the investors for portfolio adjustment.
- Index system: Based on macro data, it enables users to create self-defined indexes and then publish and share them in their own names.
- Strategy weekly: Every week, the public strategies with better performance in the week will be announced and promoted in a report.
- BBS (bulletin board system) community: For communicating and solving various problems that users come across when using the software.

Let's shift to the system architecture perspective. The whole IT system is mainly divided into four modules: data, calculation, display and trading. If Figure 1.5 is described in a technical architecture way, it will be as shown in Figure 1.6.

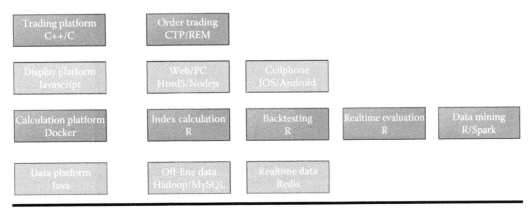

Figure 1.6 Technical architecture.

For data, there are mainly two types: historical data and realtime data. Historical data is characterized by its wide varieties and large quantity. It is used for backtesting and modeling while realtime data requires timeliness and is used for trading. The data module is mostly built with Java to ensure the system reliability, stability and high performance.

As for calculation module, it is mostly the application of various algorithms (including indexes calculation, net value valuation calculation, backtesting calculation, data mining, etc.), all of which R language does well. The calculations would be isolated with Docker, which would lead to effective computing resource distribution. Calculation is mainly completed by R, which simplifies the modeling process to just a few minutes.

Display module is used to friendly show the calculation results, i.e. returns and risk levels, to the users in a visualized way on a PC or cellphone. It is completed with Javascript and provides a better visualization experience.

Trading module is mainly used to connect to the transaction passages of exchanges, which enables us to trade with real money and to embody real value to the technology and the analysis results. It uses C++ to ensure the trading timeliness.

Overall, R language is the brain and the core part which helps us make decisions and decision making is the most complicated part in the quantitative investment process. Quantitative investment itself is an interdisciplinary field and it requires systematic and scientific methods to carry it out.

This book is just a beginning. I hope you can follow my way to start with R language, experience the fun of modeling in R and turn technology into real money. In my opinion, R is a great programming language. I will continue to propel its development in quantitative investment and make it my career. In the meantime, this book will be translated into English and published around the world. Chinese technology will make its way into the world. Let's make it.

1.2 How Algorithms Change Fate

Question
How are algorithms going to change one's fate?

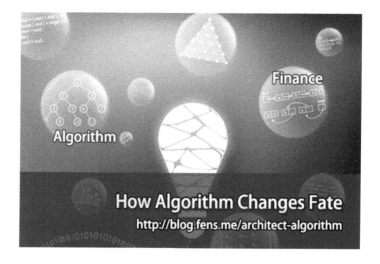

Introduction

In recent years, big data has achieved a leap in its development and it has shown more and more influence on our lives. There is data of Tencent for social life, Ali for shopping, Baidu for searching, Didi for transportation and so on. More and more data has been accumulated; and for data mining, algorithms are in need. Algorithms become more and more important especially after we've entered the era of big data.

To make dead data valuable is the power of algorithms. After we've entered the big data era, what matters is no longer the data, but the algorithms, in which lies the real productivity. The algorithm engineers will become more and more valuable. So algorithm engineers, have you really explored your value?

1.2.1 *Application of Algorithms in Different Industries*

The rising of big data has impacted every industry. It brings opportunities as well as challenges. If there's no data, there's no core value. Even if you have the data, you still need to think about how to make it valuable. In 2016, there was a pessimistic atmosphere in the investment market. Only artificial intelligence was popular. Breakthroughs have been made in algorithm areas, which can be learned from events as such: deep leaning technology was accurate in image recognition; Gradient Boosting Decision Tree (GBDT) frequently won prizes in data mining algorithm; AlphaGo beat human competitors in go; a Baidu robot named Xiaodu challenged the winner of The Brain, a Chinese TV show for brain competition; and so on.

Algorithms have already been applied in various industries and brought gradual change to people's lives and habits, as well as fields like image recognition, autopilot, user behavior, financial credit, quantitative investment, etc.

Image recognition and deep learning have developed rapidly. Now accurate face recognition, fingerprint identification and even complicated image comparison can be achieved. I remember very well that at the 2016 China Optics International A.I. Industry Forum (IAIS2016), Professor Gong Maoguo, from Xidian University, shared his research on the theme of Sparse Feature Learning of Deep Neural Networks and Space-Time Imagery Change Detection. He adopted image recognition technology to compare the satellite photos and light photos before and after the Wenchuan earthquake and found precisely the most severely hit area, i.e. the area that differed the most in landscapes before and after the earthquake. It helped to quickly locate the places most in need of help. Then the rescue team hurried there to rescue the injured and deliver relief supplies.

In autopilot field, automatic driving and parking can be realized with road surface identification. Autopilot cars of Uber have already been tested on road. They are equipped with all kinds of sensors, including radar, laser scanner and high-resolution camera, in order to plot their surrounding details. Autopilot cars are expected to improve people's lives, to save millions of lives and to provide more convenience. I was shocked 5 years ago when I saw an introduction video of autopilot in a demonstration class at Stanford University on machine learning, taught by Andrew Ng. What had happened in science fiction films was coming true.

According to user behavior analysis, human beings behave diversely and have various needs. Eating, drinking, clothing, housing, transporting, having fun, etc. are all basic behaviors of humans. Most people share common behaviors, so businessmen can collect the data and discover the rules of human behavior commonality through data mining algorithms. According to users' shopping behavior, businessmen can make recommendations of the goods that users prefer, which has brought forth the recommendation system; users' information query behavior has shown the need of information, so there are search engines; the changes of users' locations have revealed their need of transportation, hence the map apps came into being; users' personalized behavior can be tagged to identify their traits or status, and then the user profile is generated. User behavior analysis enables businessmen to know users' habits and enables users to know themselves better, which is of great commercial value.

Algorithms apply to many situations in the finance field, too.

In the area of financial credit, the traditional credit business, the core business of banks, cannot afford the high cost of serving the small-balance subscribers, for there are many of them, considering the large population of China. Therefore, non-governmental credit has risen. The outburst of P2P in the Internet finance at the end of 2014 satisfied the needs of credit demand, and yet exposed the default risks in the meantime. The lack of credit investigation system resulted in high charge-offs of many P2P companies. By the end of 2016, owners of up to thousands of P2P companies absconded. Credit investigations are urgently needed. If someone is going to buy a car but doesn't have enough money, credit is needed. Before a company makes its loans, a client should get his/her credit risk rated, so the company can have knowledge of his/her repayment ability, hence the company can decide the loan amount and repayment schedule to reduce the default risk. Zhima credit scoring of Alipay is a credit scoring model recognized by the market at present.

Quantitative investment, in my opinion, is the most complicated, challenging and interesting field. Making money with the quantification algorithm model is the easiest way of realization. In the financial investment field, there are miscellaneous kinds of data, including macro data, economic data, stock data, bond data, options data, news data, emotion data, etc. where the rules of the financial market can be found. Quants will make judgments about national economic situations and stock tendencies according to their analysis of various data and then make profits from the investment by the portfolio algorithm.

Now that we have gotten this far, I would like to ask a question. Have any smart ideas been activated by the charm of financial industry?

1.2.2 What Industry to Devote Ourselves To?

The application of algorithms in the industries mentioned above will all be promising. So as algorithm researchers, what industry is the best to devote ourselves to?

We should take different aspects into consideration and our goal is to maximize individual values. We need to choose a perfectly competitive channel of short process that we have access to, and then realize it with our algorithm technology and our understanding of the business.

In fact, the channels of individual realization are very limited. It's hard to directly make money from the market simply with an image recognition algorithm. A product is needed to carry the purpose and the product R&D process can be long. Similarly, the autopilot algorithm needs tests from car manufactures, and the user behavior algorithm would need verification of user shopping behavior from e-commerce platform.

Financial transaction is one of the channels that meet the requirement. You can open an account in the secondary financial market and complete the transaction. In this way, there are not many intermediate links. You can obtain data from the exchanges, write the algorithm model by yourselves and then deal with your own money. You are in total control. As long as the algorithm yields steadily, you will make money. The realization as such is actually quantitative investment in finance, which is one of the most reliable ways of realization.

1.2.3 Finance Is the Most Reliable

As IT engineers, we've already known programming and algorithms. We just need to learn the rules of the financial market, and then we can go and grab money from it. The secondary financial market of investment deal in China is not a very mature or stable market. Every day, there are many opportunities in the market, as well as fat fingers. Quantitative investment technology can help us find the opportunities caused by information asymmetry and make excess returns (Figure 1.7).

Finance! Finance! Finance!

Figure 1.7 Finance! Finance! Finance.

So how are we going to make quantitative investments?

For example, a private equity has raised 100 million and is ready to push into the financial market. The fund manager decides to allocate various financial assets according to the model of portfolio. Figure 1.8 shows that according to the theory of capital asset pricing model (CAPM), different assets are put into portfolios in the standard of mean-variance. Please refer to Section 2.1 for more about CAPM.

As in Figure 1.8, the x axis is the standard deviation of returns and the y axis is the mean of returns. The points in the figure have set up an area for investment. Every point stands for an investible product and the point set of every dotted line is an effective portfolio.

For nearly a hundred of points in the figure, assume that every time five assets should be allocated to make a portfolio, there would be 75,287,520 combinations; if 10 assets are needed in a portfolio, the combinations are also numerous.

We can use R language to calculate the amount of the portfolios.

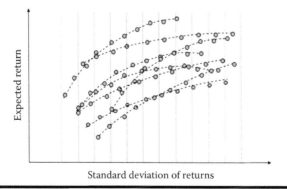

Figure 1.8 Portfolio and variance of returns.

```
# 5-combination of 100
> choose(100,5)
[1] 75287520

# 10-combination of 100
> choose(100,10)
[1] 1.731031e+13
```

In the financial market, we can choose from a lot of financial assets. In China, there are 3,000 A-shares, 2,000 funds, 3,000 bonds, 100 futures, bulk commodities, products of monetary market, exchange rate linked products, investments of overseas markets, etc. If we combine these assets, endless portfolios can be made, which results in a great amount of calculations. Therefore, we need algorithms to optimize the combinations and to find out the optimal portfolio in the market. Algorithms themselves embody the most value.

Then how do the traditional funds make their portfolios? Actually, it is done mostly by the subjective investment experience of investment managers. In the financial market, every fund is made of different assets. Randomly take one for example and see how the portfolio is made. For instance, ChinaAMC Growth Fund (000001.OF), a commingled fund. The data is from Wind Information of February 8, 2017.

As shown in Figure 1.9 Performance of ChinaAMC Growth Fund (000001.OF), 2006–2007 is the best time of this fund, with a return of 101.49% in successive 6 months; it has performed poorly in recent years, lagging behind CSI 300 and ranking low overall. In 2017, the return has been 0.58%, ranking 144 among 507 of its kind; its annualized return for 1 year is −1.45%,

	YTD	3月	6月	1年	2年	3年	5年	总回报	年化回报
华夏成长	0.58	-6.02	-3.91	-1.45	-5.15	11.67	39.96	456.40	11.99
沪深300	1.68	0.27	5.01	13.56	1.61	52.12	36.93	155.66	6.39
偏股混合型基金	-0.57	-4.30	-1.19	8.30	20.10	50.02	87.04	402.90	11.21
同类排名	144/507	353/505	365/498	400/487	398/449	378/426	290/352	-	1/2

Figure 1.9 Performance of ChinaAMC Growth Fund (000001.OF).

ranking 400 among 487; its annualized return for 3 year is 11.67%, ranking 378 among 426; and its 5-year annualized return is 39.96%, ranking 290 among 352.

Let's go to the fund constituents. It is mainly constituted with stocks and bonds.

Bond percentage (Chart 1.3):

Stock percentage (Chart 1.4):

Fund managers used to choose thousands of stocks and bonds in the market and weigh them differently. Algorithms can do this job, too. The portfolios built with algorithm modeling may perform better. If you replace the fund manager who requires a salary of several millions a year with the algorithm model, the salary can be your profit. Finally, your individual value will be

Chart 1.3 Bond Percentage

Security Name	Code	NAV Ratio (%)	Ups and Downs in Recent 3 Months (%)
China Petroleum & Chemical Corporation Bond 2012 1	122149.sh	2.34↑	−0.49
Guotai Junan Securities CP 2016 8	071602008.IB	2.12↑	−0.03
The Agricultural Development Bank of China Bond 2016 01	160401.IB	1.91↑	−0.08
Yingkou Port Group Co., Ltd Bond 2010	122049.SH	1.70↑	−1.59
Changzhou Hi-Tech Industrial Development Zone General Development Inc. (Holdings) Bond 2009 1	122956.SH	1.62↑	−0.65

Chart 1.4 Stock Percentage

Security Name	Code	NAV Ratio (%)	Ups and Downs in Recent 3 Months (%)
CAMCE	002051.SZ	4.09↑	−0.95
China Meheco	600056.SH	3.85↑	0.34
SWET	300156.SZ	3.81↑	2.56
OTMC	002175.SZ	2.89↑	−13.00
LUXSHARE-ICT	002475.SZ	1.52↑	−1.82
BGE	603588.SH	1.42↑	−14.96
SAIC MOTOR	600104.SH	1.38↑	7.88
TANAC AUTOMATION C A CNY1	300461.SZ	1.31↑	−12.28
Shanghai Pharm Hldg Co Ltd	601607.SH	1.25↑	5.39
CAHIC	600195.SH	1.21↑	−4.25

realized by algorithms and your fate is going to be changed by it, too. That is why I said finance is the most reliable way of realization.

Come and join our team, of which the fates will be changed by algorithms. Let's make money out of knowledge.

1.3 FinTech, the Booming Sector in Finance Field

Question
How do technical elites find the next booming sector?

Introduction
I've been working in IT programming for 10 years after attending university, first as a programmer then an architect. I've experienced a lot in these years, but I still insist the idea that the world can be changed with programming. I may have some insightful understanding of programming. However, I've felt that my strength falls short of my desire as I get older. In my early days, I wrote 5,000 lines of codes in Java in a day and was still energetic, but now I write 500 lines in R in a week and spend more time thinking. I write less code but accumulate more knowledge.

I am now over 30 years old. I've been through different experiences, like running my own business, raising venture capital, publishing books, teaching in a university, leading my own team and cooking at home as well. My business failed at last. However, I did not fail in technology, but the knowledge of another field and humanity.

1.3.1 Rise and Fall

To find the next booming sector is the goal of every discontented and ambitious young man. Since 2013, I have started technical preparation for my own business at home. It's been nearly 2 years. I located my business target at quantitative investment. I was assured that there would be no technical problems in R, Nodejs, Java and Hadoop; I learned theories of basic subjects, such as mathematics, statistics, finance, metrology, etc.; and I opened all transaction accounts that individuals have access to and traded with real money.

A bull market started in China at the end of 2014, with Shanghai composite index rising from 2,300 points to 5,100 within 8 months. I, kind of, caught up with the trend and smoothly got my first funding. This is the tuyere effect. As long as you stand in the tuyere, you will get great opportunities.

However, it did not last long. At the end of 2015, the stock market began to crash. After crash 1.0 and crash 2.0, a global economic crisis suddenly arrived in the beginning of 2016. Crash 3.0 stroke heavily due to two circuit breakers. Therefore, stock investors didn't dare to enter the market again. Most stock investors took heavy blows and private placements, as well, were going through a hard time under various regulations, so the stock market bubble collapsed and my business was put to an end.

Figure 1.10 shows the candlestick chart tendency of Shanghai Composite Index during the crashes.

One would fly in a tuyere. However, once the wind shifts, those who haven't prepared would fall badly. I was among those who fell.

In recent years, the Internet industry, driven by capital, has brought prosperous development to different business models. The Internet booming sectors are rotating all the time and many smart people have obtained their financial freedom from this. Social network companies have been booming since 2011 and were represented by Renren, Kaixin and Sina Weibo; then hundreds of group buying companies came into being, a great campaign was started and those surviving through it were Meituan, Dianping, Baidu Nuomi, etc.; in the following flourished the Internet finance P2P companies and just 1 year later, thousands of them slunk away; and O2O generated hype after group buying, but didn't survive the lack of capital; in 2016, VR, live broadcasting, artificial intelligence, bicycle sharing, etc. were driven unceasingly by capital.

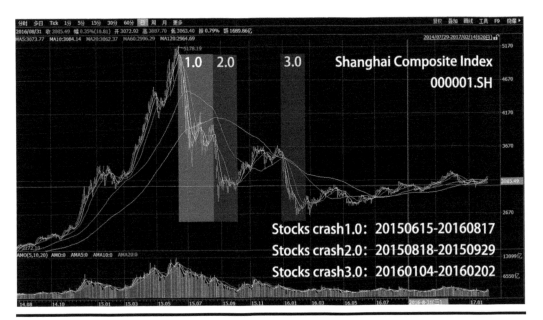

Figure 1.10   Candlestick charts of Shanghai Composite Index during the crashes.

1.3.2 Internet, at M&A Stage

In 2016, the whole business of the Internet and the mobile Internet was managed painstakingly, with either negative news reports or new names adding to the long list of shut-down companies in WeChat moments every day. The mad expansion of the Internet business stopped, and the start-up companies vanished even in the elevator advertisements. It is obvious as shown in CSI Mobile Internet Index (399970) that the market conditions of Internet business in China were appalling in 2016, compared to the skyrocketing bull market from late 2014 to 2015.

Figure 1.11 shows the weekly candlestick chart of CSI Mobile Internet Index (399970), where the area in the red box is the index tendency in 2016.

Start-up investments have obviously decreased since the latter half of 2015; instead, mergers and acquisitions increased in order to fight the capital winter. According to statistics, in the first half of 2016, 260 mergers and acquisitions were completed in China's Internet industry, including Meituan and Dianping, 58.com and ganji.com, Didi Kuaidi and Uber China, Ctrip and Qunar, Jiayuan.com and Baihe.com, Mogujie and Meilishuo, etc. These mergers and acquisitions symbolized that the whole Internet industry was shuffling and the mobile Internet, O2O, the mobile games and the mobile videos had all entered their maturation period.

1.3.3 Search for a Promising Industrial Booming Sector

Many smart people have gained their financial freedom during the Internet shuffle. How can we seize the opportunity of the next booming sector?

All the client-oriented Internet channels have been sealed by BAT, so it would be hard for start-up companies to have any breakthrough there. A big structure has been established, including the original BATJ and MI. It is more and more difficult to start a business on the Internet. One cannot run a new business individually as before, but needs to rely on big platforms and provides

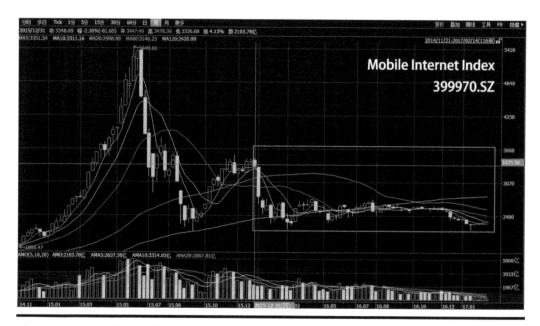

Figure 1.11 Weekly candlestick chart of CSI Mobile Internet Index (399970).

customized services for them. It will be more and more ecological in the future, i.e. every industry will provide its own specialized products.

In Dialogue, one of the CCTV financial channel programs, every guru of the Internet industry gave their directions of the next booming sector.

Liu Qiangdong said, "The booming sector may not lie in the Internet, but in traditional industries."
Li Yanhong said, "The next phase of development in the Internet is coming. The development is driven not by the big data or the cloud computing, but by the artificial intelligence."
Ma Yun said, "In our future innovative society, data will become the most important means of production and people cannot live without data, so we must invest whatever we can offer to develop data technology."

In the Dialogue program this week, Ma Huateng provided a long answer without any punctuation.

Ma Huateng: I can use just one sentence to express all their ideas – the future lies in that the traditional industries use the Internet technology to process big data with the artificial intelligence in the cloud.

All the gurus are talking about artificial intelligence, so AI is the next booming sector!

1.3.4 Gartner Hype Circle

The Hype Cycle is an instrument for enterprises to assess the visibility of a new technology and to decide the deployment of it according to its time and visibility in the market. A Hype Cycle drills down into several phases of a technology's life cycle, including Innovation Trigger, Peak of Inflated Expectations, Trough of Disillusionment, Slope of Enlightenment and Plateau of Productivity.

In July of 2016, Gartner, Inc. released the latest Hype Circle of new technologies, according to which, there are three significant trends: the era of perceptual intelligent machines is coming, the transparent and immersive experience gets optimized and a platform revolution is fermented (Figure 1.12).

The perceptual intelligent machine is a technology that efficiently integrates the computing power and the intelligence algorithm of big data processing. It enables enterprises to make full use of data and deal with complicated demands and problems unsolvable by predecessors. This field includes smart dust, machine learning, virtual personal assistant, cognitive expert advisor, intelligent data mining, smart workspace, session user interface, intelligent robot, commercial UAV, autopilot vehicle, natural-language question answering, personal analysis, enterprise taxonomy and ontology management, data broker PaaS, context broker, etc.

1.3.5 FinTech, the Booming Sector in Finance Field

In the following, we will take the financial industry as an example to illustrate how FinTech = Finance + Technology. In the innovative phase of the Internet finance, those companies driven only by technology or finance and find it hard to improve their innovation drive, may look forward to a new direction – the combination of finance and technology. All at once, every company starts to transform. Some start to do transaction, some do research and investment, some supervise orders, some do equity financing, some try live broadcasting, some turn to portfolios, some go for public opinion monitoring, etc.

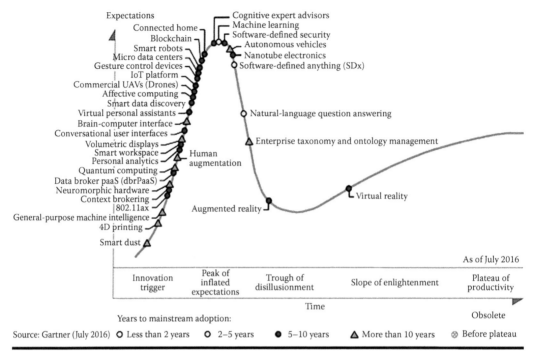

Figure 1.12   Gartner Hype Circle in 2016.

Though the overall situation was not good in 2016, everyone adjusted their attitude and tried to get prepared for the capital winter. The companies prepared themselves with more rational demand location, technology accumulation and workers' recognitive ability improvement, especially technicians' understanding of finance. The understanding of finance, a matter of great importance, is not "Internet+Finance," but "Finance+Internet," which means to improve the financial efficiency with the Internet technology, i.e. to accelerate the financial innovation by reaching to the clients with the Internet connection.

From the artificial intelligence aiding the quantitative transaction to the intelligent investment advising, we can see that FinTech has taken its root deeply in finance field.

Traditional investment advising service is provided manually, so many financial managers need to be hired and the human source cost is high. As a result, the entry threshold is high and only high net worth individuals are able to afford it. The private bank service provided in banks is an example.

While the intelligent investment advising uses the artificial intelligence algorithm to provide investment advice to the clients, it minimizes human force in the management of portfolios. The cost is low now, so it is not a dedicated service for the high net worth individuals any more. The intelligent investment advising builds models with machine learning method, draws wealth portraits for the massive clients by using the artificial intelligence technology and customizes asset management and investment plans for every client. In December of 2016, China Merchants Bank launched an application named Machine Gene Investment, which is a prelude to a campaign of artificial intelligence among banks.

Besides banks, Ant Financial, JD.com, Tencent, Baidu and many other Internet start-up companies are trying to push their ways in this direction. When large corporations enter this booming area, the new start-up companies are not competitive at all. Let's view it from another perspective.

If the large corporations are aiming at the same direction, the to B business can be an opportunity for the start-up companies.

It is like that old saying, "Do you want to be a gold digger or a water seller?" Get a clear understanding of ourselves, figure out our goals, and then seize the opportunity and work hard while we are still young.

1.4 Quantitative Investment Tools in China

Question
What investment tools are there in China?

Introduction
The Chinese market is peculiar, for it can be affected by the market conditions as well as the policies. When the stock market is rising, there can be anything: financing, leverage, high frequency trading, etc. However, if the market goes down, the new shares can stop, the stock index futures can stop, the security loans can stop and thousands of stocks can hit their limit down after two circuit breakers.

These special rules in the secondary financial market have brought forth various trading tools. In this section, we will discuss the quantitative trading tools in China's market.

1.4.1 Quantitative Trading Tools Overview

We need to, first, have knowledge of the market rules and the usage of the trading tools if we are going to trade in the secondary market. In this section, we will introduce the trading tools we can use.

Due to the independence of the financial market, the quantitative trading we usually mention can be divided into securities trading and futures trading. With securities accounts, we can trade stocks, bonds, open-ended funds, structured funds, pledges, margins, commodities, financial products of securities companies, etc. And with futures accounts, we can deal commodity futures and financial futures.

Quantification in China mainly refers to the quantitative trading of futures, stocks, bonds and funds. For individual accounts, full automation can be realized in futures quantitative trading, but hardly in security instruments trading because of its various restrictions. Usually only the institution accounts have access to securities trading, so the securities trading quantification stays mostly in stock picking and backtesting.

Figure 1.13 comprehensively introduces the tools we use to make quantitative investments, including tools for both automated trading and manual trading.

Figure 1.13 is comparatively complicated, so we divide it into several parts for illustration.

A: Let's start with the exchange data in the middle. All the data is from the exchanges and the raw data includes price, volume, bid, asked, bidding, capital inflow, capital outflow, etc.

B: Securities companies or futures companies will provide you with a client tool for securities or futures after you have opened an account, and then you can manually place an order and trade with the client tool. If the client tool is not user-friendly, you can download trade assistant tools from the third parties like Tonghuashun (10jqka.com.cn), Dazhihui (gw.com.cn), Eastmoney (eastmoney.com), Wind Information (wind.com.cn), etc. You can input the configurations of your securities company account and trade with these third-party tools. These tools usually have better operation interfaces, better stock recommendations and better personal services to help individual clients have quick knowledge of the market.

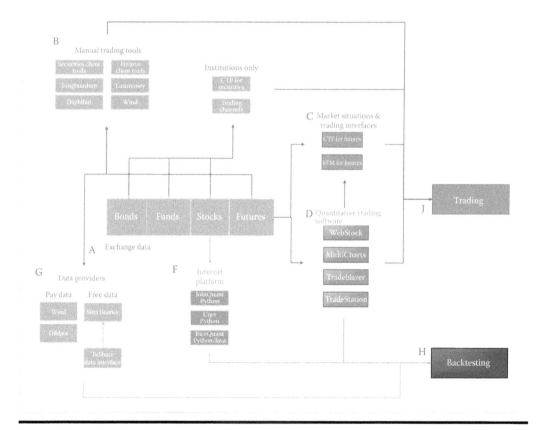

Figure 1.13   Trading tools in financial market.

However, it brings troubles as well. You will get a lot of crank calls after the third parties have your phone number.

C: Futures market is the most programmatic market at present because of CTP (Comprehensive Transaction Platform), the programmatic interface it provides. As long as you have opened a futures account, you can apply for the access to the futures CTP trading interface. Through the CTP interface, you can obtain current futures situations for free and you can also write your own codes to implement programmatic order placing, which would complete the automation for the whole trading process of futures.

D: There are five main kinds of quantitative trading software in futures trading, i.e. WebStock, MultiCharts(MC), TradeBlazer(TB), WeiStock decision-making and trading system and TradeStation(TS). These sorts of software have already packaged market quotations, trading, backtesting, account management, strategy templates, etc. and they provide independent scripting language for developers to program their own trading strategy models. Therefore, it is fast and convenient to develop scripts in a specific environment by using the software. If the script has a good backtesting result, it can go straight forward to firm bargains without any modification. Many small private equity firms and futures companies are frequently using this method to do quantitative trading and it is pretty efficient in this market of low programmatic maturity.

E: The programmatic trading in securities market is basically restricted by policies. Actually, there are some programmatic trading channels, but they are institutions only and usually those with capital over 30 million are allowed to apply for the channels. China Securities Regulatory Commission (CSRC) halted all the programmatic trading channels for new funds after the roller coaster ride of the stock market in 2015. The fund instruments which can be traded automatically in the current market must have been released before May 2015. Therefore, it is hard to realize the automated securities trading with a high threshold and under strict supervisions.

F: 2015 witnessed the rise of many Internet finance platforms, which started to take advantage of the Internet technology to seize the financial market. Besides P2P and payment, another branch of the Internet finance is quantitative investment. The mainstream platforms for programming trading models online in China are JoinQuant, Uqer and RiceQuant. These platforms mostly imitate Quantopian (https://www.quantopian.com) from abroad, which uses the Internet to break through the financial quantification, enables knowledge from outside to break down the rigid investment patterns in the finance field, gathers new strategies from the Internet communities and then makes money from the investments. I adopted this mode in my business in 2015. I took R language as the algorithm engine, while most of the other platforms were based on Python. We will not discuss programming languages here, for everyone may have his/her preference.

G: There's another way, the most professional way, i.e. using no platform. After obtaining data, you develop your own programs for backtesting, trading, accounting, risk control, etc. The raw data is mainly provided by the exchanges. However, individuals or companies cannot reach out for the exchange data, so here come the data service providers. Wind Information has become the biggest pay data service provider in finance. Gildata provides data service as well. It was taken over first by Hundsun, then by Ali, and it should be one part of Ali's layout in finance. Besides, there is another way to obtain free data – Sina Finance. There are all kinds of programs that teach you how to become a data crawler. Someone made a Python library named TuShare to collect all the data crawling methods. With this library, you can get comparatively comprehensive financial data.

What I mean is that it would be more reliable to use a professional database to accomplish professional tasks, especially those concerning money. Though the data from the Internet is free, the maintenance costs are high, for you need to program the data crawling, concern yourselves with network speed, data structure, data updates, etc. and sometime the data may get errors of missing lines or additional lines. Even if you make a strategy from the free data, you may not want to adopt it in real money trading.

H: Once we have the standardized market data, the following part will be easy. We can do a regression analysis or a tendency analysis based on MACD with R, design a hedge strategy of alpha and then you can do whatever you want. This is the most flexible way. You won't be constrained by the market or the tools. All are in control. In my opinion, as the maturity of quantification grows, more and more trading platforms would be self-built with data and this is the exact core knowledge and core technology. Moreover, the standardized common strategy models will finally become history in a perfectly competitive market. Of course, this will still take a long time.

J: Actually, there are two goals of quantitative investment: backtesting and trading. We may need to choose tools according to our goals. Trading for money, or backtesting for seeking opportunities in the market, or both? Clear goals and convenient tools will make it effective and efficient.

Now we have finished the framework idea. Next, we will explain how to use these tools in detail.

1.4.2 Client Tools for Securities and Futures

In order to trade securities or futures, we need to open an account in a securities company and get the trading client tool. The software operation interfaces differ in different securities companies.

1.4.2.1 Securities Trading Clients

When stock investors open securities accounts, securities companies are usually expected to give their customers an introduction training of stocks and futures trading and inform them of the risks. The companies also help their customers open exchange accounts, let their money from the bank get through, do bank and securities transfer, etc.

Every securities company has its own characteristics, e.g. stock trading commission discounts, plenty of securities for securities loans, remote account opening, etc. I remember the stock commission in 2010 was 0.75‰, already very low; yet in 2016, securities companies online could offer a commission of 0.2‰ due to the Internet impact. It is the bottom line and there is really no profit for securities companies.

Now some securities companies provide stock recommendation service and give advice on stock picking. Some offer good financial products with low risks and decent returns. And some feature in services like lectures on related knowledge and interaction with their customers. In the days before 2000, personal computers were not very developed and everyone went to the trading floor for communication. Now most people trade on their PC at home and young people usually on their cell phones for they don't have enough time to read stock market quotations.

As for securities account opening, Shanghai Stock Exchange (SSE) allows only one account for one customer, while one can open several accounts in Shenzhen Stock Exchange (SZSE). In the end of 2014, the stock market started to brisk up. As a result, CSRC decontrolled account opening permission, lifted the limitation of account amounts and people could open accounts in every securities company. Then after the stock market went through three risings and three downs in 2015 and continuous fluctuations in 2016, a policy came out and restricted the account number to three. This might reflect that policies override market or it could be an underlying signal. I have opened three securities accounts, one with the lowest commission, one the earliest account and the other with onsite service.

Below are the screenshots of securities client tools. Figure 1.14 is the client of China Merchants Securities and Figure 1.15 Huatai Securities.

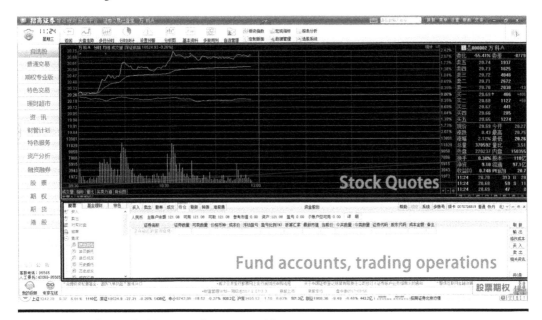

Figure 1.14   Client of China Merchants Securities.

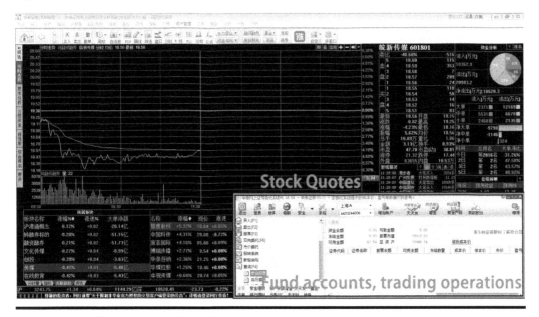

Figure 1.15   Client of Huatai Securities.

The operation interfaces of securities clients are usually divided into two parts: the stock market quotations and the account-trading. The stock market quotations are usually provided by a third-party software company. There are two main software providers in China: Tongdaxin and Tonghuashun. The market quotations interface of China Merchants Securities is from Tongdaxin, while the interface of Huatai Securities is provided by Tonghuashun. As for the account-trading part, it is customized and usually made by securities companies themselves. The trading interfaces can be completely different from each other. The trading interface of China Merchants Securities client is implanted under the quotations. However, for Huatai Securities client, the trading part is an independent software program, completely separated from the quotation interface.

The third-party securities client software includes Dazhihui, Tonghuashun, Eastmoney, Wind Information, etc. of which the former three are for individuals and the latter one is for professional customers.

- The official promotion concept of Dazhihui: Professional Strategies, Secure Risk Control Strategy and Authoritative Investment Platform.
- The official promotion concept of Tonghuashun: AI+Stock Software make investment easy.
- The official promotion concept of Wind Information: All-dimensional Financial Information, Global Financial Market Data.

From the interface designs and the operation experience, we can learn the differential positioning and focus of each software. Below are the screenshots for clients of Dazhihui (Figure 1.16), Tonghuashun (Figure 1.17) and Wind Information (Figure 1.18).

Whether they are the securities company clients or the third party clients, you will find one that suits you.

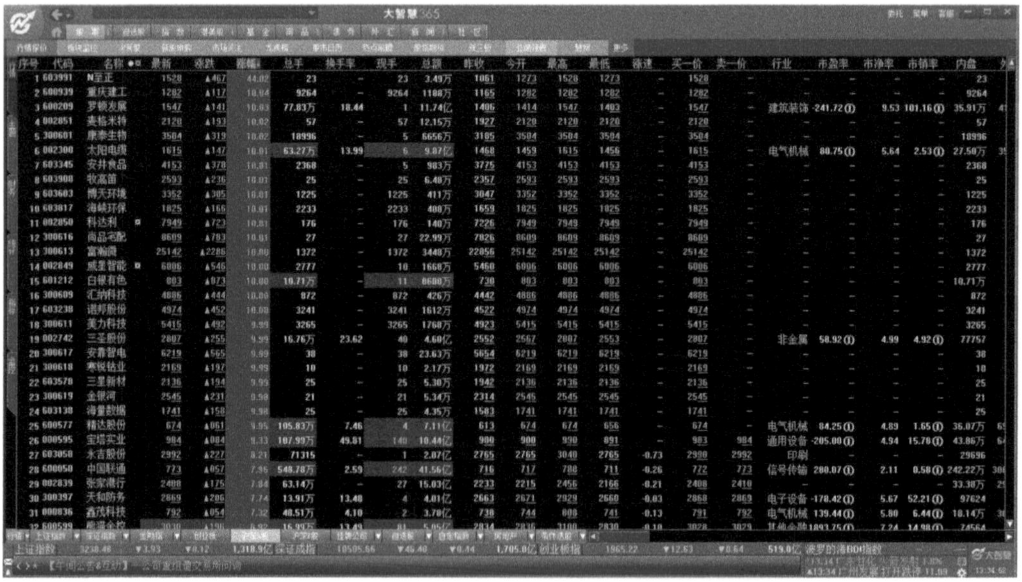

Figure 1.16 Dazhihui client.

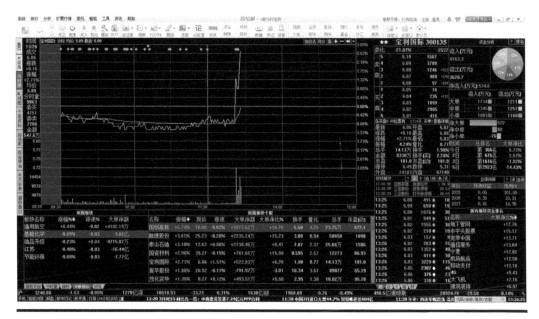

Figure 1.17   Tonghuashun client.

Figure 1.18   Wind Information client.

1.4.2.2 Futures Trading Clients

If you trade futures, you need to download a client software that futures companies provide. The interface shown in Figure 1.19 below is the client software of Guosen Futures. Its design is of typical style, with everything needed in a trade, but it's not user-friendly enough.

There is a popular third party futures client software – Kuaiqi, as shown in Figure 1.20. All the functions of Kuaiqi are set for fast trading, which is the core of Kuaiqi software. These are

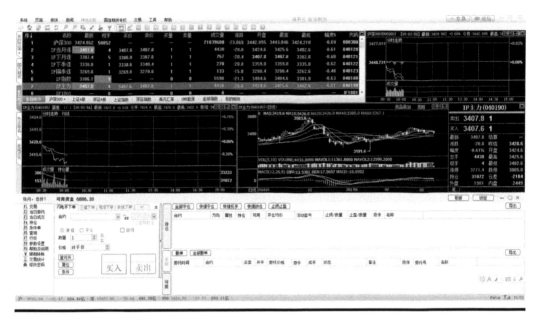

Figure 1.19   Guosen Futures client.

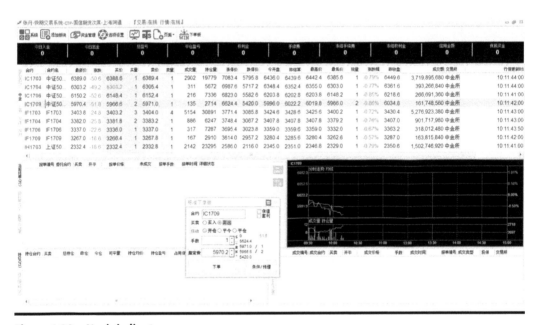

Figure 1.20   Kuaiqi client.

the main characteristics of Kuaiqi: free interface configuration, multiple placement interfaces at option, quote module placement, quick placement with mouse, quick placement with keyboard, quick withdraw, automatic position opening and closing and account trading report. With free interface configuration, users can configure the interface layout according to their own habits, which improves the comfort level.

1.4.3 Financial Database

If you want to make a quantitative investment, data is crucial. There are many data service providers in China. Some need to be paid and some are free.

1.4.3.1 Pay Data Service

Wind Information provides the most comprehensive financial database, but its charges are high. Moreover, there are databases of Gildata, Genius Finance, FinChina, SunTime, CCXE, etc. Each database has its own positioning. Some are comprehensive, some stress on macro data, some on exchange data, some on bond data, some on private equities data and some on VC/PE data. If you are still confused with these databases, you can use a third party platform like DateYes, where you can find the data you need, and then buy it and use it (Figure 1.21).

1.4.3.2 Free Data Service

There is a lot of free data on the Internet. Sina Finance has provided free data access for Internet users to download the stock quotations for a long time. The quantmod package integrated in R language can crawl data from Yahoo Finance, which is used in many cases of this book. As the popularity of the Internet, many websites provide the realtime trading quotations. Usually, any data you can view on websites can be crawled. Most websites transfer the data of JSON (JavaScript Object Notation) format with Ajax technology. If you are familiar with Javascript language, you can learn the codes through Chrome browser and crawl the data easily. However, this way you need to maintain the programs by yourselves, which is a little bit troublesome.

The problem of data crawling can be solved with TuShare package. You can simply use TuShare package to crawl data on the Internet, which saves time for the maintenance and allows you to focus on the data analysis (Figure 1.22).

At last, there is another typical Chinese way to obtain data, i.e. to buy data from Taobao. It is affordable, and can be more reliable than the data you crawl from the Internet.

1.4.4 Online Strategy Platform

JoinQuant, RiceQuant and Uqer are three online strategy R&D platforms arising from the Internet. If you are new to quantification, you can quickly learn something about it on these three platforms,

Figure 1.21   Professional financial databases.

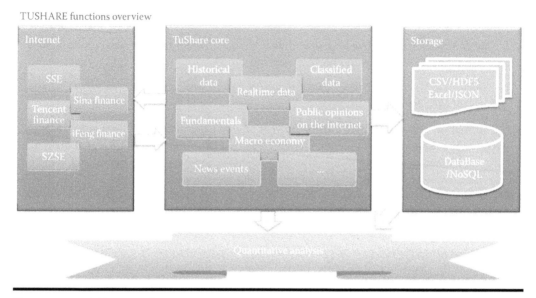

Figure 1.22   TuShare package.

where there are lots of good stuffs in their communities. You can also get to know people of your kind in various simulated quantitative trading contests, which would be beneficial for the beginners. If you are no longer a beginner and you have improved your understanding and practical experience of quantification, you will need this book for professional knowledge in your further learning.

Figures 1.23–1.25 are screenshots of backtesting on these three platforms. The designs of JoinQuant and RiceQuant are similar, just the Chinese versions of quantopian, while Uqer's design, adopting research and investment reports, is a combination of IT and finance. Which design will be more attractive to the Internet users?

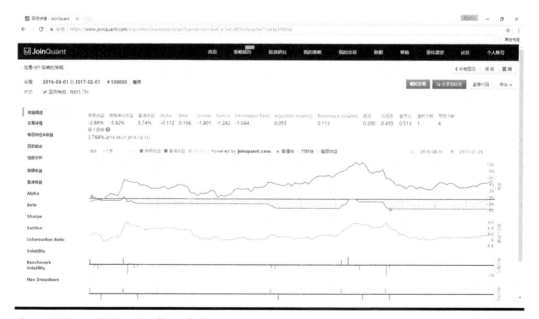

Figure 1.23   JoinQuant online platform.

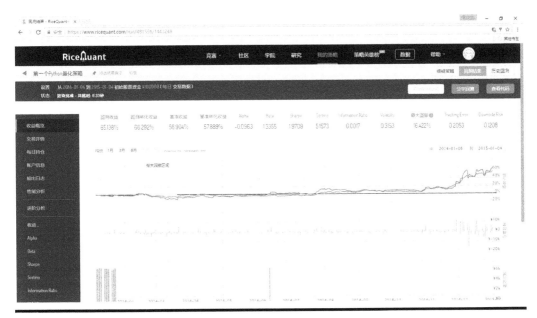

Figure 1.24   RIceQuant online platform.

Figure 1.25   Uqer online platform.

All three platforms provide Python API and automatic resource allocation schemes based on the Docker containers and they don't allow direct data downloading. Docker is an open source and lightweight virtualization technology of system level. Please refer to Section 4.3 for a detailed introduction of Docker technology.

The rise of online strategy R&D platforms have led to the decline of online strategy trading platforms. Platforms like ZLW.COM and WQUANT will be faced with huge transformation pressure.

1.4.5 Quantification Tools Software

WebStock, MultiCharts(MC), TradeBlazer(TB), WeiStock decision-making and trading system and TradeStation(TS) provide quantification client programs in the way of software. During the process of strategy development, you need to download data to your local machine and use the local resources for computing.

Therefore, the main problems of software architecture design for the local software and the Internet platforms are where to save data and where to do the computing.

For the Internet platforms, users don't have to download data; they can code online and directly submit it. The background server will launch a Docker container, allocate computing resources like memory, CPU, etc. according to the configurations and show the results to users through browsers after finishing computing. The whole process is completed on the server, so users don't need to have computers with high configurations. A computer with Internet access and a browser supporting H5 will do. Figure 1.26 shows the systematic architecture of the Internet platforms.

However, the client software has completely different architecture. As shown in Figure 1.27, all data should be locally downloaded and used for computing, which consumes local computing resources. The local computer performance affects the backtesting computing speed, so users need

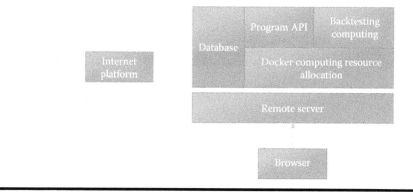

Figure 1.26 Architecture of internet platforms.

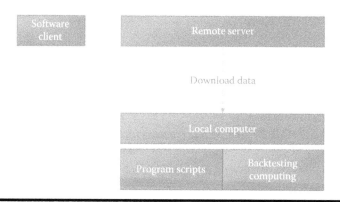

Figure 1.27 Architecture of software platforms.

highly configured computers. In this way, data can be downloaded to local machines, but it can only used by this software, for the data is encrypted without any description file or data format explanation.

By using webStock software, you can code and simultaneously intuitively view the visualized interface such as trading signals and a full report of strategy analysis. It is a very convenient design. Figures 1.28 and 1.29 are screenshots of WebStock software wh8, which contain the programming interface, the backtesting interface and the strategy report.

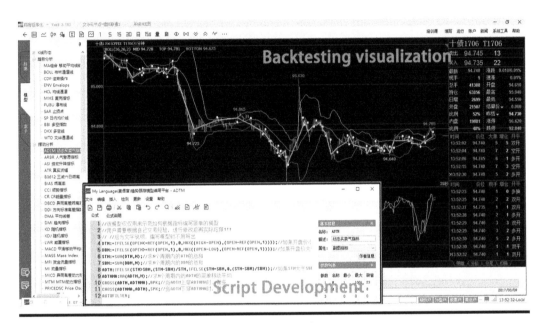

Figure 1.28 wh8 programming and backtesting interfaces.

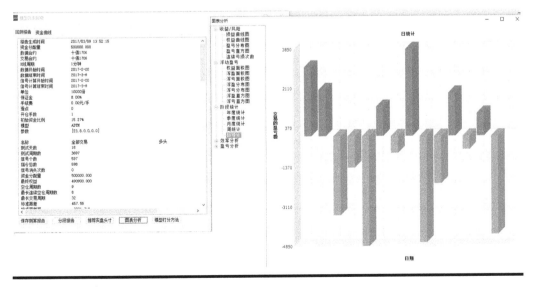

Figure 1.29 wh8 strategy report.

One main characteristic of WebStock is that it divides trading strategies into trend trading, arbitrage trading and observation trading, and it also provides special functions to support different trading modes. The other quantification software are similar to the WebStock wh8. You can download them by yourselves according to your interests or courses.

1.4.6 API Tools

In the following, I will introduce a professional way, which also shows a prominent advantage for people with a programming background. Through trading interface and quote interface, we, by ourselves, write codes to build the whole research and investment system and the trading system. It involves a lot of knowledge and information, and requires a team to carry the huge workload.

The trading interfaces available in the market are mainly CTP, REM, etc., which have their own characteristics.

CTP is a futures brokerage business management system specially developed for futures companies. It is constituted with a trading system, a risk control system and a settlement system. The trading system of CTP mainly focuses on the order processing, the market data feeding and the bank-securities transfer business; the settlement system sees to the trading management, the account management, the broker management, the capital management, the rates setting, the day end balancing, the information inquiry, the statement management, etc.; and risk control system mostly runs the high-speed realtime trial in the stock market and then reveals and controls risks. The systems connect to the four futures exchanges in China at the same time, support the trading and settlement business of commodity futures and stock index futures in China, and automatically generate and submit margin monitoring files and anti-money monitoring files.

REM refers to the rapid over-the-counter bulletin board of Shengli Financial Software. It is a system that uses FPGA hardware technology to deal with the core feed-forward risk control and the trading management. The main function of the REM system includes two parts: trading control and risk control. The system provides strict feed-forward risk control and secures ultra-low latency, and it has achieved the fastest and the most secure trading for the professional and institutional investors.

A CTP account can be opened in futures companies and is comparatively mature, while REM is only provided by some specific futures service providers and it contains only the basic API functions. CTP uses common futures trading interfaces and supports access to the four futures exchanges; however, REM supports Shanghai Futures Exchange, but not China Financial Futures Exchange. The latency of CTP is about 20 ms, and REM 5 ms, which is the data from velocity measurement in machine room. Their latencies of remote connection are both over 500 ms.

No matter what you will use, CTP or REM, they both need C++ programming. They both use asynchronous callback mechanism, so you need to know the programming design of callback as well. Using C++ for program development is comparatively complicated for common programmers. You need at least two separate processes to deal with the market data and the trading. You may need the latest quotations for biding, so data transfer is also needed between these two processes. The strategy models you develop, probably coded in R language and Matlab, will pass the trading signals to the biding application coded in C++ for trading. Besides, if different strategies go short and go long on TF1706 at the same time, you should spot them, record them but do not make any actual transaction. If you are going to place an order of over 500 round blocks, your system should help you divide it into small orders and place the orders every 5 seconds, so as to reduce the market impact on you. However, if you develop a system according to what I listed above, your system will become very complicated. Of course, if your asset size is up to several hundred billion, it is worth it.

There are many options to do quantification, so why do I prefer R language? I have already emphasized in Section 1.1 that R language is not omnipotent and I use it to solve modeling problems. You can choose Python or Matlab. They have their own advantages. Matlab is convenient, but not free. Python can achieve almost anything, but if you are going to solve a non-engineering problem with Python, you may not have any clue how.

Once you've had clear objectives and handy tools, you can exploit the advantages of yourselves and the tools to the fullest and create values.

1.5 Low-Risk Trading Strategy in China

Question
How do you make low-risk investments?

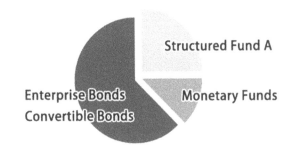

Low-Risk Trading Strategy in China
http://blog.fens.me/finance-arbitrage-strategy

Introduction
Low-risk arbitrage strategy is always the most important means of investment and financing. In the securities market, there are many low-risk arbitrage strategies that help people make use of their capital to achieve annualized returns of over 10%.

In this section, the low-risk trading strategies suitable for China's market will be introduced to make use of its trading rules.

1.5.1 *Corporate Bonds*

A corporate bond is a low-risk financial product with decent yields and comparatively low risks. At present, the annualized returns of common corporate bonds are around 6%, higher than that of the treasury bonds, the bank deposits and the bank financial products. The risks of corporate bonds mainly come from the interest rate risk, the liquidity risk, the credit risk, the reinvestment risk, the putability risk, the inflation risk, etc. For the trading practice of corporate bonds, please refer to Section 5.2 Corporate Bonds and the Arbitrage of this book.

1.5.1.1 Corporate Bond Analysis

The profit of a corporate bond comes from the coupon rate and the spread.

Coupon rate: It is usually a fixed value stated in the bond announcement. For example, if a corporate bond has a par value of 100 *yuan* and a coupon rate of 6%, you spend 10 million to buy 100 bonds and on the date payable, the interest will be paid to you after the 20% interest rate deduction, so every year you will receive an interest of 480 *yuan* = 10,000 * 6% * (1 − 20%). If the bond doesn't default, the corporate will pay you back 100 *yuan* for each bond when it is due after 7 years and besides, you will get a net income of 4.8% every year. Treasury bonds are tax free.

Spread: It refers to the bid-ask spread caused by the change of transaction prices. Corporate bonds, similar to stocks, can be traded in securities market. If the profits of most available financial products in market are lower than the coupon income of the corporate bond, people will be willing to pay more than 100 *yuan* to purchase a bond in order to get a higher profit, and the bond price will be over 100 *yuan* at the time. Likewise, if the yearly profits of the available financial products are higher than that of the corporate bond, people will sell their bonds and buy others, thus the bond price will be lower than 100 *yuan*. People trade and get spread returns from price changes.

Bonds have the characteristic of value return, which means capital with interest should be paid on the maturity date. As long as a bond doesn't default, its price will return to 100 *yuan* per bond, so bonds are more secured than stocks.

One characteristic of bond transaction is to trade at net price and to settle at full price. Full price = net price + interest. The quoted price is net price; the transaction is settled at full price, including interest. The annual interest is fixed, so every day the interest will be increased by 1/365 of the contract interest.

Factors impacting bond net price change:

- Adjustment of interest rate: There is an inverse correlation between the bond price change and the interest rate change. When the central bank raises the interest rate, the bond coupon will lose its strength and then capital will be withdrawn, so the bond price will be decreased; when the central bank reduces the interest rate, the bond coupon will show its advantage and capital will be attracted, thus the bond price will be increased. Besides, if the financial products of the same risk grade increase their annualized returns, the bond prices will fall; and vice versa.
- Spot rate: Bond prices and spot rates are negatively correlated, which is a typical seesawing relation. The increase of bond price would cause the fall of spot rate. When the spot rate is close to or lower than the average market return, capital will be withdrawn from the bonds, thus the bond price will drop; when the price declines, the spot rate will go up, capital will be flowed in the bonds again and the bond price will rise.
- Stock market: The bond market and the stock market are usually negatively correlated. If the prices rise in the stock market, capital will flow in and the bond market will tumble; on the contrary, the fall of the stock market usually encourages a bull bond market. Of course, the bull markets of both stocks and bonds in early 2015 and the bear markets of both in early 2016 broke this rule and they were caused by the capital gaming in many ways.

■ Inflation degree: In a moderate inflation, people would like to maintain the value of their assets, so they will buy more bonds, which may lead to the bond price increasing; but when hyperinflation comes and the bond interest income can no longer balance the inflation, people will transfer their assets to some place more secure, like gold, physical assets, hoarding of stock, overseas, etc., which would cause the bond price to decline.

■ Date payable: After the date payable, spot rate will rise; this will certainly lead to the increase of bond price.

■ Duration: The longer the duration of a bond, the more sensitive its price will be to the interest rate. When the interest rate increases (or decreases), the net price of the bond with longer duration will decline (or rise) more than that with shorter duration.

■ Credit spread: In a bullish bond market, the credit rating would not have great influence on the price; while in a panic market, credit spread will be enlarged and the bonds with better credit ratings will be more resistant to fall in price.

■ Repurchase conversion rate: In a bull bond market, the pledge-style repurchase bond will be more popular, for people can get more funds from the higher conversion rate.

1.5.1.2 Risk Mitigation

We should avoid two kinds of risks in corporate bonds.

1.5.1.2.1 Interest Tax

The interest of bonds is paid out once in a year, on the date payable. The bond transaction is settled at full price, i.e. an advanced interest will be paid. When you buy a bond, you need to pay the advanced interest, so you trade the bond at the settlement price, i.e. the full price. Full price = net price + interest. Because a 20% interest tax should be paid for corporate bonds and the tax will be deducted from the interest income on the date payable, the bond holder will actually receive 80% of the coupon rate. However, you have paid 100% interest in advance.

There are two ways to avoid the tax and save the 20% cost:

1. Sell the bonds before the date payable and buy them back after the ex dividend day. The bonds will be purchased by institutions, for they don't have to pay any interest tax.
2. Through the block trading of bonds, institutions can hold the bonds on behalf of you on the date payable and return them to you after the ex dividend day. The institutions will charge a fee for providing the service. There are volume thresholds for block trading: at least 10,000 bonds are required in a deal of block trading in Shanghai bond market and at least 5,000 in Shenzhen bond market.

For individuals, if your assets cannot meet the threshold, it is hard to find an institution to hold the bonds for you, so usually they sell them first and then buy them back.

1.5.1.2.2 Credit Risk

On March 4, 2014, the first material default of bond emerged. * ST Shanghai Chaori Solar Energy Science & Technology Co., Ltd announced, "Shanghai Chaori Solar Energy Science &Technology Co., Ltd Bond 2011 (112061) can only pay 4 million *yuan* instead of the full current interest of 89.8 million *yuan* on the original date payable of March 7, 2014."

Therefore, we should be careful when picking bonds to avoid defaults. It's best to pick the bonds that satisfy these conditions: credit ratings of above AA or AA+; backed with physical assets; with government background; no money loss in the recent 2 years; and with a high repurchase conversion rate.

1.5.2 Convertible Bonds

A convertible bond is a bond that can be converted into stocks. Before conversion, it is a bond with characteristic of value return and it has coupon, just like other bonds; while after conversion, it becomes a certain amount of stock shares. The rising of the stock market will bring along the price increase of the convertible bonds. A convertible bond is a hybrid security with debt- and equity-like features.

For the arbitrage of convertible bonds, please refer to Section 5.3 of this book.

1.5.2.1 Convertible Bonds Analysis

The bond market and the stock market are usually negatively correlated, but the convertible bond price and the underlying stock price are usually positively related. If the underlying stock is soaring, the convertible bond will take off in order to maintain the balance of their premium rate.

The convertible bond trading is a T+0 trading, i.e. the convertible bonds can be bought and then sold in 1 day. Some of them can be pledged. Like other bonds, a convertible bond has the characteristic of value return, so if there's no default, the capital with interest (that is, the 100 *yuan* principal and its coupon) will be paid on its maturity date.

Put provision: When the stock falls and the conversion value of the convertible bond is much lower than its par value 100 *yuan*, the investors have the right to request the issuer to buy it back at a price of its par value 100 *yuan* plus the interest compensation.

Call provision: Under certain circumstances, a company can redeem the unconverted convertible bonds at the price stated in the precedent agreement, in order to reduce the issuance cost. The call price is usually much lower than the conversion value, so this provision is mainly used to compulsorily convert the bonds to the stocks.

Whether it is a call provision or a put provision, the reason to lower the conversion price is that the underlying stock price deviates from the conversion value. The issuer is entitled to lower the conversion price, but has no right to raise the price. When the stock price jumps, the issuer may lower the conversion price to ensure that the investor's conversion profit will not suffer much and to avoid the put request from the investor. Once the put provision is invoked, the company will have to pay the debts in advance. To lower the conversion price means that a 100 *yuan* bond can be converted to more stock shares.

1.5.2.2 Risk Mitigation

Due to the call provision, the special provision of convertible bonds, the call price is usually much lower than the conversion value. Therefore, reasonable risk mitigation is needed.

During the conversion, if the price of the underlying stock is higher than 130% of the conversion price in 30 consecutive trading days, the call provision will be triggered. At this time, we need to pay attention to the company announcements and need to sell the bond or convert it before the call.

Usually, only a bull stock market can trigger the call provision.

For example:

From February 7, 2014, DHCC (DHC Software Co., Ltd.) convertible bonds can be converted to the A-share stock of the company, DHCC (002065.SZ). The closing price (15.31 *yuan* per share) of this stock has exceeded 130% of the current conversion price (11.78 *yuan* per share) in 15 trading days out of 30 consecutive trading days, i.e. from February 7, 2015 to April 24, 2015, which has triggered the call provision stated in the "Prospectus for DHCC Issuance of Convertible Bond" (the Prospectus). The 19th session of the fifth board meeting has approved "the Proposal of Implement the Call Provision of DHCC Convertible Bond." The company has decided to exercise its redemption right of DHCC Convertible Bond in advance and will redeem all the bonds that haven't been converted to stocks before the call date.

1.5.2.3 Convertible Bond Arbitrage

Because of the particularity of convertible bonds, there are many opportunities for arbitrage.

1.5.2.3.1 Negative Premium Rate Arbitrage

Negative premium rate arbitrage is the most basic and well-known arbitrage. If the stock market rises and the value of the stock that the convertible bond can be converted to is higher than the convertible bond price, there will be a negative premium rate and an arbitrage opportunity. It is a rare situation. Mostly, the conversion value is lower than the convertible bond price. If the premium rate stays negative before closing, we can buy the convertible bond and immediately convert it to stock shares, then sell the underlying stock the next day to obtain premium rate profit.

However, there is a risk here. If the stock opens low the next day, lower than the premium rate price, you are going to get held up instead of making a profit. The completely risk-free arbitrage needs to be operated together with the securities loan. But there may not be securities for lending when you want them; moreover, the annualized rate of 8% for the securities loan is not that low and it needs a higher negative premium rate and more accurate calculation.

1.5.2.3.2 Putability Arbitrage

When the underlying stock price plummets and it triggers the put provision, the holder has the right to sell the securities he/she holds back to the issuer at the agreed put price. Put option is the holders' right, not obligation. The holders can sell the bonds or continue to hold them. If there's irrational undersell of the plunged convertible bonds in market, we can buy them and then sell them back to the issuer.

It should be noted that in an interest-bearing year, the holder can execute the put option once when the terms are first met; if the holder doesn't do it the first time, he/she cannot do it again in this year.

1.5.2.3.3 Arbitrage from the Lowering of Conversion Price

If a company is faced with a put provision due to the straight falls of the underlying stock, it will usually lower the conversion price. Then the put stress is removed from the issuer. For investors, the lowering of conversion price means the increasing of conversion value. Though the convertible bond price declines, it can be converted to more underlying stock shares. Therefore, a conversion price reduction basically means to give out money.

1.5.3 Reverse Repos and Repos

In the securities market, if there are institutions or individuals that want to borrow money, they can pledge their bonds and borrow money from others in the market. Those who borrow money are buyers. When it's due, they will repay the money with interest, which will be calculated based on the realtime repo price; and the sellers, those who lend money, will receive the principal and the interest. This operation is called a reverse repo.

1.5.3.1 Reverse Repos

Reverse repos include the treasury bond reverse repo and the pledge-style reverse repo. Please refer to Section 5.4 for detailed introduction of reverse repos.

The features of reverse repos:

- Good liquidity. You can choose reverse repos of different durations according to your capital idle cycles. The principal and the interest will be paid to your account in T+1 day, and you can withdraw cash or transfer it to your bank account in T+2 days.
- Zero risk. There will be no problem as long as the treasury bond does not default. The reverse repo risk is almost the same as savings in the bank.
- Easy operation. Its trading is based on the realtime prices, the same as stock trading. It should be noted that reverse repos are about the operation of selling.

1.5.3.2 Repos

Repos are the counter parties of reverse repos. When your capital is tied up in bonds and there's not enough capital available at hand, you can pledge the bonds and borrow money, so as to seize short-term opportunities. It is the borrowers who operate repos and they must pledge their bonds.

Borrowing money through repos is of low cost, with a yearly interest rate of less than 3%. The repo price can be 2% lower when it's near closing in the afternoon. Though the cost is low, there is credit risk. If there happens to be an Internet cut off or a trading terminal breakdown, it will lead to a default for incomplete trading. Therefore, your personal credit will be impaired and the business department will be punished by CSRC. When there is low capital liquidity, i.e. there is a lack of money in the market, the repo interest rate will be high, sometimes even up to 50%. Usually, the average repo interest rate is a little bit higher than the bank financing.

When it is bullish bond market, we can add leverage and enlarge small sums of capital by repos. A bond return of four or five times can be obtained through the method of "bond purchase – repo – capital borrow – bond purchase – repo – capital borrow – bond purchase." Repo is a very good way of financing, which, if used properly, can significantly improve the profit.

Repos can be renewed. When the pledge is due, you can borrow money by repos to pay back the capital you borrowed last time, so that you don't have to take extra money or sell the bonds. But some securities companies may not allow renewal and require payback first before another repo.

1.5.4 Cash Management

Short-term cash management mainly refers to the cash operation in the monetary market, including operations of monetary funds, securities company financing, reverse repos, etc. The trading cycle is usually T+0 day.

The features of monetary funds:

■ Low-risk and with returns close to the interest rate of term deposits. Monetary funds are basically risk free, for their investment targets are usually short-term treasury bonds, repos, central bank bills, bank deposits, large NCDs, etc.
■ High capital liquidity. Monetary funds can be redeemed anytime. Usually the redemption capital will be transferred to the account in 1–4 days and there is no return between the redemption date and the capital arrival date. You can avoid weekends and holidays so as not to delay the redemption.
■ Low capital threshold. Usually, you can subscribe the funds with just 1,000 *yuan*.
■ Commission free. Usually, the fund subscription and redemption are commission free, but for some monetary funds, you may need to apply to the securities companies for free commission.
■ Tax free for gains. Monetary funds are tax free.

1.5.4.1 T+0 Monetary Funds

Here are the T+0 monetary funds subscribed and redeemed in exchanges, as shown in Chart 1.5. The data collecting date is March 1, 2017.

Take ChinaAMC Margin Money Market Fund A (519800) for example. It can be subscribed and redeemed using securities accounts, its annualized return is 2.94% and the monetary funds with codes started with 519 are commission free. If there is fund balance in the account before closing, I can use it to subscribe ChinaAMC Margin Money Market Fund A (519800) and then redeem it and get the capital back in the account in the next morning before opening, which will not affect the capital usage of other trading. However, sometimes there are restrictions of redemption. We should try to avoid the monetary funds with redemption restrictions.

Chart 1.5   Monetary Funds in Exchanges

Code	Name	Interest Calculation Rule	Return of 10,000 yuan	Annualized Rate of Return for 7 Days (%)
519858	GF Cash Pot Money Market Fund A	Start day included	0.9185	3.48
519878	China Life AMP A	Start day included	0.9084	3.31
519888	China Universal Cash Express Money Market Fund A	Start day included	0.9285	3.03
519898	Dacheng Cash Pot Money Market Fund A	Start day included	0.8185	3.01
519800	ChinaAMC Margin Money Market Fund A	Start day included	0.8346	2.94
519808	Harvest Margin Money Market Fund A	Start day included	0.6903	2.56

Besides, monetary funds bear interest at weekends. If we subscribe 519800 on Friday and redeem it next Monday, there will be interest on Friday, Saturday and Sunday. Therefore, it would be better to purchase a monetary fund than to make a reverse repo on Friday.

1.5.4.2 Arbitrage of Monetary Funds in Primary and Secondary Markets

Some of the floor monetary funds can be traded in both primary and secondary market. Subscription and redemption are dealt in the primary market, while buying and selling are dealt in the secondary market.

When there is difference between the prices in the primary and the secondary market, there is opportunity for arbitrage. If the fund price in secondary market is much lower than its net value, there is opportunity for discount arbitrage. We can buy the fund in the secondary market, then redeem it in the primary market and complete the arbitrage.

Floor funds start accruing interest immediately after the buying. For example, if you buy Fortune SG Listed Money Market Fund (511990) and then immediately redeem the same fund coded with 511991, you will get profits from two parts: the interest of the day and the fund discount income. Fortune SG Listed Money Market Fund (511990) and Fortune SG Listed Money Market Fund (511991) are two forms of one fund. 511990 is traded in the secondary market while 511991 can only be subscribed and redeemed in the primary market, thus the arbitrage opportunity in the primary and the secondary market appears.

1.5.4.3 Securities Company Financing

The financial products of securities companies such as Tiantianfa, Tiantianli, etc. can be applied for in the securities companies. Then the account balance after closing can be automatically transferred to the securities companies for financing, from which you can get an average annualized return of about 2%, not high. Different securities companies have different settlement intervals. Usually, the interest is paid every 2–3 months. Nevertheless, idle money after closing is better used in the financing of securities companies than in nothing.

For individuals, cash management is to maintain the cash value in liquidity. We can go for a reverse repo when its return is high. Usually, the returns of monetary funds, reverse repos and securities company instruments do not differ much, so you can buy any of them as you wish.

1.5.5 Structured Fund A

A structured fund refers to a fund that disintegrates a portfolio, according to fund returns or net assets, into two or more grades of fund shares with differentiated performances of risks and returns and pays returns according to the share categories.

For a structured fund, the net value of its parent fund is equal to the sum of the products of each sub fund's net value and its unit percentage. The parent fund of a structured fund is called Fund C; the sub funds are classified into Sub Fund A and Sub Fund B. Net value of parent fund = Net value of Sub Fund A * Percentage of Sub Fund A + Net value of Sub Fund B * Percentage of Sub Fund B. the risk preferences are different for Fund A and Fund B. Fund A is a fund with fixed income, similar to bonds; while Fund B, with a leverage character, would bear comparatively high market risk, for it will change as the market changes.

Parent funds belong to the primary market. The funds issued by Shenzhen Stock Exchange can only be subscribed and redeemed, while those issued by Shanghai Stock Exchange can be bought and sold as well. Fund A and Fund B belong to the secondary market and can only be bought and sold. A structured fund can be transformed in pairs. A two-share parent fund can be divided into one share of Fund A and one share of Fund B; or one share of Fund A and one share of Fund B can be integrated into a two-share parent fund. Figure 1.30 shows the transformation operations of a structured fund.

1.5.5.1 An Introduction to Fund A

Fund A is of the fixed income type, similar to bonds, and it gets profit from the interest. No matter how the market soars or falls, interest must be paid to Fund A. If the fund makes a profit, it will pay the interest of Fund A first and then pay Fund B with the rest; if it loses money, the interest of Fund A should be paid with the principal of Fund B. That is, the interest income of Fund A is pledged with the principal of Fund B. Fund A of a structured fund can be considered a bond whose default risk is much lower than those bonds in the exchanges. As long as the principal of Fund B is still there, the interest income of Fund A must be paid.

When there is a big price fluctuation for Fund B, to guarantee the interest income of Fund A, the share proportion of Fund A and Fund B will be adjusted for an ascending conversion or a descending conversion. During a regular conversion, the agreed income will be paid to the accounts of Fund A holders in the form of parent funds, which can be redeemed for cash later.

1.5.5.2 Structured Fund Arbitrage

According to the features of a structured fund, the spin-off arbitrage and the merge arbitrage can be adopted.

The spin-off arbitrage, also called the premium arbitrage, refers to the arbitrage when splitting the parent fund into Fund A and Fund B. When Current Price of Fund A * Percentage of Fund A + Current Price of Fund B * Percentage of Fund B > Net Value of Fund C, we can subscribe Fund C, then break it into Fund A and Fund B and sell them out.

The merge arbitrage, also called the discount arbitrage, refers to the arbitrage when merging Fund A and Fund B into the parent fund. When Current Price of Fund A * Percentage of Fund

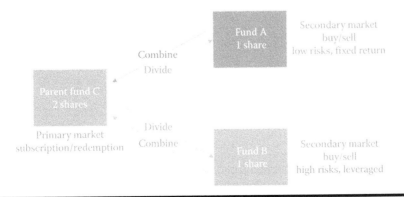

Figure 1.30   Transformation operations of A structured fund.

A + Current Price of Fund B * Percentage of Fund B < Net Value of Fund C, we can buy Fund A and Fund B, then merge them into Fund C, and redeem it.

The structured funds of Shanghai Stock Exchange and Shenzhen Stock Exchange have different arbitrage modes, so their operation terms are different (Chart 1.6).

When practicing the structured fund arbitrages, we need to accurately estimate the stability and volatility of the arbitrages, so as to smoothly complete the arbitrage operation.

1.5.6 Futures

Futures are high-risk financial derivatives. Investing futures is almost gambling. However, if practiced in a different way, futures can be low-risk and the low-risk arbitrage of futures can be realized as well.

Futures, as financial derivatives, are the most important risk mitigation tools. For example, futures can be bought for hedging. Futures are among the few products that can be traded in two-way. You can go long when there is a bullish market and go short when there is a bearish market.

For the private placements special for futures, there are three main operation approaches: arbitrage, speculation and programmatic trading of CTA.

Chart 1.6 Structured Fund Arbitrage Operation

Arbitrage Operation	Exchange	Operation Term	T Day	T+1 Day	T+2 Day	T+3 Day
Spin-off arbitrage (premium)	SZSE	T+3	Subscribe the parent fund		Split it into Fund A and Fund B	Sell Fund A and Fund B
Spin-off Arbitrage (premium)	SSE	T	Buy the parent fund, split it into Fund A and Fund B, and sell Fund A and Fund B			
Merge arbitrage (discount)	SZSE	T+2	Buy Fund A and Fund B	Merge them into the parent fund	Redeem the parent fund	
Merge arbitrage (discount)	SSE	T	Buy Fund A and Fund B, merge them into the parent fund and redeem or sell the parent fund			

For speculation, to buy futures is to gamble. Futures usually come with about 10 times of leverage, i.e. the principal is magnified by 10 times, so money can be earned fast and lost fast. The hedging operation is, on the contrary, a protection for the spot traders from the market risks, which hedges the price fluctuations of spot commodities by going long or going short. CTA analyzes the price, the volume and the technical indicators of futures with programming and then rapidly captures the market signals with computers, so as to find trading opportunities from the irrational prices. With CTA trading, we are able to make many trading strategies, such as trend tracking, volatility, period alteration, etc.

We can take the advantage of futures' two-way trading feature for arbitrage. The intertemporal arbitrage and the future-spot arbitrage are comparatively low-risk. The intertemporal arbitrage is to go long and go short at the same time. The price of two different-period contracts for the same product will be equal to the asset price when the maturity dates come, for their eventual asset targets are the same. During the transaction, when there is a big price deviation between these two contracts, we can go long on the contract with comparatively low price and go short on the one with high price. We will close the positions for arbitrage when the prices recover. The future-spot arbitrage, in a more direct way, is to trade the spot commodity and the futures of the same asset at the same time. For example, when the price of IF stock index futures contract is higher than that of Shanghai and Shenzhen 300 Index (CSI300), we will buy the CSI300 fund and meanwhile short sell the IF stock index futures; we will close the positions for arbitrage when the prices return to its normal. This trading is of low risk, for the market risks of this asset have counteracted with each other when we go long and short on the same asset; only the prices have been revealed.

The intertemporal arbitrage and the future-spot arbitrage are well-practiced in the market and there will not be many opportunities for these arbitrages. Nowadays, the mainstream way is the cross-product arbitrage, i.e. to make a cross-product trading strategy model by discovering the relation of different products. For details, please refer to Section 6.3.

The strategy will be coded into programmatic trading. Automation or half automation can be achieved, so that we can avoid error operations caused by human weaknesses like lust, greed, fear, etc.; then we can complete the trading, make money and achieve our financial freedom. The following chapters will carry out the thinking pattern of programming and implement it with R language coding. I hope everyone can cheerfully experience the fun of making money with programming.

Chapter 2

Financial Theory

In this chapter, we will make a deep interpretation of the applications of the investment theory and the statistical theory in the practical financial market with R language. The applications include four basic theoretical models, i.e. the capital asset pricing model (CAPM), the simple linear regression model, the multiple linear regression model and the autoregression model. I hope these models can help you find a way to understand the financial market.

2.1 R Interpretation of Capital Asset Pricing Model (CAPM)

Question
How do you manage portfolios with R language?

R Interpretation of
Capital Asset Pricing Model (CAPM)
http://blog.fens.me/finance-capm

Introduction
Due to the rises and falls of China's financial market in 2016, the risks have become more and more uncertain, the interest rate continues to fall and the returns of risk-free assets, e.g. some financial instruments, keep declining. Only the portfolio can maintain and increase the value of our assets. According to CAPM, the building of a portfolio after the analysis of financial data can help us control the risks and stabilize the income in an efficient market.

Here we will, in an easy approach, introduce the complicated CAPM from the theory to the modeling and then to the program implementation. CAPM reflects the relation between the asset risks and the expected return. The higher the risk, the higher the return. When the risks are the same, investors will choose those assets with the highest expected returns; if the expected returns are the same, investors will choose those with the lowest risk.*

2.1.1 Background

In 1952, Markowitz put forward the theory of portfolio selection, where he considered the intersection point of the indifference curve of the investors with risk aversion nature and the asset's effective boundary line as the best portfolio. When picking assets, investors will balance between the returns and the risks. If the risks are the same, investors will choose the assets with the highest expected returns; if the expected returns are the same, investors will choose those with the lowest risk.

In 1964, William Sharp, John Lintner and Jan Mossin came up with the single index model on the basis of Markowitz's theory. They introduced the market portfolios to the mean-variance model, which tremendously simplified the calculation. They believed that there was a linear relation between the return of any market portfolio and some common factors, which eventually evolved into the CAPM. It's been a long way from Markowitz's theory of portfolio selection to the CAPM.

In short, the core idea of CAPM lies in that the asset price is determined by the price compensation of the risks (Figure 2.1).

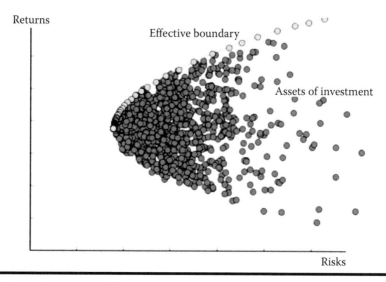

Figure 2.1 Illustration of portfolio selection.

* This is not an article from a financial textbook, so if there is any content not in accordance with the textbooks, the textbooks will prevail. Here I will try to introduce CAPM and its implementation in R language in simple terms.

2.1.1.1 Assumptions

CAPM is true based on a series of assumptions. However, these assumptions may not be true in reality, so CAPM has once been doubted.

- Assets can be divided infinitely.
- There is no transaction cost or individual income tax.
- Short-selling is unlimited.
- There is a risk-free interest rate, at which investors can lend or borrow capital of any amount without any limitation.
- Investors are price takers and the market is perfectly competitive.
- Investors are sensible; they make investment strategies by comparing the expected returns and the variances of assets and will select the assets with the lowest risks if the expected returns are the same.
- Investors make decisions in the same investment horizon; the market information is open and free of charge and it can be obtained timely.
- Investors expect the same of the economic variables in the market. They share common ideas on the expected return and the market risk of any asset.

The core assumption of CAPM is that the market is perfectly competitive, frictionless and of completely symmetrical information and the investors in the market are all, what so called in Markowitz's theory, the rational economic men. However, the assumptions are barely met in the real market. When institutions make their optimum portfolio plans by using CAPM, they do it based on their own understanding of the market and they try to get as close to the assumptions with their strengths; therefore, different institutions can make different portfolio plans.

2.1.2 Capital Market Line (CML)

We need to know several financial concepts in advance, for the financial field is involved.

- Risk assets: Refers to the assets with future profitability, but their returns are uncertain and they may cause losses. Stocks and bonds can be classified in this category.
- Risk-free assets: Refers to the assets with no risk or very low risks. The returns are fixed and there is no default risk.
- Rate of return: Refers to the rate of return on investment we get from the beginning of the investment till the end.
- Risk-free return rate: Refers to the rate of return on investment of risk-free assets.
- Portfolio: Refers to the combination of stocks, bonds, funds, financial derivatives, etc. that investors or financial institutions hold, aiming at risk diversification.
- Leverage: Refers to the trade in which a small capital is used to make an investment of multiple times the original amount, expecting to get a profit or a loss of multiple times the original return of the investment target.

2.1.2.1 Risk Assets

We can describe risk assets with expected returns and risks in two-dimensional coordinates. See Figure 2.2.

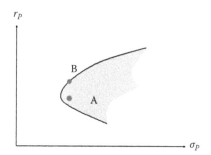

Figure 2.2   Risk assets.

Explanation of Figure 2.2:

- *X*-axis refers to the risks.
- *Y*-axis refers to the rates of returns.
- The grey area refers to the investible area of the financial assets.
- The black line is the effective investment boundary.
- Point A and Point B refer to two risk assets.

A and B share the same *x* value, which means they have the same risks. Point B is above Point A, which means B's return is higher than A's. For rational investors, if they have to choose between A and B, they all will invest their money in B rather than A.

2.1.2.2 Risk-Free Assets

We add risk-free assets to figure out the relation between risk-free assets and risk assets.
Explanation of Figure 2.3:

- Point B refers to a risk asset on the effective investment boundary.
- Point C refers to a risk-free asset on the *Y*-axis.
- *X*-axis refers to the risks.
- *Y*-axis refers to the rates of returns.
- The grey area refers to the investible area of the financial assets.
- The black line is the effective investment boundary.

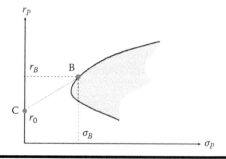

Figure 2.3   Risk-free assets.

Point C, the risk-free asset, locates on *Y*-axis, which means its *x* value is 0, i.e. the risk is 0. We can allocate the investment capital on C or B. If we invest all on C, we will get the risk-free return of r_0; if we invest all on B, we will get the return of r_B and take the risk of σ_B. However, if we invest partly in the risk asset of Point B and partly in the risk-free asset of Point C, we will get a portfolio constituted of asset B and asset C and the risk and return that we will get will be reflect in the straight line connecting B and C.

2.1.2.3 The Optimal Portfolio

So is there an optimal portfolio with the best return and the lowest risk? Let's find out this optimal portfolio M.

Explanation of Figure 2.4:

■ Point M refers to the optimal portfolio of risk assets.
■ Point B refers to a risk asset on the effective investment boundary.
■ Point C refers to a risk-free asset on the *Y*-axis.
■ *X*-axis refers to the risks.
■ *Y*-axis refers to the rates of returns.
■ The grey area refers to the investible area of the financial assets.
■ The black line is the effective investment boundary.

Assume that there is an optimal portfolio, represented by Point M in Figure 2.4. We connect C and M and make the connecting line tangent to the grey area. We can see from the figure that the slope of CM is always bigger than that of any line connecting C and the points in the investible area, e.g. the connecting line of C and B. Let's extend the line connecting C and B and find Point B' on the extension line which shares the same *x* value as M. M and B' have the same risk, but M gets higher return, for it is above B'. That is, when the risks are the same, we will all choose the asset with the highest return.

The slope of M is the biggest in the investible area, so we consider M the optimal portfolio of assets in the market.

No matter what risk and return preferences the investors have, we can make the best investment plans for them as long as we find out the optimal portfolio and know the risk-free asset. For rational investors, they will only invest in the optimal portfolio, if found, regardless their return and risk preferences (Figure 2.5).

The portfolio of M is usually constituted with all the investible securities, and the proportion of each security asset is the relative market value of the security. The risk-free asset C is not

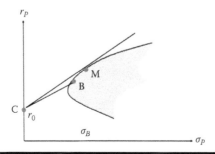

Figure 2.4 The optimal portfolio.

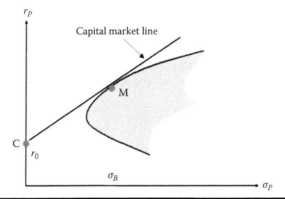

Figure 2.5 Capital market line.

included in M. People will always use CM connecting line for investment and for building the optimal portfolio.

In a practical trading, the price of the financial assets may deviate for the impact of supply and demand in the market. When there's price deviation, the market will automatically repair and restore the price to its equilibrium level.

2.1.2.4 Capital Market Line (CML)

The CM connection line fits the theory of portfolio selection which Markowitz put forward, where he considered the intersection point of the indifference curve of investors with risk aversion nature and the asset's effective boundary line as the best portfolio. And the CM connection line is called the Capital Market Line (CML).

The CML shows the simple linear relation between the expected return and the standard deviation of efficient portfolios.

The CML determines the prices of securities, because the CML is the balance of risk and return in an efficient portfolio. If this balance is thrown off, another relation of risk and return will be formed beyond the CML.

2.1.2.5 Portfolio Construction

The CML is the optimal portfolio. If we find this portfolio, all of our money will be invested in it. We can get the risk and return we want by assigning different weights of investment to the risk-free asset C and the risk asset M; we can also use financial instruments to add leverage and magnify the risk and return, as shown in Figure 2.6.

Investors can be classified into risk aversion investors and risk radical investors.

For risk aversion investors, they value the capital safety, they do not go for high return but principal safety and their capital is usually used for living. If we are making capital allocation plans for them, we can invest most part of the capital in risk-free asset and small part in the optimal portfolio on M, so as to assure low risks, but still get small returns.

In Figure 2.6, if 50% of the capital is allocated in the risk assets and another 50% in risk-free assets to make a portfolio, the formula will be:

```
CM1 = 0.5C + 0.5M
```

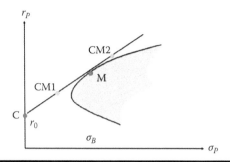

Figure 2.6   Portfolio management.

For risk radical investors, they value the returns of their capital; the principal can be partly or completely lost, for it is usually idle money specially for investment. When making capital allocation plans for them, we can invest all the money in M. Or more, we can even, by borrowing money or raising capital, add leverage and magnify the capital for investment. This method can enlarge the risks by the lever magnification, as well as the returns.

In Figure 2.6, Point CM2 is located on the extension line of CM. We can allocate 150% of the capital in risk assets M, meanwhile pledge 50% of the capital and the risk-free assets return C to borrow money. The formula will be:

```
CM2 = -0.5C + 1.5M
```

2.1.2.6 Relation between Risk and Return

In the above, we have qualitatively introduced the relation between risks and returns in the way of thought train, but we did not describe it quantitatively. So how can we define the relation between risks and returns in mathematical way?

Explanation of Figure 2.7:

- Point M refers to the optimal portfolio of risk assets
- Point C refers to the risk-free asset on the *Y*-axis.
- r_0 refers to the return of the risk-free asset.
- r_M refers to the return of Point M.
- σ_p on x-axis refers to the return variances of risk-free assets.
- r_p on y-axis refers to the returns.

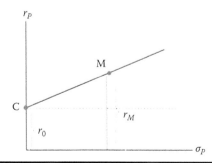

Figure 2.7   Relationship between risks and returns.

The mean–variance model, brought in by William Sharp, significantly simplifies the calculation, i.e. it has solved the problem of formula calculation. It uses variance to depict risks and establishes a simple linear relation between returns and risks, which can be represented with the formula below:

Formula:

```
E(rm) - r0 = A * σM^2
```

Explanation:

- $E(rm)$: The expected return of a market portfolio.
- $r0$: The risk-free return rate.
- $E(rm) - r0$: The risk premium of the market portfolio.
- $σM^2$: The variance of the market portfolio, $Var(rM)$.
- A: The level of risk aversion.

With this formula, we can clearly understand the quantitative relation of risks and returns, and we can use data for calculation as well.

2.1.3 Capital Asset Pricing Model (CAPM)

For a market portfolio, its risk premium and its variance show a linear relation. However, if for one single asset, its return and risk are part of the portfolio; the asset and the portfolio are both influenced by the market situation.

2.1.3.1 The Risk Premium of Single Asset

In CAPM, the risk of one single asset is represented by β. β evaluates the change consistence degree of a single financial asset and its market return and it is calculated by covariance. The risk of a single asset is to divide the covariance of the current asset and the portfolio return by the variance of the portfolio return.

The formula of the single asset risk:

```
βi = Cov(ri, rm) / Var(rm)
   = Cov(ri, rm) / σm^2
```

The formula of the single asset risk premium:

```
E(ri) - rf = (Cov(ri, rm) / σm^2)*[E(rm) - rf]
           = βi  *  [E(rm) - rf]
```

Explanation:

- $E(ri)$ refers to the expected return of the risk asset i.
- $E(rm)$ refers to the expected return of the market portfolio.
- rf refers to the return of the risk-free asset.
- $Cov(ri, rm)$ refers to the covariance of the risk asset return and the portfolio return.
- $Var(rm)$ refers to the variance of the market portfolio return.

We can see from the formula that the risk premium of a single asset is proportional to the risk premium of market portfolio M and they are both impacted by β.

2.1.3.2 Capital Asset Pricing Model (CAPM)

The CAPM is the cornerstone theory in modern finance. Specific formulas of CAPM can be deduced under the assumptions stated in the above. And the introduction above is the whole deduction process. From a learner's perspective, this theory is comparatively simple, for just simple statistics is involved; but it did take the predecessors a long time to research and explore.

To assess the risk of a single asset, if $\beta = 1$, it means the tendencies of the current asset and the whole market are consistent; if $\beta = 2$, the risk is high and its return changes more significantly than the market; if $\beta = 0.5$, it is a low-risk asset allocation.

2.1.3.3 Two Types of Risks

Two types of risks are defined in CAPM, i.e. systematic risks and non-systemic risks.

Systematic risks are the risks caused by the external factors, such as inflation, GDP, major political events, etc. The influence of these factors on the asset return cannot be eliminated by portfolios, so these risks are unavoidable to investors.

Non-systemic risks are the risks caused by the internal structure of portfolios. For example, if Share A and Share B are highly relevant, once there is big fluctuation of Share A's return, Share B will present a similar fluctuation; the meeting of two crests or troughs will increase the risk of the portfolio. Conversely, if A and B are negatively correlated, their fluctuations will counteract with each other. This kind of risks is determined by the asset types in the portfolio, so theoretically or practically, it can be minimized or even eliminated by diversification of the investment. In the process of risk elimination, the return of the portfolio will not decline.

2.1.3.4 Two Types of Returns

Returns are what correspond to risks. We bear two types of risks, so we will receive two returns they bring along. One type of return, beta(β), is completely relevant to the market; another type, alpha(α), is irrelevant to the market.

The return of beta can be obtained easily. For instance, if you are positive about a market, you can buy its corresponding index fund at a low cost and wait for its rising. The return of alpha is comparatively hard to get. It is a strategy return that reflects the investment capability. Alpha is the difference between the actual return of a portfolio and its expected return. Here is the alpha calculation formula:

```
E(ri) - rf = αi + βi  *  [E(rm) - rf]
αi          = [E(ri) - rf] -  βi * [E(rm) - rf]
```

Alpha can be used to evaluate the investment capability of investors. Let's illustrate it with an example. When the market return is 14%, β of Security A is 1.2 and the interest rate of the short-term treasury bond is 6%, the investor trades this stock and gets an actual return of 17%. So how should we assess the investment capability of this investor?

First, solve the expected return of Security A = 6% + 1.2 * (14 − 6)% = 15.6%; then subtract the expected return of Security A from its actual return, i.e. 17% − 15.6% = 1.4%. The result 1.4% is the alpha, which means the investor is able to get an extra return of 1.4%.

2.1.3.5 Application Scenarios of Capital Asset Pricing Model (CAPM)

Risk diversification of portfolio: Investors can diversify their holdings of multiple risk assets in the proportion of market portfolio structure, so as to get their risk assets portfolio as close as to the market portfolio and eliminate the non-systemic risks.

Return-risk ratio adjustment: Recombine the market portfolio of risk-free assets and risk assets to get a customized return-risk portfolio.

Indexing investment: Refers to the investment which allocates the weights of the assets according to an index. The weights or the components can be fine tuned to achieve a higher alpha than the index.

Asset pricing: The CAPM can be used to judge whether the market prices of securities or other financial assets are at a balanced level, not overestimated or underestimated, so as to arbitrage for excess returns.

Fund purchase: When we are going to purchase funds, the CAPM can help us with analysis. We can take an example in the market. Fund A has an expected return of 12% and its risk of $\beta = 1$; while Fund B has an expected return of 13% and its $\beta = 1.5$. The market expected return is 11% and the return of the risk-free asset $r_0 = 5\%$. So which one will you buy?

When you are faced with various fund recommendations of Alipay every day, you will realize this is an actual question. If you understand this section, the answer will be to solve alpha according to the CAPM. Buy whatever fund with a higher alpha.

Apply the formula to solve alpha.

```
αA = 12 - 5 - 1 * [11 - 5] = 1%
αB = 13 - 5 - 1.5* [11 -5 ] = -1%
```

Fund A has an alpha of 1%, while Fund B −1%. Obviously, the conclusion is that the manager of Fund A, with excess return of 1%, is more capable and the manager of Fund B, with a return lower than the market by 1%, is less capable. Therefore, we will invest in Fund A, not Fund B.

2.1.4 Portfolio Modeling with R

A large part of this section has been devoted to the introduction of CAPM theory, but it is very simple to implement it with program, because R language has encapsulated the related calculation functions of CAPM. To complete the calculation, what we need to do is just call the functions (Figure 2.8).

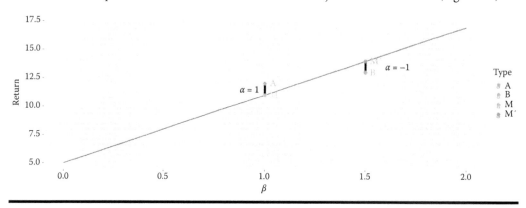

Figure 2.8 Calculation of the fund alphas.

Mostly, we will use two packages in the programming implementation of R language: quant-mod and PerformanceAnalytics. For quantitative investment packages of R language, please refer to Section 1.1.

- Quantmod is used for data downloading.
- PerformanceAnalytics is used for the calculation of different evaluation indexes.

We can design an application scenario. Assume that we have 100,000 USD and we want to invest it in the American stock market and gain better return than S&P 500 Index. So how should we buy stocks?

First, we need to make clear that our final goal is to gain "better return than S&P 500 Index." Second, we need to, based on the CAPM theory, figure out the investment strategy from the perspective of portfolio, not technical indicator. Aiming at better return than S&P 500 Index, we need to consider S&P 500 Index an ideal portfolio. Then we need to pick some stocks in the market, calculate their return, beta, alpha, etc. to see if they meet the expectation and then test them repeatedly until we find the suitable stocks or stock portfolios.

This article is just a case description used to illustrate the investment ideas and methods. It does not constitute any stock recommendation.

The system environment used here:

- Win10 64bit
- R: 3.2.3 x86_64-w64-mingw32/x64 b4bit

Download from Yahoo the daily market data of these three stocks, IBM, GE and YHOO, since January 1, 2010 and the daily data of S&P 500 Index as well.

Run the R language programs.

```
# Load the package
> library(quantmod)
> library(PerformanceAnalytics)

# Download the data of three stocks and S&P 500 Index from Yahoo
> getSymbols(c('IBM','GE','YHOO','^GSPC'), from = '2010-01-01')

# Print the first six lines and the last six lines
> head(GSPC)
              open     high      low    close      volume adjusted
2010-01-04 1116.56 1133.87 1116.56 1132.99 3991400000  1132.99
2010-01-05 1132.66 1136.63 1129.66 1136.52 2491020000  1136.52
2010-01-06 1135.71 1139.19 1133.95 1137.14 4972660000  1137.14
2010-01-07 1136.27 1142.46 1131.32 1141.69 5270680000  1141.69
2010-01-08 1140.52 1145.39 1136.22 1144.98 4389590000  1144.98
2010-01-11 1145.96 1149.74 1142.02 1146.98 4255780000  1146.98

> tail(GSPC)
              open     high      low    close      volume adjusted
2016-12-20 2266.50 2272.56 2266.14 2270.76 3298780000  2270.76
2016-12-21 2270.54 2271.23 2265.15 2265.18 2852230000  2265.18
2016-12-22 2262.93 2263.18 2256.08 2260.96 2876320000  2260.96
2016-12-23 2260.25 2263.79 2258.84 2263.79 2020550000  2263.79
```

```
2016-12-27 2266.23 2273.82 2266.15 2268.88 1987080000  2268.88
2016-12-28 2270.23 2271.31 2249.11 2249.92 2392360000  2249.92

# Draw the candlestick chart of S&P 500 Index, as shown in Figure 2.9
> barChart(GSPC)
```

Merge the adjusted prices of these four assets.

```
> # Change the column names
> names(IBM)<-c("open","high","low","close","volume","adjusted")
> names(GE)<-c("open","high","low","close","volume","adjusted")
> names(YHOO)<-c("open","high","low","close","volume","adjusted")
> names(GSPC)<-c("open","high","low","close","volume","adjusted")

# Merge the data
> dat=merge(IBM$adjusted,GE$adjusted,YHOO$adjusted,GSPC$adjusted)
> names(dat)<-c('IBM','GE','YHOO','SP500')

# Print the first six lines
> head(dat)
                IBM       GE    YHOO    SP500
2010-01-04 112.2859 12.27367  17.10  1132.99
2010-01-05 110.9295 12.33722  17.23  1136.52
2010-01-06 110.2089 12.27367  17.17  1137.14
2010-01-07 109.8274 12.90920  16.70  1141.69
2010-01-08 110.9295 13.18724  16.70  1144.98
2010-01-11 109.7680 13.31435  16.74  1146.98
```

Calculate the daily returns and merge the returns to dat_ret.

```
> dat_ret=merge(IBM_ret,GE_ret,YHOO_ret,SP500_ret)
> names(dat_ret)<-c('IBM','GE','YHOO','SP500')
> head(dat_ret)
```

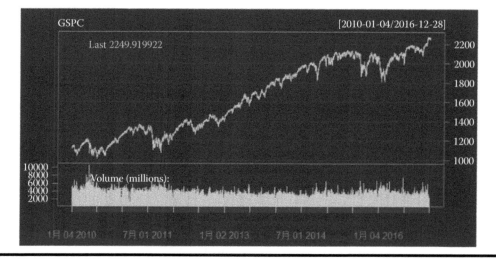

Figure 2.9   Candlestick chart of S&P 500 index.

```
                  IBM           GE          YHOO        SP500
2010-01-04   0.009681385   0.015111695   0.009445041 0.0147147759
2010-01-05  -0.012079963   0.005177994   0.007602339 0.0031156762
2010-01-06  -0.006496033  -0.005151320  -0.003482298 0.0005455205
2010-01-07  -0.003461515   0.051779935  -0.027373267 0.0040012012
2010-01-08   0.010034759   0.021538462   0.000000000 0.0028817272
2010-01-11  -0.010470080   0.009638554   0.002395150 0.0017467554
```

Define the risk-free return as 4% and calculate the average annualized returns of these four assets

```
# Risk-free return
> Rf<-0.04/12

# Calculate the average annualized returns, the average annualized
standard deviations and the average annualized Sharpes
> results<-table.AnnualizedReturns(dat_ret,Rf=Rf)
> results
                            IBM      GE    YHOO    SP500
Annualized Return        0.0345  0.1108  0.1257  0.1055
Annualized Std Dev       0.1918  0.2180  0.3043  0.1555
Annualized Sharpe (Rf=84%) -2.8892 -2.3899 -1.6911 -3.3659
```

We can see that during this period of time, S&P 500 Index gains an average annualized return of 10%, which is quite good. It will be difficult to surpass it.

Let's run the statistical index analysis and view the data from another dimension.

```
# Calculate the statistical indexes
> table.Stats(dat_ret)
                       IBM        GE       YHOO       SP500
Observations      1760.0000 1760.0000 1760.0000 1760.0000
NAs                  0.0000    0.0000    0.0000    0.0000
Minimum             -0.0828   -0.0654   -0.0871   -0.0666
Quartile 1          -0.0060   -0.0065   -0.0098   -0.0039
Median               0.0002    0.0004    0.0005    0.0005
Arithmetic Mean      0.0002    0.0005    0.0007    0.0004
Geometric Mean       0.0001    0.0004    0.0005    0.0004
Quartile 3           0.0067    0.0077    0.0112    0.0053
Maximum              0.0567    0.1080    0.1034    0.0474
SE Mean              0.0003    0.0003    0.0005    0.0002
LCL Mean (0.95)     -0.0004   -0.0001   -0.0002    0.0000
UCL Mean (0.95)      0.0008    0.0012    0.0015    0.0009
Variance             0.0001    0.0002    0.0004    0.0001
Stdev                0.0121    0.0137    0.0192    0.0098
Skewness            -0.5876    0.3084    0.0959   -0.3514
Kurtosis             4.6634    4.7294    2.9990    4.0151
```

According to statistical index analysis, every asset has 1760 sample points and no NA value. The minimum daily return of YHOO is −0.0871. The maximum daily return of GE is 0.1080. YHOO has the biggest arithmetic mean, geometric mean, variance and standard deviation.

Draw the plots of the daily returns of IBM (Figure 2.10), the monthly returns of IBM (Figure 2.11) and the cumulative returns of the four assets (Figure 2.12), and then run correlation analysis for these four assets.

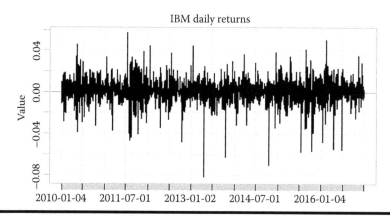

Figure 2.10   IBM daily returns.

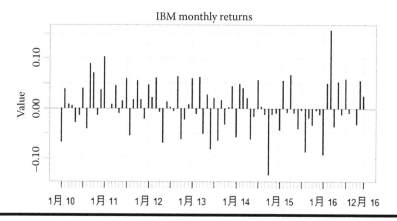

Figure 2.11   IBM monthly returns.

Figure 2.12   Cumulative returns of the four assets.

According to Figure 2.12, the GE tendency is basically in line with S&P 500 Index, which means that GE has sustainably developed by following the American economy since 2010. YHOO rose dramatically from early 2013 to early 2015, far ahead of S&P 500 Index, which means the Internet industry where YHOO belongs has brought huge market dividend during the period; but it plummeted from 2015 to 2016. Its rises and falls were significantly influenced by the market. For IBM, it's below S&P 500 Index for most of the time, which indicates that the rapid development of American economy in these years has not brought about much developing space for IBM. With the purpose of better return than S&P 500 Index, we have to choose GE or YHOO.

2.1.4.1 Correlation Analysis

After the correlation analysis for these four assets, we find out that the correlation coefficient of GE and S&P 500 Index is 0.78, which means GE is the most related stock to S&P 500 Index among the other three. And YHOO is the most uncorrelated among the others (Figure 2.13).

Finally, take S&P 500 Index as the market portfolio and respectively calculate the alphas and betas of the other three stocks.

```
# Calculate the alphas
> CAPM.alpha(dat_ret[,1:3],dat_ret[,4],Rf=Rf)

                      IBM           GE           YHOO
Alpha: SP500 -0.000752943 0.0003502332 0.0003944279

# Calculate the betas
> CAPM.beta(dat_ret[,1:3],dat_ret[,4],Rf=Rf)

                 IBM       GE       YHOO
Beta: SP500 0.8218135 1.098877 1.064844
```

IBM has the smallest alpha of the other three stocks, a minus alpha, which means IBM is behind the market and it's better to buy S&P 500 Index than IBM. The beta of GE is the largest. The larger the beta on its rising, the better the market return we will get. YHOO has an alpha and a

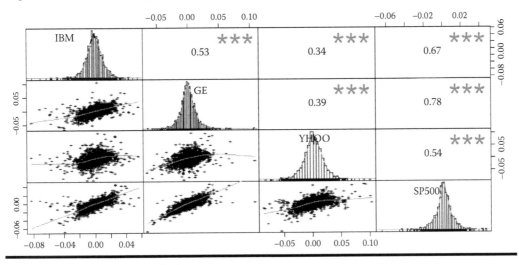

Figure 2.13   Correlation analysis.

beta close to GE, but its standard deviation and maximum drawdown, etc. are too large, which indicates big fluctuations.

In all, if we allocate our capital partly in GE and partly in YHOO, we will get better return than S&P 500 Index. However, the betas of both GE and YHOO are higher than S&P 500 Index, so their risks will be higher than S&P 500 Index as well. We need to add new stocks to diversify the risks. Details of quantitative analysis will be introduced in the following articles.

2.1.5 Beta versus Alpha

At the end of this chapter, we will supplement some knowledge about Alpha and Beta. The concepts of Alpha and Beta originated from the financial market to solve the problem of portfolio return decomposition.

- Alpha: Usually refers to the excess return of a portfolio, i.e. the capability of the manager.
- Beta: Refers to the market risks. Initially, it mostly referred to the systematic risks of the stock market.

Alpha is the difference of the average actual return and the average expected return.

- If $\alpha > 0$, it means the price of the fund or stock may be underestimated, so buy-in is recommended.
- If $\alpha < 0$, it means the price of the fund or stock may be overestimated, so short selling is recommended.
- If $\alpha = 0$, it means the price of the fund or stock reflects accurately its intrinsic value, not overestimated or underestimated.

Beta reflects the co-movement of a single security and the market portfolio.

- If $\beta > 1$, it means the security is aggressive and will rise drastically when the market goes up.
- If $\beta < 1$, it means the security is defensive and will drop slightly when the market falls.
- If $\beta = 1$, it means the security is neutral and will go in line with the market.

It's been long since the CAPM started. The financial theory is developing all the time. Another important theory breakthrough after the CAPM is the arbitrage pricing theory (APT), which I will introduce in my blog in the future.

Detailed introductions to the CAPM financial theory, its deducing process and its realization with R language have been given in this section. They are explicated in my own understanding. Hope these will give some help and instructions to those beginner friends of quantification. I also hope to find people like me who cross boundary from IT to finance. Let's use our thinking pattern of IT + finance and keep fighting in the financial market.

2.2 R Interpretation of Simple Linear Regression Model

Question
How do you run regression analysis with R language?

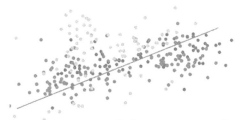

R Interpretation of
Simple Linear Regression Model
http://blog.fens.me/r-linear-regression

Introduction

In our daily lives, there are many correlated events. Take atmospheric pressure and altitude for example. The higher the altitude, the lower the atmospheric pressure. Take human heights and weights for another example. Usually, taller people weigh more. There may be other correlated events as well, like the higher the intellectual level, the higher their income; the better the economy of a marketized country, the stronger its currency; or if there is a global economic crisis, the safe-haven assets like gold, etc. will go stronger.

If we are going to study these events and find out the relation of different variables, we need to use regression analysis. Simple linear regression analysis is the simplest model to deal with the relation of two variables. Its variables are linearly correlated. Let's discover the rules in life.

2.2.1 An Introduction to Simple Linear Regression

Regression Analysis is a statistical analysis technique used to define the relationship of two or more than two variables. If the regression analysis concerns only one independent variable X and one dependent variable Y, which show a linear relation, it is called simple linear regression analysis.

Regression analysis is one of the basic statistical models, which involves the fundamentals of statistics. It is required to master plenty of terms and knowledge to understand it.

There are two types of variables in regression analysis: dependent variables and independent variables. A dependent variable, represented with Y, usually refers to the target concerned in practical problems. An independent variable, represented with X, is a variable that influences the value of the dependent variable. If there are more than one independent variables, $X1, X2, ..., Xn$ will be used for representations.

The followings are the main steps of a regression analysis research:

1. To determine the quantitative expression of the relation between the dependent variable Y and the independent variables $X1, X2, ..., Xn$, i.e. the regression equation.
2. To inspect the confidence coefficient of the regression equation.
3. To assess the impact of the independent variables Xn ($n = 1, 2, ..., m$) on the dependent variable.
4. To predict the outcome with the regression equation.

Next we will, according to the main steps of regression analysis (see Figure 2.14), sort out the structure and introduce the usage of the simple linear regression model.

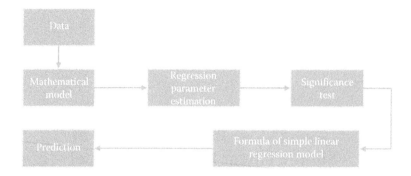

Figure 2.14 Main steps of regression analysis

The system environment used here:

■ Win10 64bit
■ R: 3.2.3 x86_64-w64-mingw32/x64 b4bit

2.2.2 *Dataset and Mathematical Model*

Let's start with an example. A simple set of data will be used to illustrate the mathematical model principle and formula of simple linear regression analysis. Please find the quantitative relation of *Y* and *X* in the following dataset.

The dataset is the trading data of the day opening on March 1, 2016. It's the price data of two zinc future contracts by minutes, which is saved to a file zn.csv. The dataset includes three columns, with time as Index, zn1.Close the price data of ZN1604 by minutes and zn2.Close the price data of ZN1605 by minutes.

Open the zn.csv file to get the dataset as follow:

```
"Index","zn1.Close","zn2.Close"
2016-03-01 09:01:00,14075,14145
2016-03-01 09:02:00,14095,14160
2016-03-01 09:03:00,14095,14160
2016-03-01 09:04:00,14095,14165
2016-03-01 09:05:00,14120,14190
2016-03-01 09:06:00,14115,14180
2016-03-01 09:07:00,14110,14170
2016-03-01 09:08:00,14110,14175
2016-03-01 09:09:00,14105,14170
2016-03-01 09:10:00,14105,14170
2016-03-01 09:11:00,14120,14180
2016-03-01 09:12:00,14105,14170
2016-03-01 09:13:00,14105,14170
2016-03-01 09:14:00,14110,14175
2016-03-01 09:15:00,14105,14175
2016-03-01 09:16:00,14120,14185
2016-03-01 09:17:00,14125,14190
2016-03-01 09:18:00,14115,14185
2016-03-01 09:19:00,14135,14195
2016-03-01 09:20:00,14125,14190
```

```
2016-03-01 09:21:00,14135,14205
2016-03-01 09:22:00,14140,14210
2016-03-01 09:23:00,14140,14200
2016-03-01 09:24:00,14135,14205
2016-03-01 09:25:00,14140,14205
2016-03-01 09:26:00,14135,14205
2016-03-01 09:27:00,14130,14205
2016-03-01 09:28:00,14115,14180

# Omitted
```

We take zn1.Close as X and zn2.Close as Y. Let's try to find the relation expression of the independent variable X and the dependent variable Y.

To be intuitive, we can first draw a scatter plot, with X as abscissa and Y ordinate. Every point corresponds to an X and a Y (Figure 2.15).

```
# Load the class library
> library(zoo)
> library(xts)

# Set the environment variables
> options(stringsAsFactors = FALSE)

# Read data
> dat<-read.csv(file="zn.csv",sep=",")

# Convert to xts format
> df<-xts(dat[,-1],order.by=as.POSIXct(dat[,1]))

# Dataset already exists in the df variables
> head(df)
                    zn1.Close zn2.Close
2016-03-01 09:01:00     14075     14145
2016-03-01 09:02:00     14095     14160
2016-03-01 09:03:00     14095     14160
2016-03-01 09:04:00     14095     14165
2016-03-01 09:05:00     14120     14190
2016-03-01 09:06:00     14115     14180

# Assign values to x and y respectively
> x<-as.numeric(df[,1])
> y<-as.numeric(df[,2])
```

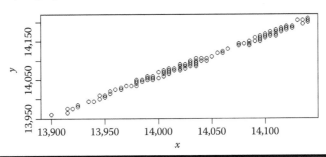

Figure 2.15   Scatter plot of zn1.Close and zn2.Close.

```
# Draw the plot
> plot(y~x+1)
```

From the scatter plot, we find that X and Y almost align around a straight line, so we can assume that there is a linear relation between X and Y, with the formula as:

```
Y = a + b * X + c
```

- ▪ Y, the dependent variable
- ▪ X, the independent variable
- ▪ a, the Y-intercept
- ▪ b, the independent variable coefficient
- ▪ $a + b * X$, the part of Y that changes linearly as X varies
- ▪ c, the residual or random error, which is the sum of the impact from all the uncertain factors. Its value is unobservable. Suppose c accords with normal distribution, with the mean value as 0 and the variance as σ^2. It is recorded as $c \sim N(0,\sigma^2)$

For the formula above, the function $f(X) = a + b * X$ is called a simple linear regression function, where a is the regression constant, b is the regression coefficient and both a and b are called the regression parameters. X is the independent variable of regression or the regressor, while Y is the dependent variable of regression or the response variable. If $(X1,Y1)$, $(X2,Y2)$, ..., (Xn,Yn) is a group of observed values for (X,Y), the simple linear regression model is:

```
Yi = a + b * X + ci,      i= 1,2,...n
Where E(ci)=0, var(ci)=σ^2, i=1,2,...n
```

Now a mathematical definition of simple linear regression model is given and next we will estimate the parameters of the regression model by using the dataset.

2.2.3 Regression Parameter Estimation

The regression parameters a and b of the formula above are unknown for us, so we need to use parameter estimation to figure out the values of a and b, so that we can acquire the quantitative relation of X and Y in the dataset. Our goal is to work out a straight line and minimize the sum of squares of difference between the value of Y for each point in the line and its actual value of Y, that is, minimize the value of $(Y1\ Actual - Y1\ Predicted)^2 + (Y2\ Actual - Y2\ Predicted)^2 + \cdots + (Yn\ Actual - Yn\ Predicted)^2$. To estimate the parameters, we merely take into account the part of Y that linearly changes as X varies. The residuals are unobservable and will not be used in the parameter estimation. The analysis of residuals will be introduced later.

Deform the formula into an a and b function $Q(a,b)$, i.e. deform the sum of squares $(Y\ Actual - Y\ Predicted)$ into the sum of squares $(Y\ Actual - (a + b * X))$.

$$Q(a,b) = \sum_{i-1}^{n} \left(Yi - (aXi + b) \right)^2$$

Formula 2.1 Deformation Formula of Regression Parameters

Deduce the solution formula of a and b by using the least square estimation.

$$a = \bar{y} - b\bar{x}, \quad b = \frac{\sum_{i=1}^{n}(x_i - \bar{x})(y_i - \bar{y})}{\sum_{i=1}^{n}(x_i - \bar{x})^2}$$

Formula 2.2 Formula of Regression Parameters
where the mean values of *x* and *y* can be calculated as follow:

$$\bar{x} = \frac{1}{n}\sum_{i=1}^{n}x_i \quad \bar{y} = \frac{1}{n}\sum_{i=1}^{n}y_i$$

Formula 2.3 Mean Value Formula
We can solve the two regression parameters *a* and *b* by using this formula.

Next, we will use R language to implement the regression model parameter estimation for the dataset above. lm() function of R language can be used to perform the modeling process of simple linear regression (Figure 2.16).

```
# Build the linear regression model
> lm.ab<-lm(y~1+x)

# Print the parameter estimation result
> lm.ab

Call:
lm(formula = y~1 + x)

Coefficients:
(Intercept)              x
   -349.493          1.029
```

If you want to calculate manually, you can perform the formula by yourself.

```
# Mean value of x
> Xm<-mean(x);Xm
[1] 14034.82

# Mean value of y
> Ym<-mean(y);Ym
[1] 14096.76

# Calculate the regression coefficient
> b <- sum((x-Xm)*(y-Ym)) / sum((x-Xm)^2) ;b
[1] 1.029315

# Calculate the regression constant
> a <- Ym - b * Xm;a
[1] -349.493
```

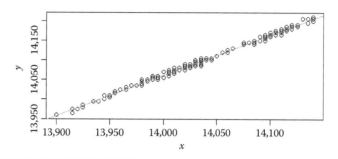

Figure 2.16 Linear regression.

The results of the regression parameters a and b from manual calculation are the same as those from lm() function. With the values of *a* and *b*, we can draw a proximate direct line.

Here is the calculation formula:

```
Y= a + b * X = -349.493 + 1.029315 * X
```

Draw the regression line.

```
> plot(y~x+1)
> abline(lm.ab)
```

This is a straight line fitted for the data. They are approximate values. We can see that some points are in the line while some are not. To evaluate how the regression line fits, we need to take a significance test for the regression model.

2.2.4 Significance Test of Regression Equation

According to Formula 2.2 Formula of Regression Parameters, we don't have to know if Y and X are linearly correlated in the calculating process. If there is no correlation, the regression equation will be meaningless; if Y and X are correlated, i.e. Y changes linearly as X varies, the simple linear regression will matter. Therefore, we need to verify the correlation validity with hypothesis testing.

Usually, three significance test methods will be used.

- T test: It is used to test the significance of a certain independent variable Xi of the model in Y. P-value is usually used to evaluate the significance. If its P-value is smaller than 0.01, it shows that this independent variable Xi is significantly correlated with Y.
- F test: It is used to test the linear significance of all the independent variables in Y. it is also evaluated with P-value. If its P-value is smaller than 0.01, it means that overall, the independent variables X and Y show a significant correlation.
- R^2 (R-squared) correlation coefficient test: It is used to assess the fit degree of the regression equation. The R^2 value ranges from 0 to 1. The closer the value to 1, the higher the fit degree.

All these three test methods have been implemented with R language. All we need to do is to interpret the result. We've already built the simple linear regression model with lm() function in the above. Now we can extract the model calculating result of summary() function.

```
> summary(lm.ab)        # Calculate the result


Call:
lm(formula = y~1 + x)


Residuals:
Min            1Q     Median      3Q       Max
-11.9385   -2.2317   -0.1797   3.3546   10.2766


Coefficients:
Estimate Std. Error t value Pr(>|t|)
(Intercept) -3.495e+02   7.173e+01   -4.872 2.09e-06 ***
x            1.029e+00   5.111e-03 201.390  < 2e-16 ***
---
Signif. codes: 0 '***' 0.001 '**' 0.01 '*' 0.05 '.' 0.1 ' ' 1


Residual standard error: 4.232 on 223 degrees of freedom
Multiple R-squared: 0.9945,    Adjusted R-squared: 0.9945
F-statistic: 4.056e+04 on 1 and 223 DF,  p-value: < 2.2e-16
```

Model interpretation:

- Call presents the formula of regression model.
- Residuals list the minimal point, $Q1$ point, median point, $Q3$ point and the maximal point of the residuals.
- Coefficients represent the result of parameter estimation calculation.
- Estimate represents the parameter estimation column. The intercept line stands for the estimated value of constant parameter of a, while the x line represents the estimated value of b, the parameter of the independent variable x.
- Std. Error represents the standard deviation of the parameters, i.e. sd(a) and sd(b).
- t value is the value of T test.
- Pr(>|t|) means P-value, used to assess T test and match the significance marks.
- For significance marks, *** means extremely significant, ** highly significant, * significant, less significant, and no mark means not significant.
- Residual standard error stands for the residual standard deviation; the degree of freedom is $n - 2$.
- Multiple R-squared is for the correlation coefficient R^2 test. The closer its value to 1, the more significant.
- Adjusted R-squared is the adjusted coefficient of the correlation coefficient, used to solve the problem that the determination coefficient R^2 grows bigger with more independent variables of multiple regression.
- For F-statistic represents the statistics of F; the degree of freedom is $(1, n - 2)$; P-value is used to assess F test and match the significance marks.

By checking the model result data, we find out that the y-intercept and the independent variable x by T test are both extremely significant, the model independent variable by F test is extremely significant and the independent variable and the dependent variable are highly relevant by R^2 correlation coefficient test.

Finally, after the regression parameter test and the regression equation test, we conclude the simple linear regression equation as:

```
Y = -349.493 + 1.029315 * X
```

2.2.5 Residual Analysis and Outlier Detection

After the significance test of the regression model, the residual analysis (the difference of predicted values and actual values) still need to be run to verify the correctness of the model. The residuals should be distributed normally.

We can calculate the data residuals by ourselves and test if it complies with normal distribution.

```
# Residuals
> y.res<-residuals(lm.ab)

# Print the first six lines of data
> head(y.res)
          1         2         3         4         5         6
6.8888680 1.3025744 1.3025744 6.3025744 5.5697074 0.7162808

# Normal distribution test
> shapiro.test(y.res)

    Shapiro-Wilk normality test

data: y.res
W = 0.98987, p-value = 0.1164

# Draw the scatter plot of residuals
> plot(y.res)
```

See Figure 2.17 for the result.

Test the residuals with Shapiro-Wilk normality. W is close to 1 and *P*-value > 0.05, which means the dataset obeys normal distribution. For introduction to normal distribution, please refer to Section 1.4 Introduction to Frequently Used Continuous Distributions and Their R Implementations of *R for Programmers: Advanced Techniques*.

Meanwhile, we can use tools of R language to generate four types of plots for model diagnose, so that we can simplify our coding process for calculation.

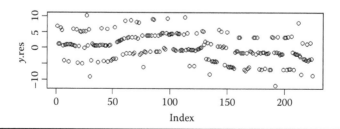

Figure 2.17   Scatter plot of residuals.

```
# Draw the plot and hit Enter to show the next one
> plot(lm.ab)
Hit  to see next plot: # Residual fitness plot
Hit  to see next plot: # Residual QQ plot
Hit  to see next plot: # Standardized residual pair fitted value
Hit  to see next plot: # Standardized residual pair leverage value
```

Draw the plot of residuals and fitted values, with the fitted values as abscissa and the residuals as ordinate. Data points are uniformly distributed on two sides of $y = 0$, showing a random distribution. The red line is a steady curve without any significant shape characteristic, which means the residual data performs well.

As shown in Figure 2.19, the residual QQ plot is used to describe whether the residuals are normally distributed. The data points in the figure are aligned along the diagonal, tending to a straight line and passed through by the diagonal, which intuitively obeys the normal distribution. The standardized residuals approximately comply with normal distribution, with 95% of the sample points located in the range of $[-2, 2]$.

Draw the plot of standardized residual square roots and fitted values (see Figure 2.20), with the fitted values as abscissa and the standardized residual square roots as ordinate. Similar to the assessment of residuals and fitted values comparison (see Figure 2.18), data is randomly distributed and the red line is a steady curve without any shape characteristic.

Draw the plot of standardized residuals and leverages (see Figure 2.21). The dotted line represents the contour of cook's distance which is usually used to measure the regression influential points. There is no red contour in the plot, which means there is no outlier in the data that significantly influences the regression result.

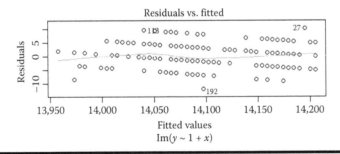

Figure 2.18   Residuals vs. fitted values.

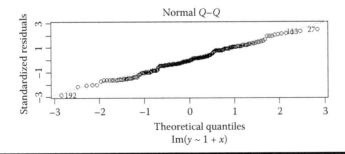

Figure 2.19   Residual QQ plot.

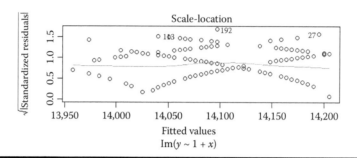

Figure 2.20   Standardized residual square roots vs. fitted values.

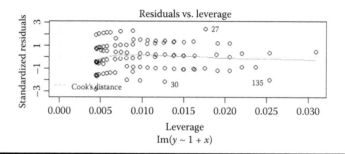

Figure 2.21   Standardized residuals vs. leverages.

We can change the layout and present these four plots together, as shown in Figure 2.22.

```
> par(mfrow=c(2,2))
> plot(lm.ab)
```

In every plot of Figure 2.22, there are some marked points, which may be the outliers. If the model needs optimization, we can start with these marked points. This residual analysis has a very good result, so it may not improve the model a lot to optimize the outliers.

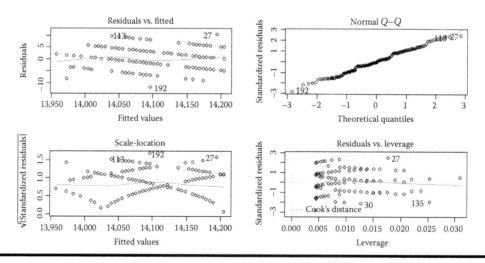

Figure 2.22   Combination of four plots.

We find out in Figure 2.22 that these two points (No. 27 and No. 192) appear in several plots. Suppose these two points are outliers. Let's remove these two points and run the significance test and residual analysis again.

```
# Check No.27 and No.192
> df[c(27,192),]
                   zn1.Close zn2.Close
2016-03-01 09:27:00    14130     14205
2016-03-01 14:27:00    14035     14085


# Rebuild the dataset without No.27 and No.192
> df2<-df[-c(27,192),]
```

Regression modeling and significance test.

```
> x2<-as.numeric(df2[,1])
> y2<-as.numeric(df2[,2])
> lm.ab2<-lm(y2~1+x2)
> summary(lm.ab2)


Call:
lm(formula = y2~1 + x2)


Residuals:
Min      1Q  Median      3Q     Max
-9.0356 -2.1542 -0.2727  3.3336  9.5879


Coefficients:
Estimate Std. Error t value Pr(>|t|)
(Intercept) -3.293e+02  7.024e+01  -4.688 4.83e-06 ***
x2           1.028e+00  5.004e-03 205.391  < 2e-16 ***
---
Signif. codes:
0 '***' 0.001 '**' 0.01 '*' 0.05 '.' 0.1 ' ' 1


Residual standard error: 4.117 on 221 degrees of freedom
Multiple R-squared: 0.9948,    Adjusted R-squared: 0.9948
F-statistic: 4.219e+04 on 1 and 221 DF,  p-value: < 2.2e-16
```

Comparing the results of significance test, T test, F test and R^2 test of this time with those of before, there is no significant improvement, just as I predicted. Therefore, after the residual analysis and outlier analysis, I think the model is valid.

2.2.6 Model Prediction

Finally, we've got the simple linear regression equation and now we can start the data prediction. For example, given $X = x0$, calculate the value of $y0 = a + b * x0$ and the prediction interval of confidence coefficient $1 - \alpha$.

So when $X = x0$ and $Y = y0$, the prediction interval of confidence coefficient $1 - \alpha$ is:

$$\left[\hat{y}_0 - l, \hat{y}_0 + l \right],$$

$$l = t_{\alpha/2} (n-2) \hat{\sigma} \sqrt{1 + \frac{1}{n} + \frac{(\bar{x} - x_0)^2}{S_{xx}}}.$$

Formula 2.4 Prediction Interval

$$P\left\{ \hat{y}_0 - l < y_0 < \hat{y}_0 + l \right\} = 1 - \alpha.$$

Formula 2.5 Prediction Interval Adjustment

We can use predict() function in R language to calculate the predicted value $y0$ and its prediction interval. The programmed algorithm is as follow:

```
> newX<-data.frame(x=14040)
> lm.pred<-predict(lm.ab,newX,interval="prediction",level=0.95)

# Predict the results
> lm.pred
       fit       lwr       upr
1 14102.09 14093.73 14110.44
```

So when $x0 = 14,040$, the confidence coefficient is 0.95, and the value of $y0$ is 14,102, the prediction interval will be [14,093.73, 14,110.44].

Let's present it with a plot.

```
> plot(y~x+1)
> abline(lm.ab,col='red')
> points(rep(newX$x,3),y=lm.pred,pch=19,col=c('red','blue','green'))
```

Where the red point is the value of $y0$, the blue point is the minimum value of the prediction interval and the green point is the maximum value of the prediction interval.

The core of using statistical model is the result interpretation. In this section, we have introduced the basic modeling process of simple regression and its detailed interpretation method. With grasp of the method, we can easily learn and understand the models of multiple regression, nonlinear regression, etc. and apply these models in the practical work. In the next section, we will introduce the R interpretation of multiple linear regression model (Figure 2.23).

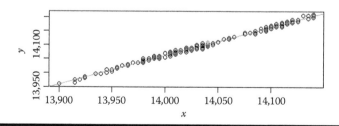

Figure 2.23   Regression prediction plot.

2.3 R Interpretation of Multiple Linear Regression Model

Question
How do you run multiple linear regression analysis with R language?

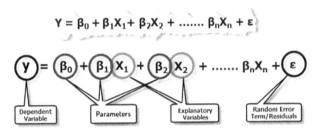

R Interpretation of
Multiple Linear Regression Model
http://blog.fens.me/r-multi-linear-regression

$$Y = \beta_0 + \beta_1 X_1 + \beta_2 X_2 + \dotsb \beta_n X_n + \varepsilon$$

Introduction
In practical problems of life and work, there can be more than one factor that will influence the dependent variable. For example, for the saying that people with higher intellectual level will get better pay, the factors that impact the better pay can be better family conditions, therefore better education, that the first-tier cities which the people work in provide with better job opportunities, and that the industries the people work for are in a big economy-rising cycle, etc. Multiple regression analysis is fit for interpreting these complicated and multi-dimensional life rules.

2.3.1 An Introduction to Multiple Linear Regression

Compared with simple linear regression, multiple linear regression is a statistical analysis technique used to define the relationship of two or more than two variables. The basic analysis method of multiple linear regression is similar to that of simple linear regression. We need to select a multivariate dataset, define its mathematical model, estimate the parameters, run the significance test for the estimated parameters, run the residual analysis, run the outlier detection, determine the regression equation and finally run the model prediction.

What's different from the simple linear regression equation is that the multiple regression equation includes multiple independent variables, so there is one very important operation, i.e. the independent variable optimization, i.e. to pick the independent variables with the most significant correlation and meanwhile eliminating those with the least. There are functions in R language for convenient optimization, which will be a great help for the model improvement. Let's begin the modeling of multiple linear regression.

The system environment used here:

- Win10 64bit
- R: 3.2.3 x86_64-w64-mingw32/x64 b4bit

2.3.2 Modeling of Multiple Linear Regression

Those who have studied commodity futures would know that black futures are related to the upstreams and downstreams of their industry chains. Iron ores are the raw materials for steel-making; coking coal and coke are the energy resources of steelmaking; HC (Hot Coils) are steel plates made by taking slabs as raw materials and getting them heated; deformed steel bars are bars with ribs.

Taking the industry chain into consideration, if we are going to predict the price of the deformed steel bar, we need to think about the factors that would influence its price, such as raw materials, energy resources, the materials alike, etc. For instance, intuitively, we would think, if the iron ore price is going up, the price of deformed steel bar will rise. Then if the iron ore price has increased by 1000 *yuan* per ton, how much would 1 ton of deformed steel bar will rise? It would be inaccurate to predict it with experience. We need quantitative calculation here.

2.3.2.1 Dataset and Mathematical Model

Let's start with data introduction. I select the black commodity futures for dataset, including JM(coking coal), J(coke) and I(iron ore) from Dalian Commodity Exchange and RU(deformed steel bar) and HC(hot coil) from Shanghai Futures Exchange.

The dataset is the trading data of the day opening on Tuesday, March 15, 2016. It's the price data of five black future contracts by minutes, which is saved to a file future.csv.

```
# Load the library
> library(xts)
> library(reshape2)
> library(ggplot2)

# Set the environment variables
> options(stringsAsFactors = FALSE)

# Read data
> dat<-read.csv(file="future.csv",sep=",")

# Convert to xts format
> df<-xts(dat[,-1],order.by=as.Date(dat[,1]))

# Dataset exists in the df variables
> head(df,20)

                       x1    x2    x3   x4    y
2016-03-15 09:01:00 754.5 616.5 426.5 2215 2055
2016-03-15 09:02:00 752.5 614.5 423.5 2206 2048
2016-03-15 09:03:00 753.0 614.0 423.0 2199 2044
2016-03-15 09:04:00 752.5 613.0 422.5 2197 2040
2016-03-15 09:05:00 753.0 615.5 424.0 2198 2043
2016-03-15 09:06:00 752.5 614.5 422.0 2195 2040
2016-03-15 09:07:00 752.0 614.0 421.5 2193 2036
2016-03-15 09:08:00 753.0 615.0 422.5 2197 2043
2016-03-15 09:09:00 754.0 615.5 422.5 2197 2041
2016-03-15 09:10:00 754.5 615.5 423.0 2200 2044
2016-03-15 09:11:00 757.0 616.5 423.0 2201 2045
2016-03-15 09:12:00 756.0 615.5 423.0 2200 2044
```

```
2016-03-15 09:13:00 755.5 615.0 423.0 2197 2042
2016-03-15 09:14:00 755.5 615.0 423.0 2196 2042
2016-03-15 09:15:00 756.0 616.0 423.5 2200 2045
2016-03-15 09:16:00 757.5 616.0 424.0 2205 2052
2016-03-15 09:17:00 758.5 618.0 424.0 2204 2051
2016-03-15 09:18:00 759.5 618.5 424.0 2205 2053
2016-03-15 09:19:00 759.5 617.5 424.5 2206 2053
2016-03-15 09:20:00 758.5 617.5 423.5 2201 2050
```

The dataset includes six columns:

- The index is time.
- $x1$ is the quoted price of J(j1605) contract by minutes.
- $x2$ is the quoted price of JM(jm1605) contract by minutes.
- $x3$ is the quoted price of I(i1605) contract by minutes.
- $x4$ is the quoted price of HC(hc1605) contract by minutes.
- y is the quoted price of RU(rb1605) contract by minutes.

Assume that the price of RU is linearly correlated to the price of the other four commodities. Then we build the multiple linear regression model with RU price as the dependent variable and the price of J, JM, I and HC as the independent variables. The formula is:

```
y = a + b * x1 + c * x2 + d * x3 + e * x4 + ε
```

- y, the dependent variable, stands for RU(Deformed Steel Bar).
- $x1$, the independent variable, stands for J(Coke).
- $x2$, the independent variable, stands for JM(Coking Coal).
- $x3$, the independent variable, stands for I(Iron Ore).
- $x4$, the independent variable, stands for HC(Hot Coil).
- a, the Y-intercept.
- b, c, d and e, the independent variable coefficients.
- ε, the residual or random error, which is the sum of the impact from all the uncertain factors. Its value is unobservable. Assume that ε complies with the normal distribution $N(0, \sigma^2)$.

Now a mathematical definition of multiple linear regression model is given and next we will estimate the parameters of the regression model by using the dataset.

2.3.2.2 Regression Parameters Estimation

We don't know the regression parameters of a, b, c, d and e in the formula above. Parameter estimation is used to estimate these parameters and then determine the relation of independent variables and the dependent variable. Our goal is to work out a straight line and minimize the sum of squares of difference between the value of Y for each point in the line and its actual value of Y, that is, minimize the value of ($Y1$ Actual − $Y1$ Predicted) ^2+($Y2$ Actual − $Y2$ Predicted) ^2+ ⋯ + (Yn Actual − Yn Predicted) ^2. To estimate the parameters, we merely take into account the part of Y that linearly changes as X varies. The residuals are unobservable and will not be used in the parameters estimation.

Similar to simple linear regression, we will use R language to implement the regression model parameter estimation for the dataset. lm() function can be used to perform the modeling process of the multiple linear regression.

```
# Build the multiple linear regression model
> lm1<-lm(y~x1+x2+x3+x4,data=df)

# Print the parameter estimation result
> lm1

Call:
lm(formula = y~x1 + x2 + x3 + x4, data = df)

Coefficients:
(Intercept)            x1            x2            x3            x4
   212.8780        0.8542        0.6672       -0.6674        0.4821
```

Then we've got the equation of *x* and *y* relation.

```
y = 212.8780 + 0.8542 * x1 + 0.6672 * x2 - 0.6674 * x3 + 0.4821 * x4
```

2.3.2.3 Significance Test of Regression Equation

Similar to simple linear regression, the significance test of multiple linear regression also includes *T* test, *F* test and R^2 (*R*-squared) correlation coefficient test. All these three test methods have been implemented with R language. All we need to do is to interpret the result and we can use summary() function to extract the model calculation result.

```
> summary(lm1)

Call:
lm(formula = y~x1 + x2 + x3 + x4, data = df)

Residuals:
Min       1Q  Median      3Q      Max
-4.9648 -1.3241 -0.0319  1.2403  5.4194

Coefficients:
Estimate Std. Error t value Pr(>|t|)
(Intercept) 212.87796     58.26788    3.653 0.000323 ***
x1            0.85423      0.10958    7.795 2.50e-13 ***
x2            0.66724      0.12938    5.157 5.57e-07 ***
x3           -0.66741      0.15421   -4.328 2.28e-05 ***
x4            0.48214      0.01959   24.609  < 2e-16 ***
---
Signif. codes: 0 '***' 0.001 '**' 0.01 '*' 0.05 '.' 0.1 ' ' 1

Residual standard error: 2.028 on 221 degrees of freedom
Multiple R-squared: 0.9725,    Adjusted R-squared: 0.972
F-statistic: 1956 on 4 and 221 DF,   p-value: < 2.2e-16
```

- *T* test: All independent variables are extremely significant***
- *F* test: Extremely significant as well, with *P*-value < 2.2e-16
- Adjusted R^2: The value is 0.972, very correlated.

Finally, after the regression parameter test and the regression equation test, we conclude the multiple linear regression equation as:

```
y = 212.87796 + 0.85423 * x1 + 0.66724 * x2 - 0.66741 * x3 + 0.48214 * x4
```

That is:

```
The price of deformed steel bar = 212.87796 + 0.85423 * the price of
coking coal + 0.66724 * the price of coke - 0.66741 * the price of iron
ore + 0.48214 * the price of hot coil
```

2.3.2.4 Residual Analysis and Outlier Detection

After the significance test of the regression model, the residual analysis (the difference of predicted values and actual values) still need to be run to verify the correctness of the model. The residuals should comply with the normal distribution $N(0, \sigma^2)$. For intuitive analysis, we can use plot() function to generate four types of plots for model diagnosis.

```
> par(mfrow=c(2,2))
> plot(lm1)
```

- Data points of residuals and fitted values (upper left) are uniformly distributed on two sides of $y = 0$, showing a random distribution. The red line is a steady curve without any significant shape characteristic.
- The data points of residual QQ plot (upper right) are aligned along the diagonal, tending to a straight line and passed through by the diagonal, which intuitively obeys the normal distribution.
- Data points of standardized residual square roots and fitted values (lower left) are uniformly distributed on two sides of $y = 0$, showing a random distribution. The red line is a steady curve without any significant shape characteristic.
- There is no red contour in the plot of standardized residuals and leverage values (lower right), which means there is no outlier in the data which significantly influences the regression result (Figure 2.24).

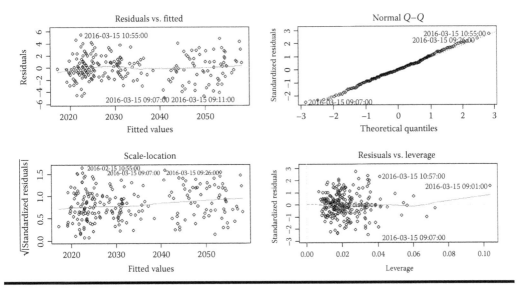

Figure 2.24   Residual tests for multiple regression.

In conclusion, there is no significant outlier, so the residuals match the assumptions.

2.3.2.5 Model Prediction

We've got the multiple linear regression equation and now we can start the data prediction. We can use predict() function in R language to calculate the predicted value $y0$ and its prediction interval and visualize them together.

```
> par(mfrow=c(1,1))   #Set up layout

# Prediction calculation
> dfp<-predict(lm1,interval="prediction")

# Print the predicted values
> head(dfp,10)
                fit       lwr       upr
2014-03-21 3160.526 3046.425 3274.626
2014-03-24 3193.253 3078.868 3307.637
2014-03-25 3240.389 3126.171 3354.607
2014-03-26 3228.565 3114.420 3342.710
2014-03-27 3222.528 3108.342 3336.713
2014-03-28 3262.399 3148.132 3376.666
2014-03-31 3291.996 3177.648 3406.344
2014-04-01 3305.870 3191.447 3420.294
2014-04-02 3275.370 3161.018 3389.723
2014-04-03 3297.358 3182.960 3411.755

# Merge data
> mdf<-merge(df$y,dfp)

# Draw the plot as shown in Figure 2.25
> draw(mdf)
```

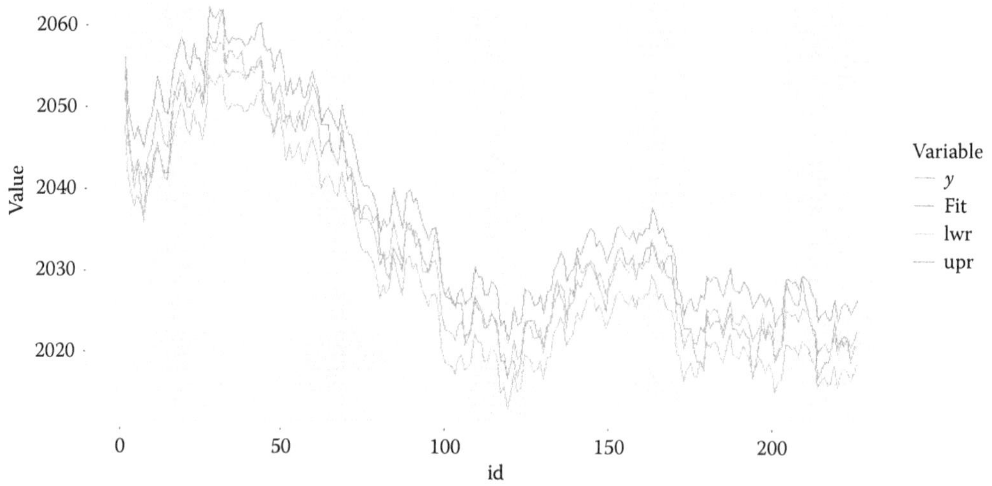

Figure 2.25   Model prediction analysis.

Plot explanation

- *y*, the actual price, represented with the red line.
- Fit, the predicted price, represented with the green line.
- lwr, the lowest predicted price, represented with the blue line.
- upr, the highest predicted price, represented with the purple line.

According to the plot, the actual price *y* and the predicted price fit are close for most of the time. Now we have already built a model.

2.3.3 Model Optimization

We've already find a good model in the above. If we are going to optimize the model, we can use update() function of R language for the model adjustment. We can first check the relation between each of the independent variables *x*1, *x*2, *x*3, *x*4 and the dependent variable *y*. See Figure 2.26 for the pair display plot.

```
# The relation of variables
> pairs(as.data.frame(df))
```

According to Figure 2.26, *x*2 and *y* show the least linear relation. So we can try to adjust the multiple linear regression model by removing the variable *x*2 from the original model.

```
# Model adjustment
> lm2<-update(lm1, .~. -x2)

> summary(lm2)

Call:
lm(formula = y~x1 + x3 + x4, data = df)
```

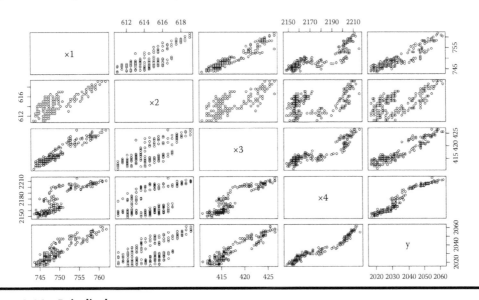

Figure 2.26   Pair display.

```
Residuals:
Min      1Q  Median     3Q      Max
-6.0039 -1.3842  0.0177  1.3513  4.8028

Coefficients:
Estimate Std. Error t value Pr(>|t|)
(Intercept) 462.47104    34.26636    13.50   < 2e-16 ***
x1            1.08728     0.10543    10.31   < 2e-16 ***
x3           -0.40788     0.15394    -2.65   0.00864 **
x4            0.42582     0.01718    24.79   < 2e-16 ***
---
Signif. codes:
0 '***' 0.001 '**' 0.01 '*' 0.05 '.' 0.1 ' ' 1

Residual standard error: 2.142 on 222 degrees of freedom
Multiple R-squared: 0.9692,    Adjusted R-squared: 0.9688
F-statistic: 2330 on 3 and 222 DF,  p-value: < 2.2e-16
```

After removing the independent variable $x2$, the T test value of independent variable $x3$ gets larger and meanwhile the Adjusted R-squared gets smaller, so there can be problems in this adjustment.

If we analyze the internal logic of the production and material supply, coking coal and coke are of upstream and downstream relationship. Coking coal is one of the raw materials to produce coke; charred coal and other coking coal are blended and made into coke. Usually, it requires 1.33 ton of coking coal, of which at least 30% is charred coal, to produce 1 ton of coke.

We can change the relation of coking coal and coke. Add the relation of $x1 * x2$ to the model and see what happens.

```
# Model adjustment
> lm3<-update(lm1, .~. + x1*x2)
> summary(lm3)

Call:
lm(formula = y~x1 + x2 + x3 + x4 + x1:x2, data = df)

Residuals:
Min      1Q  Median     3Q      Max
-4.8110 -1.3501 -0.0595  1.2019  5.3884

Coefficients:
Estimate Std. Error t value Pr(>|t|)
(Intercept) 7160.32231 7814.50048   0.916    0.361
x1           -8.45530   10.47167   -0.807    0.420
x2          -10.58406   12.65579   -0.836    0.404
x3           -0.64344    0.15662   -4.108 5.63e-05 ***
x4            0.48363    0.01967   24.584   < 2e-16 ***
x1:x2         0.01505    0.01693    0.889    0.375
---
Signif. codes:
0 '***' 0.001 '**' 0.01 '*' 0.05 '.' 0.1 ' ' 1

Residual standard error: 2.029 on 220 degrees of freedom
Multiple R-squared: 0.9726,    Adjusted R-squared: 0.972
F-statistic: 1563 on 5 and 220 DF,  p-value: < 2.2e-16
```

We can see from the result that after adding the column of $x1 * x2$, the T test values of the original $x1$, $x2$ and Intercept are not significant. Continue to adjust the model by removing two independent variables $x1$ and $x2$ from the model.

```
# Adjust the model
> lm4<-update(lm3, .~. -x1-x2)
> summary(lm4)

Call:
lm(formula = y~x3 + x4 + x1:x2, data = df)

Residuals:
Min      1Q  Median      3Q     Max
-4.9027 -1.2516 -0.0167  1.2748  5.8683

Coefficients:
Estimate Std. Error t value Pr(>|t|)
(Intercept)  6.950e+02  1.609e+01  43.183  < 2e-16 ***
x3          -6.284e-01  1.530e-01  -4.108 5.61e-05 ***
x4           4.959e-01  1.785e-02  27.783  < 2e-16 ***
x1:x2        1.133e-03  9.524e-05  11.897  < 2e-16 ***
---
Signif. codes: 0 '***' 0.001 '**' 0.01 '*' 0.05 '.' 0.1 ' ' 1

Residual standard error: 2.035 on 222 degrees of freedom
Multiple R-squared: 0.9722,    Adjusted R-squared: 0.9718
F-statistic: 2588 on 3 and 222 DF,  p-value: < 2.2e-16
```

The result of this adjustment seems good. However, it does not improve much, comparing to the original model.

If we adjust and test the model manually, we will usually do it according to our professional knowledge. If we calculate according to the data indicators, we can use the optimization method of stepwise regression in R language and make a judgment with AIC indicator on whether the parameter optimization is needed.

```
#Run stepwise regression on lm1 model
> step(lm1)
Start: AIC=324.51
y~x1 + x2 + x3 + x4

        Df Sum of Sq    RSS    AIC
              908.8 324.51
- x3     1    77.03  985.9 340.90
- x2     1   109.37 1018.2 348.19
- x1     1   249.90 1158.8 377.41
- x4     1  2490.56 3399.4 620.65

Call:
lm(formula = y~x1 + x2 + x3 + x4, data = df)

Coefficients:
(Intercept)          x1          x2          x3          x4
  212.8780      0.8542      0.6672     -0.6674      0.4821
```

By calculation, the smallest value of the lm1 model AIC indicator is 324.51. The AIC value grows larger every time an independent variable is removed, so we'd better keep this model without any adjustment.

Adjust the lm3 model with stepwise regression.

```
#Run stepwise regression on lm3 model
> step(lm3)
Start: AIC=325.7                    #Current AIC
y~x1 + x2 + x3 + x4 + x1:x2

          Df Sum of Sq     RSS    AIC
- x1:x2    1      3.25   908.8 324.51
                  905.6 325.70
- x3       1     69.47   975.1 340.41
- x4       1   2487.86  3393.5 622.25

Step: AIC=324.51                   #AIC with x1*x2 removed
y~x1 + x2 + x3 + x4

          Df Sum of Sq     RSS    AIC
                  908.8 324.51
- x3       1     77.03   985.9 340.90
- x2       1    109.37  1018.2 348.19
- x1       1    249.90  1158.8 377.41
- x4       1   2490.56  3399.4 620.65

Call:
lm(formula = y~x1 + x2 + x3 + x4, data = df)

Coefficients:
(Intercept)           x1           x2           x3           x4
   212.8780       0.8542       0.6672      -0.6674       0.4821
```

The value of AIC is the smallest if $X1 * X2$ is removed. The final result tells us that the original model is the best.

2.3.4 Case: Candlestick Chart Data Verification of Black Futures

At the end of this section, we are going to test the five future contracts with their candlestick chart data and find their multiple regression relation. Load the data of these five black futures, which are saved in different files. The correspondences of futures varieties and their file names are shown in Chart 2.1.

Next, we will read data and convert the data format with R language.

```
# Define the function to read files
> dailyData <- function(file) {
+    df <- read.table(file = file, header = FALSE,sep = ',', na.strings =
'NULL')
+    names(df)<-c('date','price')
+    return(df)
+ }

# Convert to xts format
> toXts<-function(data,format='%Y-%m-%d %H:%M:%S'){
```

Chart 2.1 Correspondences of Futures Varieties and File Names

Futures Varieties	File Name
J(j1605)	j_daily.csv
JM(jm1605)	jm_daily.csv
I(i1605)	i_daily.csv
HC(hc1605)	hc_daily.csv
RU(rb1605)	rb_daily.csv

```
+    df<-subset(data,select=-c(date))
+    xts(df,order.by=strptime(data$date, format))
+ }

# Read data respectively from files
> x1<-toXts(dailyData(file='j_daily.csv'),'%Y%m%d')
> x2<-toXts(dailyData(file='jm_daily.csv'),'%Y%m%d')
> x3<-toXts(dailyData(file='i_daily.csv'),'%Y%m%d')
> x4<-toXts(dailyData(file='hc_daily.csv'),'%Y%m%d')
> y<-toXts(dailyData(file='rb_daily.csv'),'%Y%m%d')

# Merge dataset
> df<-na.omit(merge(x1,x2,x3,x4,y))

# Rename the columns
> names(df)<-c('x1','x2','x3','x4','y')
```

Build the multiple regression analysis model for the candlestick chart data of the five varieties above.

```
# Multiple regression analysis modeling
> lm9<-lm(y~x1+x2+x3+x4,data=df)  # Candlestick chart data
> summary(lm9)

Call:
lm(formula = y~x1 + x2 + x3 + x4, data = df)

Residuals:
Min          1Q      Median      3Q        Max
-173.338   -37.470    3.465    32.158    178.982

Coefficients:
Estimate Std. Error t value Pr(>|t|)
(Intercept) 386.33482   31.07729   12.431  < 2e-16 ***
x1            0.75871    0.07554   10.045  < 2e-16 ***
x2           -0.62907    0.14715   -4.275 2.24e-05 ***
x3            1.16070    0.05224   22.219  < 2e-16 ***
x4            0.46461    0.02168   21.427  < 2e-16 ***
---
Signif. codes: 0 '***' 0.001 '**' 0.01 '*' 0.05 '.' 0.1 ' ' 1
```

```
Residual standard error: 57.78 on 565 degrees of freedom
Multiple R-squared: 0.9844,    Adjusted R-squared: 0.9843
F-statistic: 8906 on 4 and 565 DF,  p-value: < 2.2e-16
```

Check the basic statistical information of the dataset.

```
> summary(df)
     Index                         x1              x2
Min.   :2014-03-21 00:00:00  Min.   : 606.5  Min.   :494.0
1st Qu.:2014-10-21 06:00:00  1st Qu.: 803.5  1st Qu.:613.1
Median :2015-05-20 12:00:00  Median : 939.0  Median :705.8
Mean   :2015-05-21 08:02:31  Mean   : 936.1  Mean   :695.3
3rd Qu.:2015-12-16 18:00:00  3rd Qu.:1075.0  3rd Qu.:773.0
Max.   :2016-07-25 00:00:00  Max.   :1280.0  Max.   :898.0

       x3              x4              y
Min.   :284.0   Min.   :1691   Min.   :1626
1st Qu.:374.1   1st Qu.:2084   1st Qu.:2012
Median :434.0   Median :2503   Median :2378
Mean   :476.5   Mean   :2545   Mean   :2395
3rd Qu.:545.8   3rd Qu.:2916   3rd Qu.:2592
Max.   :825.0   Max.   :3480   Max.   :3414
```

Finally, generate the pair display plot of the candlestick chart data for these five future varieties. See Figure 2.27.

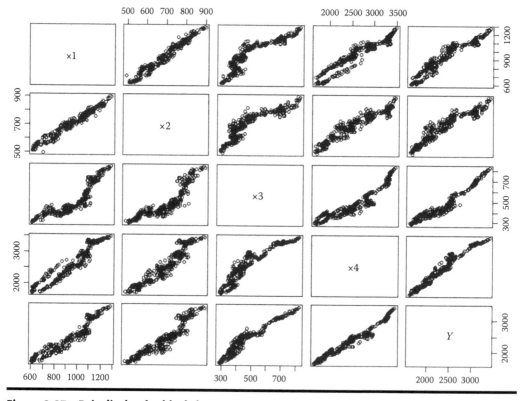

Figure 2.27   Pair display for black futures.

```
> pairs(as.data.frame(df))
```

The candlestick chart data of these five black futures shows very significant correlation. Therefore, we can apply this conclusion in the actual tradings.

With the statistical analysis techniques of multiple regression, we have introduced the basic application of multiple regression in the financial market in this section. We have found some complicated rules in life by building models of multiple independent variables and a dependent variable and we have also established effective indicators to test the model. Let's find more rules of the financial market with our technical advantages.

2.4 R Interpretation of Autoregression Model

Question
How do you analyze time series with R language?

AR(*P*)

R Interpretation of Autoregression Model
http://blog.fens.me/r-ar

Introduction
Time series is a common data format in financial analysis. Autoregression model is a basic method to analyze time series data. Finding periodic rules of the data by autoregression modeling would help us understand the changes in the financial market.

There are many frequently used models in time series analysis, including AR, MA, ARMA, ARIMA, ARCH and GARCH. The main difference of these models is their different application conditions. Moreover, these models are the outcomes of different progress stages and the models that came after have solved the inherent problems of the previous ones. In this section, we will start with AR model to introduce the whole system of time series analysis and then implement the model with R language.

2.4.1 An Introduction to Autoregression Model

Autoregressive model, AR model for short, is a statistical method to deal with time series. It is used to describe the relation between the current values and the historical values. A variable can be self-predicted by using its own historical data. Autoregression model should meet the requirement of stationarity. For example, assume that the historical data X_1 to X_{t-1} of a time series dataset X are linearly correlated. We can predict the performance of the current value X_t. The current value of X is equal to the sum of one or multiple backward linear combinations, the constant term and the random error.

The formula of the p-order autoregressive process

$$X_t = c + \sum_{i=1}^{p} \varphi_i X_{t-i} + \varepsilon_t$$

Formula 2.6 Autoregression Formula
  Field interpretation:

- X_t is the performance of the current value of X.
- c is the constant term.
- p is the order. i is a value from 1 to p.
- φ_i is the coefficient of autocorrelation.
- t is the time period.
- ε_t is the random error with mean as 0 and standard deviation as δ. δ is independent of t.

The first-order autoregression model can be represented with AR(1). The simplified formula is:

$$X_t = c + \varphi X_{t-1} + \varepsilon_t$$

Formula 2.7 First-order Autoregression Formula
  Autoregression is developed from the linear regression analysis. It analyzes the independent variable x itself instead of the impact of independent variable x on the dependent variable y. For knowledge about linear regression, please refer to Section 2.2.

2.4.1.1 Restrictions on the Autoregression Model

An autoregression model uses its own data for prediction, but there are some restrictions on it:

- It should be of stationarity, which requires that the random character should not change during the random process as time goes.
- It should be of autocorrelation. If the autocorrelation coefficient (φ_i) is smaller than 0.5, the model should not be adopted, or the prediction result will be very inaccurate.
- Autoregression only applies to the prediction of the phenomenon related to its prior stage, i.e. the phenomenon influenced much by its own historical factors. For the phenomenon impacted by other factors, the vector autoregression model instead of the autoregression should be used.

2.4.1.2 Stationarity of Time Series

If the random character of the random process where time series Y is produced does not change as time goes, we say that the process is stationary; if the random character of the random process changes over time, we say that the process is non-stationary.

Stationarity is the fitted curve generated by the sample time series. It will continue the current shape in future within a period of time. If the data is non-stationary, it means the shape of the sample fitted curve will not continued, i.e. the fitted curve shape does not fit the current curve.

- The mean and the variance of the random variable Yt are irrelevant to the time t.
- The covariance of random variables Yt and Ys is only related to the time difference (step length) $t – s$.
- For a stationary time series, there are abundant mathematical methods to deal with it; for a non-stationary time series, it will be transformed to a stationary time series by means of difference, etc. and then will get dealt with as a stationary time series.

2.4.2 Autoregression Modeling with R language

With understanding of the autoregression model definition, we can simulate the processes of autoregression modeling and calculation with R language.

The system environment used here:

- Win10 64bit
- R: 3.2.3 x86_64-w64-mingw32/x64 b4bit

First, let's generate a stationary random walk dataset.

```
# A random walk dataset
> set.seed(0)
> x<-w<-rnorm(1000)          # Generate data that obeys normal distribution
N(0,1)
> for(t in 2:1000) x[t]<-x[t-1]+w[t]
> tsx<-ts(x)                 # Generate the data of ts time series

# View the dataset
> head(tsx,15)
 [1] 1.2629543 0.9367209 2.2665202 3.5389495 3.9535909 2.4136409
 [7] 1.4850739 1.1903534 1.1845862 3.5892396 4.3528331 3.5538238
[13] 2.4061668 2.1167053 1.8174901

> plot(tsx)                  # generate a visulized plot
> a<-ar(tsx);a               # Autoregression modeling

Call:
ar(x = tsx)

Coefficients:
1
0.9879

Order selected 1  sigma^2 estimated as  1.168
```

The autocorrelation coefficient is 0.9879, so this is a very significantly autocorrelation and the series above complies with the autocorrelation character.

Ar() function of R language provides with various estimation methods of autocorrelation coefficient, including "yule-walker", "burg", "ols", "mle" and "yw". Yule-walker is the default method. The least square (ols) and the maximum likelihood (mle) are also frequently used (Figure 2.28).

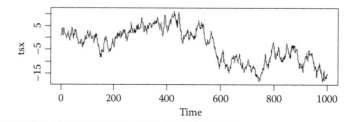

Figure 2.28   Random walk dataset.

Let's estimate the parameters by the least square method.

```
> b<-ar(tsx,method = "ols");b

Call:
ar(x = tsx, method = "ols")

Coefficients:
1
0.9911

Intercept: -0.017 (0.03149)

Order selected 1  sigma^2 estimated as  0.9906
```

The calculation result of the least square method is that the autocorrelation coefficient is 0.9911 and the intercept is −0.017. Only when the parameters are estimated by the least square method is there an intercept.

Let's estimate the parameters by the maximum likelihood method.

```
> d<-ar(tsx,method = "mle");d

Call:
ar(x = tsx, method = "mle")

Coefficients:
1
0.9904

Order selected 1  sigma^2 estimated as  0.9902
```

The calculation result of the maximum likelihood method is that the autocorrelation coefficient is 0.9904. The autocorrelation coefficients generated from the three above methods are very close.

2.4.3 ACF/PACF Model Identification

In the above case, the first-order autoregression model AR(1), by default, is implemented with the programs. However, in practical applications, the order p of a time series is unknown in the autoregression model AR. It should be determined by actual data. The autocorrelation function (ACF) and the partial autocorrelation function (PACF) are frequently used ways to determine the order of an autoregression model. If ACF/PACF cannot do, we would need the information criterion function like AIC (Aikaike info Criterion) and BIC (Bayesian information criterion) to determine the order.

The process of autoregression modeling includes the order determination, the parameter estimation and the order re-determination. The autocorrelation function ACF is used to assess whether the autoregression model fits. If the autocorrelation function has a tail, the AR model is an appropriate model. The partial autocorrelation function PACF is used to determine the order of a model. If the value of the partial autocorrelation function is close to 0 after a certain order, take this order as the order of the model. The partial autocorrelation function determines the order by its truncation.

2.4.3.1 Autocorrelation Function ACF

To compare an ordered sequence of random variables with itself is the statistical definition of autocorrelation function. Every sequence without any phase difference is similar to itself, i.e. the value of the autocorrelation function is the largest under this circumstance. If the components of the sequence are correlated (not random any more), the value of the correlation function below would not be zero and these components would be autocorrelated.

The autocorrelation function reflects the correlation degree of the values in a sequence in different times.

The formula of ACF:

$$P_k = \frac{\sum_{t=k+1}^{n} (Y_t - \mu)(Y_{t-k} - \mu)}{\sum_{t=1}^{n} (Y_t - \mu)^2}$$

Formula 2.8 Formula of Autocorrelation Function
Field interpretation:

■ P_k is the standard error of ACF.
■ t is the dataset length.
■ k is the lag, with its value range from 1 to $t-1$, meaning the correlation of sequence values in a time distance of k.
■ Y_t is the sample value in t time.
■ Y_{t-k} is the sample value in $t-k$ time.
■ M is the mean of the samples.

So the autocorrelation value P_k ranges [−1, 1], where 1 is its maximum positive correlation value, −1 is its maximum negative correlation value and 0 means irrelevant.

According to the formula above, we can manually calculate the ACF value of the tsx dataset.

```
> u<-mean(tsx)   #Mean
> v<-var(tsx)    #Variance

> # First-order lag
> p1<-sum(((x[1:length(tsx)-1]-u)*(x[2:length(tsx)]-u)))/
((length(tsx)-1)*v);p1
[1] 0.9878619
> # Second-order lag
```

```
> p2<-sum((x[1:(length(tsx)-2)]-u)*(x[3:length(tsx)]-u))/
((length(tsx)-1)*v);p2
[1] 0.9760271
> # Third-order lag
> p3<-sum((x[1:(length(tsx)-3)]-u)*(x[4:length(tsx)]-u))/
((length(tsx)-1)*v);p3
[1] 0.9635961
```

Meanwhile, we can calculate with acf() function of R language and print the first 30 lagging ACF values.

```
> acf(tsx)$acf
, , 1

             [,1]
 [1,]  1.0000000
 [2,]  0.9878619
 [3,]  0.9760271
 [4,]  0.9635961
 [5,]  0.9503371
 [6,]  0.9384022
 [7,]  0.9263075
 [8,]  0.9142540
 [9,]  0.9024862
[10,]  0.8914740
[11,]  0.8809663
[12,]  0.8711005
[13,]  0.8628609
[14,]  0.8544984
[15,]  0.8462270
[16,]  0.8384758
[17,]  0.8301834
[18,]  0.8229206
[19,]  0.8161523
[20,]  0.8081941
[21,]  0.8009467
[22,]  0.7942255
[23,]  0.7886249
[24,]  0.7838154
[25,]  0.7789733
[26,]  0.7749697
[27,]  0.7709313
[28,]  0.7662547
[29,]  0.7623381
[30,]  0.7604101
[31,]  0.7577333
```

Compare the first three results of the manual calculation with those of the function calculation and they are all the same. We can also visualize the results with R language.

```
> acf(tsx)
```

According to Figure 2.29, the data ACF is tailing and it is significantly autocorrelated. Next, we will determine the order of AR with the partial autocorrelation function.

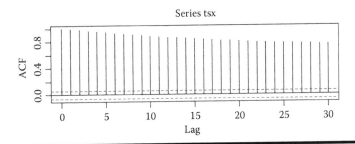

Figure 2.29   ACF test.

2.4.3.2 *Partial Autocorrelation Function (PACF)*

The partial autocorrelation function is derived from the autocorrelation function. Actually, what we get from solving the autocorrelation coefficient $p(k)$ of the lag k for a stationary AR(p) model is not a simple correlation between $x(t)$ and $x(t-k)$. $X(t)$ will be impacted by the $k-1$ variables, i.e. $x(t-1), x(t-2), \ldots, x(t-k+1)$, which is correlated with $x(t-k)$, so the autocorrelation coefficient $p(k)$ involves the influence of other variables on $x(t)$ and $x(t-k)$.

To simply measure the impact of $x(t-k)$ on $x(t)$, the concept of partial autocorrelation coefficient should be introduced. For a stationary time series $\{x(t)\}$, the partial autocorrelation coefficient of the lag k refers to the influence or correlation degree of $x(t-k)$ on $x(t)$ after giving the $k-1$ random variables of $x(t-1), x(t-2), \ldots, x(t-k+1)$ or eliminating the interference of the $k-1$ random variables of $x(t-1), x(t-2), \ldots, x(t-k+1)$.

In simple terms, the autocorrelation coefficient of ACF involves the impacts of other variables, while the partial autocorrelation coefficient of PACF reflects the correlation of just two variables. There are linear relations and nonlinear relations in ACF. The partial autocorrelation function is what eliminates the linear relations from the autocorrelation. When PACF is proximate to 0, it means the relation of the two timing values is caused completely by the linear relation.

Let's calculate the partial autocorrelation function with the pacf() function of R language.

```
> pacf(tsx)$acf
, , 1

            [,1]
 [1,]   0.987861891
 [2,]   0.006463542
 [3,]  -0.030541593
 [4,]  -0.041290415
 [5,]   0.047921168
 [6,]  -0.009774246
 [7,]  -0.006267004
 [8,]   0.002146693
 [9,]   0.028782423
[10,]   0.014785187
[11,]   0.019307564
[12,]   0.060879259
[13,]  -0.007254278
[14,]  -0.004139848
[15,]   0.015707900
[16,]  -0.018615370
[17,]   0.037067452
```

```
[18,]   0.019322565
[19,]  -0.048471479
[20,]   0.023388065
[21,]   0.027640953
[22,]   0.051177900
[23,]   0.028063875
[24,]  -0.003957142
[25,]   0.034030631
[26,]   0.004270416
[27,]  -0.029613088
[28,]   0.033715973
[29,]   0.092337583
[30,]  -0.031264028
```

```
# Visualize the output
> pacf(tsx)
```

According to the result analysis of Figure 2.30, when the lag value is 1, the AR model is significant; when the lag has a value other than 1, the PACF value is close to 0 and the model is not significant. Therefore, the data of dataset tsx satisfies the autoregression model of AR(1). The first-order autocorrelation coefficient value estimated in the above is usable.

2.4.4 Model Prediction

With the model identification, we haveonfirmed that the dataset tsx meets the modeling condition of AR(1); and we have already c built the AR(1) model. Next, we can predict with this backtesting model and then discover some rules and find value. We can use the predict() function of R language to run the prediction calculation.

Predict with the AR(1) model and preserve the first 5 predicted positions.

```
> predict(a,10,n.ahead=5L)
$pred
Time Series:
Start = 2
End = 6
Frequency = 1
[1] 9.839680 9.681307 9.524855 9.370303 9.217627

$se
Time Series:
Start = 2
```

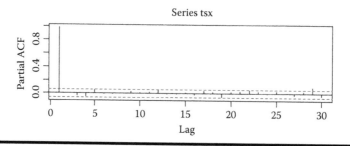

Figure 2.30   PACF test.

```
End = 6
Frequency = 1
[1] 1.080826 1.519271 1.849506 2.122810 2.359189
```

In the above result, the variable $pred means the predicted value and $se the error. We can generate a visualized plot for intuitive predicted results.

```
# Generate 50 predicted values
> tsp<-predict(a,n.ahead=50L)

# Draw a plot with the original data
> plot(tsx)

# Draw the predicted values and errors
> lines(tsp$pred,col='red')
> lines(tsp$pred+tsp$se,col='blue')
> lines(tsp$pred-tsp$se,col='blue')
```

In Figure 2.31, the black line is the original data, the red line on the right side is the predicted value and the blue line is the range of the predicted values. Now we have implemented the prediction calculation with AR(1) model.

The processes of prediction and visualization are completed with the native predict() function and the plot() function. The forecast package of R language can simplify the above processes with less codes and more convenient operations (Figure 2.32).

```
# Load the forecast package
> library('forecast')

# Generate the AR(1) model
> a2 <- arima(tsx, order=c(1,0,0))
> tsp2<-forecast(a2, h=50)
> plot(tsp2)
```

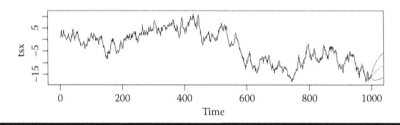

Figure 2.31   **Autoregression model prediction.**

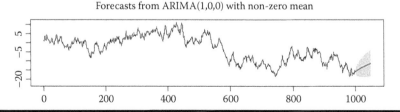

Figure 2.32   **Prediction and visualization with forecast package.**

Check the prediction result of the forecast() calculation.

```
> tsp2
     Point Forecast      Lo 80      Hi 80      Lo 95       Hi 95
1001      -15.71590 -16.99118 -14.440628 -17.66627 -13.7655369
1002      -15.60332 -17.39825 -13.808389 -18.34843 -12.8582092
1003      -15.49181 -17.67972 -13.303904 -18.83792 -12.1456966
1004      -15.38136 -17.89579 -12.866932 -19.22685 -11.5358726
1005      -15.27197 -18.06994 -12.474000 -19.55110 -10.9928432
1006      -15.16362 -18.21425 -12.112996 -19.82915 -10.4980922
1007      -15.05631 -18.33593 -11.776682 -20.07206 -10.0405541
1008      -14.95001 -18.43972 -11.460312 -20.28705  -9.6129750
1009      -14.84474 -18.52891 -11.160567 -20.47919  -9.2102846
1010      -14.74046 -18.60591 -10.875013 -20.65216  -8.8287673
1011      -14.63718 -18.67257 -10.601802 -20.80877  -8.4655994
1012      -14.53489 -18.73030 -10.339486 -20.95121  -8.1185723
1013      -14.43357 -18.78024 -10.086905 -21.08123  -7.7859174
1014      -14.33322 -18.82333  -9.843112 -21.20026  -7.4661903
1015      -14.23383 -18.86034  -9.607319 -21.30947  -7.1581923
1016      -14.13538 -18.89190  -9.378864 -21.40985  -6.8609139
```

The forecast() function directly generates the Forecast value, the predicted value ranges of 80% possibility and the predicted value ranges of 95% possibility.

We've already gone through the whole autoregression model processes of design pattern, modeling, test conditions, prediction calculation and visualized display, now we can apply the autoregression model in the actual business. To discover rules, to find value!

The autoregression model is just a beginning. For more, please subscribe my blog (http://fens.me). Only with thorough study of time series can we interpret the profound economy of China.

DATA PROCESSING AND HIGH PERFORMANCE COMPUTING OF R

II

Chapter 3

Data Processing of R

In this chapter which is cored with the R language data processing technology, we will introduce different ways to process various data with R language, including the standard structured dataset processing and the character dataset processing and meanwhile, we will introduce in simple terms how R language processes data, including the frequently used data processing operations like looping, grouping, merging, pipeline, word segmentation, etc.

3.1 Family of apply Functions

Question
How do you do a loop with R language?

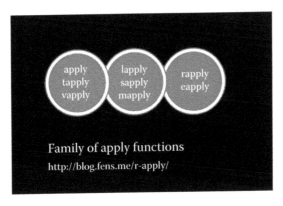

Introduction
When we begin to learn the R language, we may be exposed to different tips and tricks, the most important one of which would be that we should use the vector computing other than the cycle computing, for the cycle computing is inefficient.

Why is that? Because R's loop operations of *for* and *while* are both implemented based on R language itself, while the vector operation is implemented based on the underlying C functions, thus their performances will be significantly different. So how can we do efficient loop computing?

Here we need to introduce the family of apply functions, which includes apply, sapply, tapply, mapply, lapply, rapply, vapply and eapply.

3.1.1 Family of apply Functions

The family of apply functions is a group of core functions for data processing in R language. With this function family, we can implement the data operations of looping, grouping, filtering, type control, etc. However, the loop body processing pattern of the apply function family in R language is different from other languages, so the family is always not easy for its users.

Many beginners of R language would rather write *for* loops than learn the apply function family, so they write codes similar to C language. I have serious contempt for the programmers who can only write *for* loops.

The family of apply functions is created to deal with the data looping problems. It is formed to deal with different data types and different return values and it includes eight functionally similar functions. Some of them look similar, while some are not (Figure 3.1).

Usually, the most frequently used functions are apply, sapply and lapply. In the following, these eight functions will be introduced respectively.

The system environment used here:

- Win10 64bit
- R: 3.2.3 x86_64-w64-mingw32/x64 b4bit

3.1.2 apply Function

The apply function is the most frequently used function to replace the *for* loops. It can loop compute matrix, data frame and array (two-dimension and multi-dimension) by rows or columns, iterate the subelements, pass the subelements in the way of parameters to the self-defined FUN function and then return the computing result.

Function definition:

```
apply(X, MARGIN, FUN, ...)
```

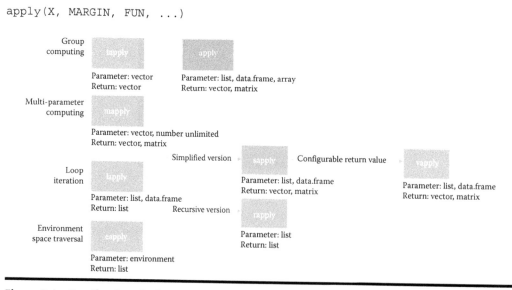

Figure 3.1   Family of apply functions.

Parameter list:

- *X*: array, matrix, data frame.
- MARGIN: Means calculating by rows or columns. 1 stands for calculating by rows and 2 by columns.
- FUN: The self-defined call function.
- ...: More parameters, optional.

For example, if we are going to sum every row of the matrix, the apply function will be used to do loops.

```
> x<-matrix(1:12,ncol=3); x
     [,1] [,2] [,3]
[1,]    1    5    9
[2,]    2    6   10
[3,]    3    7   11
[4,]    4    8   12

> apply(x,1,sum)  # Sum by rows
[1] 15 18 21 24
```

In the following, let's take an example with a bit complicated calculation. Do loops by rows, add 1 to column $x1$ of the data frame and calculate the mean of $x1$ and $x2$.

```
# Generate data.frame
> x <- cbind(x1 = 3, x2 = c(4:1, 2:5)); x
     x1 x2
[1,]  3  4
[2,]  3  3
[3,]  3  2
[4,]  3  1
[5,]  3  2
[6,]  3  3
[7,]  3  4
[8,]  3  5

# Self-define function myFUN, with the first parameter as data
# The second and third parameters are self-defined parameters, which
can be introduced by "..." of the apply function.
> myFUN<- function(x, c1, c2) {
+    c(sum(x[c1],1), mean(x[c2]))
+ }

# Loop the data frame by rows, pass every row separately to the function
myFUN and set c1 as the second parameter and c2 as the third
# Transpose the rows and columns of the result with function t()
> t(apply(x,1,myFUN,c1='x1',c2=c('x1','x2')))
     [,1] [,2]
[1,]    4  3.5
[2,]    4  3.0
[3,]    4  2.5
[4,]    4  2.0
```

```
[5,]    4    2.5
[6,]    4    3.0
[7,]    4    3.5
[8,]    4    4.0
```

A common loop computing has been implemented by the above self-defined function myFUN.
    If we use a *for* loop to achieve the above result, the codes will be:

```
# Define a data frame of the result
> df<-data.frame()

# Define the for loop
> for(i in 1:nrow(x)){
+     row<-x[i,]                                   # Values of
every row
+     df<-rbind(df,rbind(c(sum(row[1],1), mean(row))))    # Calculate and
assign the values to the result data frame
+ }

# Print the result data frame
> df
  V1  V2
1  4 3.5
2  4 3.0
3  4 2.5
4  4 2.0
5  4 2.5
6  4 3.0
7  4 3.5
8  4 4.0
```

It's easy to implement the calculation process above with a *for* loop, but there are some extra operations, such as building the loop bodies, defining the result dataset and merging every loop result to the result dataset.

    And there is another way to meet the demand above, i.e. to use R's characteristic and implement it with the vector computing.

```
> data.frame(x1=x[,1]+1,x2=rowMeans(x))
  x1   x2
1  4  3.5
2  4  3.0
3  4  2.5
4  4  2.0
5  4  2.5
6  4  3.0
7  4  3.5
8  4  4.0
```

Therefore, only one line of codes is enough to complete the calculation. Next, let's compare the performance consumptions of these three ways.

```
# Clear the environment variables
> rm(list=ls())
```

```
# Encapsulate fun1
> fun1<-function(x){
+    myFUN<- function(x, c1, c2) {
+       c(sum(x[c1],1), mean(x[c2]))
+    }
+    apply(x,1,myFUN,c1='x1',c2=c('x1','x2'))
+ }

# Encapsulate fun2
> fun2<-function(x){
+    df<-data.frame()
+    for(i in 1:nrow(x)){
+       row<-x[i,]
+       df<-rbind(df,rbind(c(sum(row[1],1), mean(row))))
+    }
+ }

# Encapsulate fun3
> fun3<-function(x){
+    data.frame(x1=x[,1]+1,x2=rowMeans(x))
+ }

# Generate the dataset
> x <- cbind(x1=3, x2 = c(400:1, 2:500))

# Count respectively the CPU time consumptions of the three ways
> system.time(fun1(x))
User System Elapsed
0.01 0.00 0.02

> system.time(fun2(x))
User System Elapsed
0.19 0.00 0.18

> system.time(fun3(x))
User System Elapsed
   0    0    0
```

As for the CPU time consumption, it takes the longest for loop to implement the calculation, short for the apply function and nearly no time consumption for the inherent vector computing of R language. After the test above, we conclude that for the same calculation, the inherent vector computing of R language should be a priority; if looping is a must, the apply function can be used; and try to avoid the direct use of the operations like *for, while*, etc.

3.1.3 lapply Function

lapply function is one of the most basic loop operation functions. It is used to loop the datasets of list and data.frame and to return the list structure with the same length as X to be the result set. The result set type can be recognized by "l," the first letter of lapply.

Function definition:

```
lapply(X, FUN, ...)
```

Parameter list:

- *X*: The data of list and data.frame.
- FUN: The self-defined call function.
- ...: More parameters, optional.

For example, we are going to calculate the quantile of the data corresponding to every KEY in list.

```
# Build a list dataset x, which includes three KEY values: a, b and c.

> x <- list(a = 1:10, b = rnorm(6,10,5), c = c(TRUE,FALSE,FALSE,TRUE));x
$a
[1]  1  2  3  4  5  6  7  8  9 10
$b
[1]   0.7585424 14.3662366 13.3772979 11.6658990  9.7011387 21.5321427
$c
[1]  TRUE FALSE FALSE  TRUE

# Calculate the quantile of the data corresponding to every KEY
> lapply(x,fivenum)
$a
[1]  1.0  3.0  5.5  8.0 10.0
$b
[1]  0.7585424  9.7011387 12.5215985 14.3662366 21.5321427
$c
[1] 0.0 0.0 0.5 1.0 1.0
```

Then the lapply function can easily loop the list dataset and it can also loop the dataset of data. frame by columns. However, if the incoming dataset is a vector object or a matrix object, simply using lapply would not achieve what we want.

For example, sum the matrix columns.

```
# Generate a matrix
> x <- cbind(x1=3, x2=c(2:1,4:5))
> x; class(x)
     x1 x2
[1,]  3  2
[2,]  3  1
[3,]  3  4
[4,]  3  5
[1] "matrix"

# Sum
> lapply(x, sum)
[[1]]
[1] 3
[[2]]
[1] 3
[[3]]
[1] 3
[[4]]
[1] 3
[[5]]
[1] 2
```

```
[[6]]
[1] 1
[[7]]
[1] 4
[[8]]
[1] 5
```

lapply would respectively loop every value of the matrix, rather than group compute it by rows or columns.

Sum the columns of the data frame.

```
> lapply(data.frame(x), sum)
$x1
[1] 12

$x2
[1] 12
```

3.1.4 sapply Function

The sapply function is a simplified lapply. Two parameters of simplify and USE.NAMES are added to sapply, in order to make its output more user-friendly. The return value is t a vector, not a list object.

Function definition:

```
sapply(X, FUN, ..., simplify=TRUE, USE.NAMES = TRUE)
```

Parameter list:

- *X*: array, matrix, data frame.
- FUN: The self-defined call function.
- ...: More parameters, optional.
- simplify: Whether to be arrayed. If the value is array, the output result will be grouped by array.
- USE.NAMES: If X is a string, TRUE means to set string as the data name and FALSE means not.

We will still illustrate it with the above computing requirements of the lapply.

```
> x <- cbind(x1=3, x2=c(2:1,4:5)); x
     x1 x2
[1,]  3  2
[2,]  3  1
[3,]  3  4
[4,]  3  5

# Calculate the matrix and the calculation process is the same as the
lapply function
> sapply(x, sum)
[1] 3 3 3 3 2 1 4 5

# Calculate the data frame
> sapply(data.frame(x), sum)
```

```
x1 x2
12 12

# Check the result type. The return type of sapply is vector and the
return type of lapply is list
> class(lapply(x, sum))
[1] "list"
> class(sapply(x, sum))
[1] "numeric"
```

If simplify=FALSE and USE.NAMES=FALSE, the sapply function would be totally equal to the lapply function.

```
> lapply(data.frame(x), sum)
$x1
[1] 12
$x2
[1] 12

> sapply(data.frame(x), sum, simplify=FALSE, USE.NAMES=FALSE)
$x1
[1] 12
$x2
[1] 12
```

When simplify is an array, we can refer to the following example and build a three-dimensional array, two dimensions of which are square matrices.

```
> a<-1:2

# Group by array
> sapply(a,function(x) matrix(x,2,2), simplify='array')
, , 1

     [,1] [,2]
[1,]   1    1
[2,]   1    1

, , 2

     [,1] [,2]
[1,]   2    2
[2,]   2    2

# By default, auto-merge and group
> sapply(a,function(x) matrix(x,2,2))
     [,1] [,2]
[1,]   1    2
[2,]   1    2
[3,]   1    2
[4,]   1    2
```

For a string vector, the data name can be auto-generated.

```
> val<-head(letters)
# Set the data name by default
> sapply(val,paste,USE.NAMES=TRUE)
  a   b   c   d   e   f
"a" "b" "c" "d" "e" "f"

# If USE.NAMES=FALSE, do not set the data name
> sapply(val,paste,USE.NAMES=FALSE)
[1] "a" "b" "c" "d" "e" "f"
```

3.1.5 vapply Function

vapply, similar to sapply, provides the parameter of FUN.VALUE to control the row names of the return values, which makes the program stronger.

Function definition:

```
vapply(X, FUN, FUN.VALUE, ..., USE.NAMES = TRUE)
```

Parameter list:

- *X*: array, matrix, data frame.
- FUN: The self-defined call function.
- FUN.VALUE: Define the row names (row.nemes) of the return values.
- ...: More parameters, optional.
- USE.NAMES: If X is a string, TRUE means to set string as the data name and FALSE means not.

For example, accumulatively sum the data of data frame and set the row.names of every row.

```
# Generate the dataset
> x <- data.frame(cbind(x1=3, x2=c(2:1,4:5)));
  x1 x2
1  3  2
2  3  1
3  3  4
4  3  5

# Set the row names respectively as a, b, c and d
> vapply(x,cumsum,FUN.VALUE=c('a'=0,'b'=0,'c'=0,'d'=0))
  x1 x2
a  3  2
b  6  3
c  9  7
d 12 12

# If no row name is set for it, it will be the index value by default
> a<-sapply(x,cumsum);a
     x1 x2
[1,]  3  2
[2,]  6  3
[3,]  9  7
[4,] 12 12
```

```
# Set the row names manually
> row.names(a)<-c('a','b','c','d')
> a
   x1  x2
a   3   2
b   6   3
c   9   7
d  12  12
```

The row names of the return values can be directly set with vapply, which actually saves one line of codes and makes the codes smoother. If you do not want to remember one more function, you can just ignore this. It will be enough to just use sapply.

3.1.6 mapply Function

mapply function is the deforming function of sapply. It is similar to the sapply function with multiple variables but different in the parameter definition. Its first parameter is the self-defined FUN function and the second one "..." can receive multiple data, which can be called as the parameter of function FUN.

Function definition:

```
mapply(FUN, ..., MoreArgs = NULL, SIMPLIFY = TRUE,USE.NAMES = TRUE)
```

Parameter list:

- FUN: The self-defined call function.
- ...: Receives multiple data.
- MoreArgs: Parameter list.
- SIMPLIFY: Whether to be arrayed. If the value is array, the output result will be grouped by array.
- USE.NAMES: If X is a string, TRUE means to set string as the data name and FALSE means not.

For example, compare the sizes of three vectors and take the largest value according to the index order.

```
> set.seed(1)

# Define three vectors
> x<-1:10
> y<-5:-4
> z<-round(runif(10,-5,5))

# Take the largest value according to the index order.
> mapply(max,x,y,z)
 [1]   5   4   3   4   5   6   7   8   9  10
```

Take another example. Generate four datasets which obey normal distribution, and their respectively means and variances are $c(1, 10, 100, 1000)$.

```
> set.seed(1)
```

```
# The length is 4
> n<-rep(4,4)

# m is mean and v is variance
> m<-v<-c(1,10,100,1000)

# Generate four groups of data, grouped by columns
> mapply(rnorm,n,m,v)
          [,1]       [,2]      [,3]        [,4]
[1,]  0.3735462 13.295078 157.57814   378.7594
[2,]  1.1836433  1.795316  69.46116 -1214.6999
[3,]  0.1643714 14.874291 251.17812  2124.9309
[4,]  2.5952808 17.383247 138.98432   955.0664
```

mapply can receive multiple parameters, so we don't have to merge data into data.frame when we are operating data and we can calculate the result by just one operation.

3.1.7 tapply Function

Tapply function is used to do loop computing for groups. Dataset X can be grouped by the INDEX parameter, functioned the same as the frequently used group by operation of SQL.

Function definition:

```
tapply(X, INDEX, FUN = NULL, ..., simplify = TRUE)
```

Parameter list:

- *X*: The vector.
- INDEX: Used as the index for grouping.
- FUN: The self-defined call function.
- ...: Receives multiple data.
- simplify: Whether to be arrayed. If the value is array, the output result will be grouped by array.

For example, calculate the mean of the petal length of different types of irises.

```
# Group by iris$Species
> tapply(iris$Petal.Length,iris$Species,mean)
    setosa versicolor  virginica
     1.462      4.260      5.552
```

Calculate the vectors x and y, group them with vector t as the index and sum them.

```
> set.seed(1)

# Define the vectors x and y
> x<-y<-1:10;x;y
 [1]  1  2  3  4  5  6  7  8  9 10
 [1]  1  2  3  4  5  6  7  8  9 10
# Set the grouping index t
> t<-round(runif(10,1,100)%%2);t
 [1] 1 2 2 1 1 2 1 0 1 1
```

```
# Group and sum x
> tapply(x,t,sum)
 0  1  2
 8 36 11
```

tapply receives only one vector parameter, so we can pass other parameters to the self-defined FUN function by "...". If we want to sum the vector *y*, we can pass *y* to tapply as the fourth parameter and then do the calculation.

```
> tapply(x,t,sum,y)
 0  1  2
63 91 66
```

The result does not meet our expectation. It did not sum *x* and *y* which are grouped by *t*; it came out with other result. When the fourth parameter is passed to the function, it is not passed to it one by one according to the loop. It passes through all the vector data every time. So for sum(*y*) = 55, when *t* = 0, *x* = 8 and *y* = 55, the final result should be 63. Therefore, if we are going to pass another parameter with "...", we should make clear the pass process and avoid the algorithm mistake.

3.1.8 rapply Function

rapply function is a recursive version of lapply function. It deals with the data of list type only. It will perform the recursive traversal for every element of list, including the subelements of list if there are some.

Function definition:

```
rapply(object, f, classes = "ANY", deflt = NULL, how = c("unlist",
"replace", "list"), ...)
```

Parameter list:

- object: The data of list.
- *f*: The self-defined call function.
- classes: The match type. ANY means all types.
- deflt: The default value for non-matching types.
- how: Three operations: if the value is replace, the result after calling *f* will replace the original element in the original list; if the value is list, a new list will be established, where the function *f* will be called if the type matches or the value will be deflt if the type is not matched; if the its value is unlist, the unlist(recursive = TRUE) operation will be performed once.
- ...: More parameters, optional.

For example, filter the data of list and then sort all the data of numeric type from the smallest to the largest.

```
> set.seed(1)
> x=list(a=12,b=1:4,c=c('b','a'))
> y=pi
> z=data.frame(a=rnorm(10),b=1:10)
> a <- list(x=x,y=y,z=z)
```

```
# Sort the data and replace the values of the original list
> rapply(a,sort, classes='numeric',how='replace')
$x
$x$a
[1] 12
$x$b
[1] 4 3 2 1
$x$c
[1] "b" "a"

$y
[1] 3.141593

$z
$z$a
 [1] -0.8356286 -0.8204684 -0.6264538 -0.3053884  0.1836433  0.3295078
 [7]  0.4874291  0.5757814  0.7383247  1.5952808
$z$b
 [1] 10  9  8  7  6  5  4  3  2  1
> class(a$z$b)
[1] "integer"
```

We can see in the result that only the data of za is sorted. The zb type is integer, not numeric, so it did not get sorted.

Next, let's process the string type data. Add a string of "++++" to all the string type data and set the non-string type data as NA.

```
> rapply(a,function(x) paste(x,'++++'),classes="character",deflt=NA, how
= "list")
$x
$x$a
[1] NA
$x$b
[1] NA
$x$c
[1] "b ++++" "a ++++"

$y
[1] NA

$z
$z$a
[1] NA
$z$b
[1] NA
```

Only xc is the string vector. Merge it into a new string. Therefore, the data of list type can be filtered conveniently with rapply.

3.1.9 eapply Function

Perform a traversal for all the variables in an environment space. If we have acquired a good habit of saving the self-defined variables into a self-defined environment space according to certain rules, this function will make your operations really convenient. Many of us may be unfamiliar with the operations of environment spaces, so please refer to Section 3.2 Uncovering the Mystery of R Environments and Section 3.3 Revealing the Function Environments in R of *R for Programmers: Advanced Techniques*.

Function definition:

```
eapply(env, FUN, ..., all.names = FALSE, USE.NAMES = TRUE)
```

Parameter list:

- env: The environment space.
- FUN: The self-defined call function.
- ...: More parameters, optional.
- all.names: The match type. ANY means all types.
- USE.NAMES: If X is a string, TRUE means to set string as the data name and FALSE means not.

In the following, we will define an environment space and loop the variables in it.

```
# Define an environment space
> env <- new.env(hash = FALSE)

# Save three variables to this environment space
> env$a <- 1:10
> env$beta <- exp(-3:3)
> env$logic <- c(TRUE, FALSE, FALSE, TRUE)
> env
<environment: 0x0000000005eccc00>

# View the variables in env
> ls(env)
[1] "a"      "beta"  "logic"

# View the string structure of variables in env
> ls.str(env)
a :   int [1:10] 1 2 3 4 5 6 7 8 9 10
beta :   num [1:7] 0.0498 0.1353 0.3679 1 2.7183 ...
logic :   logi [1:4] TRUE FALSE FALSE TRUE
```

Calculate the mean of all the variables in env.

```
> eapply(env, mean)
$logic
[1] 0.5
$beta
[1] 4.535125
$a
[1] 5.5
```

Then calculate the memory size that all the variables in the current environment space occupy.

```
# View the variables in the current environment space
> ls()
 [1] "a"      "df"      "env"     "x"      "y"      "z"      "X"

# Check the memory size that all variables occupy
> eapply(environment(), object.size)
```

```
$a
2056 bytes

$df
1576 bytes

$x
656 bytes

$y
48 bytes

$z
952 bytes

$X
1088 bytes

$env
56 bytes
```

The eapply function is not a very commonly used function, but the usage of the environment space should be grasped for the R package development. Especially when R is used as the industrialized tools, it is necessary to accurately control and manage the variables.

In this article, we have comprehensively introduced the family of apply functions in R language for the data loop processing, which should be enough to deal with almost all the loop situations. Meanwhile, when discussing the apply functions, we have compared the performance of three data processing methods. The inherent vector computing of R is better than the apply loop, and is much better than the *for* loop. Therefore, we should use the apply function more in our future development and application of R.

Leave behind the thinking pattern of a programmer, take in that of data and you may be suddenly enlightened.

3.2 Ultra-High Performance Data Processing Package: data.table

Question
How do you efficiently process data?

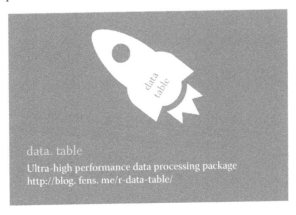

Introduction

The most frequently used data type in R language is data.frame. Most of the data processing operates in the structure of data.frame. With data.frame, it is convenient to store and query data, to loop compute data along with the family of apply functions and to implement data operations such as segmentation, grouping, merging, etc. by using packages like plyr, reshape2, melt, etc. It is really convenient when the data size is not big. However, it is not very efficient to process data with data.frame structure. Especially when the data size is slightly bigger, the process will slow down.

data.table provides functions similar to data.frame and has an extra setting of indexing, which makes its data processing pretty efficient, maybe 1 or 2 orders of magnitude faster than data. frame. In the following of this section, the usage of data.table package will be introduced.

3.2.1 An Introduction to data.table Package

data.table package is an extended toolkit of data.frame. It can set index by its self-defined keys and can efficiently implement the data operations like indexing and querying, quick grouping, quick connecting, quick valuation, etc. data.table significantly improves the efficiency of data operations mainly by the binary search method and it is also compatible with the vector retrieval method of data.frame. Meanwhile, data.table is also good for rapid aggregation of big data. According to the official introduction, data.table can efficiently process the memory data of 100 GB size. So let's test it.

data.table project address: https://cran.r-project.org/web/packages/data.table/

The system environment used here:

- Win10 64bit
- R: 3.2.3 x86_64-w64-mingw32/x64 b4bit

data.table package can be found in the standard library of CRAN (Comprehensive R Archive Network). Its installation is very simple, just two commands to go.

```
~ R
> install.packages("data.table")
> library(data.table)
```

3.2.2 Usage of data.table Package

Next, we will start using data.table package and get familiar with its basic operations.

1. Create a dataset with data.table

Usually, to create a dataset with data.frame, we can use the syntax below.

```
# Create a data frame with data.frame
> df<-data.frame(a=c('A','B','C','A','A','B'),b=rnorm(6))
> df
  a          b
1 A   1.3847248
2 B   0.6387315
3 C  -1.8126626
4 A  -0.0265709
5 A  -0.3292935
6 B  -1.0891958
```

For data.table, the syntaxes of creating a dataset and a data frame are the same.

```
# Create an object of data.table
> dt = data.table(a=c('A','B','C','A','A','B'),b=rnorm(6))
> dt
   a         b
1: A  0.09174236
2: B -0.84029180
3: C -0.08157873
4: A -0.39992084
5: A -1.66034154
6: B -0.33526447
```

In the types of df and dt objects, we can see that data.table is the extended type of data.frame.

```
# data.frame type
> class(df)
[1] "data.frame"
```

```
# data.table type
> class(dt)
[1] "data.table" "data.frame"
```

If data.table is just an S3 extended type of data.frame, data.table would not improve greatly in efficiency. To verify it, we need to check the structure definition of the data.table codes.

```
# Print the function definition of data.table
> data.table
function (..., keep.rownames = FALSE, check.names = FALSE, key = NULL)

{
    x <- list(...)
    if (!.R.listCopiesNamed)
        .Call(CcopyNamedInList, x)
    if (identical(x, list(NULL)) || identical(x, list(list())) ||
        identical(x, list(data.frame(NULL))) || identical(x,
        list(data.table(NULL))))
        return(null.data.table())
    tt <- as.list(substitute(list(...)))[-1L]

    # Code omitted

    setattr(value, "names", vnames)
    setattr(value, "row.names", .set_row_names(nr))
    setattr(value, "class", c("data.table", "data.frame"))
    if (!is.null(key)) {
        if (!is.character(key))
            stop("key argument of data.table() must be character")
        if (length(key) == 1L) {
            key = strsplit(key, split = ",")[[1L]]
        }
        setkeyv(value, key)
    }
```

```
    else {
        if (length(ckey) && !recycledkey && !any(duplicated(ckey)) &&
            all(ckey %in% names(value)) && !any(duplicated(names(value)
[names(value) %in%
            ckey])))
            setattr(value, "sorted", ckey)
    }
    alloc.col(value)
}
<bytecode: 0x0000000013c1edf0>
<environment: namespace:data.table>
```

According to the codes above, data.table does not use any dependent codes of the data.frame structure as its code definition and it processes data in its own function definitions, so we can confirm that the underlying structures of data.table and data.frame are different.

But why did we find that there was extension relation between data.table and data.frame just now when we checked the object of data.table with the class function? Here we need to know the S3 object-oriented system architecture design of R language. If you are not familiar with the S3 type, please refer to Section 4.2 The S3-Based OOP in R of *R for Programmers: Advanced Techniques*.

Find the line setattr(value, "class", c("data.table", "data.frame")) from the above codes, and we can see that the extension definition is initiatively set by the developer. That is, the developer of the data.table package hoped that data.table should be used as data.frame, so he/she packages it so to achieve zero switching costs for the users.

2. Switch between data.table and data.frame

If you want to switch between the data.frame objects and the data.table objects, the code is easy. You can switch them directly.

Switch the type of data.frame object to data.table object.

```
# Create an object of data.frame
> df<-data.frame(a=c('A','B','C','A','A','B'),b=rnorm(6))

# Check the type
> class(df)
[1] "data.frame"

# Switch the type to data.table object
> df2<-data.table(df)

# Check the type
> class(df2)
[1] "data.table" "data.frame"
```

Switch the type of data.table object to data.frame object.

```
# Create an object of data.table
> dt <- data.table(a=c('A','B','C','A','A','B'),b=rnorm(6))

# Check the type
> class(dt)
[1] "data.table" "data.frame"
```

```
# Switch to data.frame object
> dt2<-data.frame(dt)

# Check the type
> class(dt2)
[1] "data.frame"
```

3. Query with data.table

data.table are expected to be operated similarly to data.frame, so those query methods of data. frame are basically fit for data.table as well. Moreover, data.table has its own features and provides with some self-defined keys for high-efficient query.

Let's first have a look at the basic data query methods of data.table.

```
# Create an object of data.table
> dt = data.table(a=c('A','B','C','A','A','B'),b=rnorm(6))
> dt
     a         b
1:  A   0.7792728
2:  B   1.4870693
3:  C   0.9890549
4:  A  -0.2769280
5:  A  -1.3009561
6:  B   1.1076424
```

Query by rows or columns.

```
# Take the data of second row
> dt[2,]
     a         b
1:  B 1.487069

# It will work without commas
> dt[2]
     a         b
1:  B 1.487069

# Take the values of column a
> dt$a
[1]  "A"  "B"  "C"  "A"  "A"  "B"

# Take the rows whose values are B in column a
> dt[a=="B",]
     a         b
1:  B 1.487069
2:  B 1.107642

# Take the judgment if the row values of column a are equal to B
> dt[,a=='B']
[1] FALSE   TRUE FALSE FALSE FALSE   TRUE

# Take the index of the rows whose values are B in column a
> which(dt[,a=='B'])
[1] 2 6
```

The operations above, such as the uses of index value, == and $, are all the same as those of data. frame. Next, let's query with the specially designed keys of data.table.

```
# Set column a as the index column
> setkey(dt,a)

# Print dt objects and find that the data has been sorted according to
the ASCII values of the column a letters.
> dt
     a          b
1:  A   0.7792728
2:  A  -0.2769280
3:  A  -1.3009561
4:  B   1.4870693
5:  B   1.1076424
6:  C   0.9890549
```

Query with the self-defined index.

```
# Take the rows whose values are B in column a
> dt["B",]
     a          b
1:  B 1.487069
2:  B 1.107642

# Take the rows whose values are B in column a and keep the first line
> dt["B",mult="first"]
     a          b
1:  B 1.487069

# Take the rows whose values are B in column a and keep the last line
> dt["B",mult="last"]
     a          b
1:  B 1.107642

# Take the rows whose values are B in column a. Take NA when there is
none
> dt["b"]
     a  b
1:  b NA
```

We can see in the above code testing that after defining keys, we don't need to define columns for query. The first place in the square brackets is, by default, for the keys, which are the matching conditions of indexing. For this, we can save the codes of variable definition. Meanwhile, the filter conditions can be added to the dataset by the mult parameter, which improves the code efficiency. If the query value is not included in the index column, it will go to NA.

4. Add, remove and revise the data.table object

To add a column to the data.table object, we can use this format: data.table[, colname := var1].

```
# Create an object of data.table
```

```
> dt = data.table(a=c('A','B','C','A','A','B'),b=rnorm(6))
> dt
   a          b
1: A  1.51765578
2: B  0.01182553
3: C  0.71768667
4: A  0.64578235
5: A -0.04210508
6: B  0.29767383

# Add a column named c
> dt[,c:=b+2]
> dt
   a          b        c
1: A  1.51765578 3.517656
2: B  0.01182553 2.011826
3: C  0.71768667 2.717687
4: A  0.64578235 2.645782
5: A -0.04210508 1.957895
6: B  0.29767383 2.297674

# Add two columns named respectively c1 and c2
> dt[,`:=`(c1 = 1:6, c2 = 2:7)]
> dt
   a          b        c c1 c2
1: A  0.7545555 2.754555  1  2
2: B  0.5556030 2.555603  2  3
3: C -0.1080962 1.891904  3  4
4: A  0.3983576 2.398358  4  5
5: A -0.9141015 1.085899  5  6
6: B -0.8577402 1.142260  6  7

# Add two columns named d1 and d2
> dt[,c('d1','d2'):=list(1:6,2:7)]
> dt
   a          b        c c1 c2 d1 d2
1: A  0.7545555 2.754555  1  2  1  2
2: B  0.5556030 2.555603  2  3  2  3
3: C -0.1080962 1.891904  3  4  3  4
4: A  0.3983576 2.398358  4  5  4  5
5: A -0.9141015 1.085899  5  6  5  6
6: B -0.8577402 1.142260  6  7  6  7
```

To remove a column from the data.table object, i.e, to assign null to this column, we can use this format: data.table[, colname := NULL]. We will continue using the dt object created just now.

```
# Delete column c1
> dt[,c1:=NULL]
> dt
   a          b        c c2 d1 d2
1: A  0.7545555 2.754555  2  1  2
2: B  0.5556030 2.555603  3  2  3
3: C -0.1080962 1.891904  4  3  4
4: A  0.3983576 2.398358  5  4  5
```

```
5:  A -0.9141015 1.085899   6  5  6
6:  B -0.8577402 1.142260   7  6  7

# Delete column d1 and d2 at the same time
> dt[,c('d1','d2'):=NULL]
> dt
     a          b          c c2
1:  A   0.7545555 2.754555   2
2:  B   0.5556030 2.555603   3
3:  C  -0.1080962 1.891904   4
4:  A   0.3983576 2.398358   5
5:  A  -0.9141015 1.085899   6
6:  B  -0.8577402 1.142260   7
```

To revise the value of a data.table object, i.e. to locate the data with index and then replace its value, we can use this format: data.table[condition, colname := 0]. We will continue using the dt object created just now.

```
# Assign 30 to b
> dt[,b:=30]
> dt
     a  b        c c2
1:  A 30 2.754555   2
2:  B 30 2.555603   3
3:  C 30 1.891904   4
4:  A 30 2.398358   5
5:  A 30 1.085899   6
6:  B 30 1.142260   7

# For those rows with the values of column a as B and the values of
column c2 larger than 3, assign 100 to their column b
> dt[a=='B' & c2>3, b:=100]
> dt
     a   b        c c2
1:  A  30 2.754555   2
2:  B  30 2.555603   3
3:  C  30 1.891904   4
4:  A  30 2.398358   5
5:  A  30 1.085899   6
6:  B 100 1.142260   7

# Another expression
> dt[,b:=ifelse(a=='B' & c2>3,50,b)]
> dt
     a  b        c c2
1:  A 30 2.754555   2
2:  B 30 2.555603   3
3:  C 30 1.891904   4
4:  A 30 2.398358   5
5:  A 30 1.085899   6
6:  B 50 1.142260   7
```

5. Group computing of data.table

To group compute the data.frame objects, we either adopt the apply function for its auto-management, or use the group computing function of plyr package. The data.table package itself supports group computing, similar to the group by function of SQL, which is the main advantage of the data.table package.

For example, group by column *a* and sum column *b* by groups.

```
# Create data
> dt = data.table(a=c('A','B','C','A','A','B'),b=rnorm(6))
> dt
   a          b
1: A   1.4781041
2: B   1.4135736
3: C  -0.6593834
4: A  -0.1231766
5: A  -1.7351749
6: B  -0.2528973


# Sum column b
> dt[,sum(b)]
[1] 0.1210455


# Group by column a and sum column b by groups
> dt[,sum(b),by=a]
   a          V1
1: A  -0.3802474
2: B   1.1606763
3: C  -0.6593834
```

6. Join operation of multiple data.table

When operating data, we usually need to join two or more datasets through an index. Our algorithm will merge multiple data and process them together, where join operation will be needed. Join operation is similar to the LEFT JOIN of the relational database.

Take the students' exam scenario for example. According to the ER design pattern, we usually partition the data by entities. There are two entities here: students and scores. The student entity includes the basic information of students such as their names, while the score entity includes the subjects of the tests and the scores.

Assume that there are six students all taking the exams of A and B and scoring differently.

```
# Six students
> student <- data.table(id=1:6,name=c('Dan','Mike','Ann','Yang','Li',
'Kate'));student
   id name
1: 1   Dan
2: 2   Mike
3: 3   Ann
4: 4   Yang
5: 5   Li
6: 6   Kate
```

```
# All take the exams of A and B
> score <- data.table(id=1:12,stuId=rep(1:6,2),score=runif(12,60,99),
class=c(rep('A',6),rep('B',6)));score
     id stuId    score class
 1:  1     1 89.18497     A
 2:  2     2 61.76987     A
 3:  3     3 74.67598     A
 4:  4     4 64.08165     A
 5:  5     5 85.00035     A
 6:  6     6 95.25072     A
 7:  7     1 81.42813     B
 8:  8     2 82.16083     B
 9:  9     3 69.53405     B
10: 10     4 89.01985     B
11: 11     5 96.77196     B
12: 12     6 97.02833     B
```

Join operate the datasets of students and scores by the student ID.

```
# Set the dataset of scores with stuID as key
> setkey(score,"stuId")

# Set the dataset of students with id as key
> setkey(student,"id")

# Merge the data of these two datasets
> student[score,nomatch=NA,mult="all"]
     id name i.id    score class
 1:  1  Dan    1 89.18497     A
 2:  1  Dan    7 81.42813     B
 3:  2 Mike    2 61.76987     A
 4:  2 Mike    8 82.16083     B
 5:  3  Ann    3 74.67598     A
 6:  3  Ann    9 69.53405     B
 7:  4 Yang    4 64.08165     A
 8:  4 Yang   10 89.01985     B
 9:  5   Li    5 85.00035     A
10:  5   Li   11 96.77196     B
11:  6 Kate    6 95.25072     A
12:  6 Kate   12 97.02833     B
```

Finally, we will see that the results of two datasets are merged into one. And the join operation of data is completed. From the perspective of coding, one line of codes is more convenient than the jointing of data.frame.

3.2.3 Performance Comparison of data.table Package

Nowadays, we need to process data of very large size, maybe millions of lines. If we are going to analyze or process the data with R language and without adding any new hardware, we need to operate it with high-performance packages. And data.table would be a good choice here.

1. Index query performance comparison of data.table and data.frame

Let's generate a dataset of slightly large size, including two columns *x* and *y*, to which we assign English letters. The dataset contains 100,000,004 lines and takes a memory of 600 mb. Compare the performances and the time consumptions of the index query operations for data.frame and data.table.

Create a dataset with data.frame.

```
# Clear the environment variables
> rm(list=ls())

# Set the size
> size = ceiling(1e8/26^2)
[1] 147929

# Calculate the time of data.frame object generating
> t0=system.time(
+    df <- data.frame(x=rep(LETTERS,each=26*size),y=rep(letters,each=s
ize))
+ )

# Print the time
> t0
User System Elapsed
7.53 1.22 8.93

# Row numbers of the df object
> nrow(df)
[1] 100000004

# Memory taken
> object.size(df)
800004440 bytes

# Run the condition query
> t1=system.time(
+    val1 <- df[df$x=="R" & df$y=="h",]
+ )

# Time query
> t1
```

User System Elapsed

```
13.98   2.30 16.43
```

Create a dataset with data.table.

```
# Clear the environment variables
> rm(list=ls())

# Set the size
> size = ceiling(1e8/26^2)
[1] 147929

# Calculate the time of data.table object generating
> t2=system.time(
```

```
+    dt <- data.table(x=rep(LETTERS,each=26*size),y=rep(letters,each=s
ize))
+ )

# Time of the object generating
> t2
User System Elapsed
2.96 0.51 3.52

# Row numbers of the object
> nrow(dt)
[1] 100000004

# Memory taken
> object.size(dt)
1600003792 bytes

# Run the condition query
> t3=system.time(
+ val2 <- dt[x=="R" & y=="h",]
+ )

# Time query
> t3
User System Elapsed
6.68 0.10 6.93
```

From the test above, data.table is much more efficient than data.frame in creating and querying objects. Let's set an index for the data.table dataset and see how does index query work.

```
# Set x and y as the key indexes
> setkey(dt,x,y)

# Condition query
> t4=system.time(
+    val3  <- dt[list("R","h")]
+ )

# View the time
> t4
User System Elapsed
0.00 0.00 0.05
```

With the index columns, when we run the index query, the CPU time consumption is zero. Shocked!

2. Assignment performance comparison of data.table and data.frame

There are usually two actions for assignment operation: query and replace, which will be implemented by both data.frame and data.table. According to the previous section, data.table is much faster than data.frame in index query. However, we'd better avoid complicated query in the assignment operation test.

For the rows whose value of *x* column is R, assign value to their *y*s. First, let's test the computing time for data.frame.

```
> size = 1000000
> df <- data.frame(x=rep(LETTERS,each=size),y=rnorm(26*size))
> system.time(
+    df$y[which(df$x=='R')]<-10
+ )
User System Elapsed
0.75 0.01 0.77
```

Calculate the assignment time of data.table.

```
> dt <- data.table(x=rep(LETTERS,each=size),y=rnorm(26*size))
> system.time(
+    dt[x=='R', y:=10]
+ )
User System Elapsed
0.11 0.00 0.11
> setkey(dt,x)
> system.time(
+    dt['R', y:=10]
+ )
User System Elapsed
0.01 0.00 0.02
```

Comparing the assignment tests of data.table and data.frame, data.table with index shows great advantage in its performance. Let's increase the data size and run the assignment test again.

```
> size = 1000000*5
> df <- data.frame(x=rep(LETTERS,each=size),y=rnorm(26*size))
> system.time(
+    df$y[which(df$x=='R')]<-10
+ )
User System Elapsed
3.22 0.25 3.47

> rm(list=ls())
> size = 1000000*5
> dt <- data.table(x=rep(LETTERS,each=size),y=rnorm(26*size))
> setkey(dt,x)
> system.time(
+    dt['R', y:=10]
+ )
User System Elapsed
0.08 0.01 0.08
```

After increasing the data size, data.table is several times faster than data.frame in assignment. The advantage of data.table is even more significant.

3. Group computing performance comparison of data.table and tapply

Let's compare the group computing performances of data.table and tapply. The test will be run in a simple computing setting, e.g. to group the dataset by *x* column and then sum the *y* column by the grouping.

```
# Set the dataset size
> size = 100000
> dt <- data.table(x=rep(LETTERS,each=size),y=rnorm(26*size))

# Set column x as key
> setkey(dt,x)

# Calculate the time of grouping column x and summing column y by the
grouping
> system.time(
+ r1<-dt[,sum(y),by=x]
+ )
User System Elapsed
0.03 0.00 0.03

# Calculate the sum time of the tapply implementation
> system.time(
+ r2<-tapply(dt$y,dt$x,sum)
+ )
User System Elapsed
0.25 0.05 0.30

# View the dataset size, 40mb
> object.size(dt)
41602688 bytes
```

For data of 40 mb size, tapply is faster than data.table in group computing. What if I increase the dataset size? Let's increase the size by 10 times and test it again.

```
> size = 100000*10
> dt <- data.table(x=rep(LETTERS,each=size),y=rnorm(26*size))
> setkey(dt,x)
> val3<-dt[list("R")]

> system.time(
+    r1<-dt[,sum(y),by=x]
+ )
User System Elapsed
0.25 0.03 0.28

> system.time(
+    r2<-tapply(dt$y,dt$x,sum)
+ )
User System Elapsed
2.56 0.36 2.92

# Data of 400mb
> object.size(dt)
416002688 bytes
```

For data of 400 mb, data.table is obviously better than tapply at the computing performance. Let's increase the data by five more times and see what will happen.

```
> size = 100000*10*5
> dt <- data.table(x=rep(LETTERS,each=size),y=rnorm(26*size))
> setkey(dt,x)

> system.time(
+       r1<-dt[,sum(y),by=x]
+ )
User System Elapsed
1.50 0.11 1.61

> system.time(
+       r2<-tapply(dt$y,dt$x,sum)
+ )
 User System Elapsed
13.30   3.58 16.90

# Data of 2G
> object.size(dt)
2080002688 bytes
```

For data of 2G size, it takes tapply 16 seconds and data.table 1.6 seconds. According to these two tests above, if data size is above 400 mb, the CPU time consumption is linear.

Put the test data together, as shown in Figure 3.2.

According to the comparison above, the data.table package is 10 times faster than tapply in group computing, 30 times faster than data.frame in assignment operation, and 100 times faster than data.frame in index query. Therefore, the data.table package is definitely a package worth your hard study.

Hurry up and get your programs optimized by the data.table package.

Item for comparison	Log row number	Data size (mb)	Data. table	Data. frame	tapply
Data creating time	100,000,004	1,600	3.63	3.8	
Index query performance	100,000,004	1,600	0.06	9.42	
Assignment performance	26,000,000	416	0.02	0.69	
Assignment performance	130,000,000	2,000	0.08	3.47	
Group computing performance	2,600,000	40	0.03		0.3
Group computing performance	26,000,000	400	0.28		2.92
Group computing performance	130,000,000	2,000	1.61		16.9

Figure 3.2 Data processing performance comparison of data.table.

3.3 Efficient Pipe Operation of R: magrittr

Question
How do you simplify the codes?

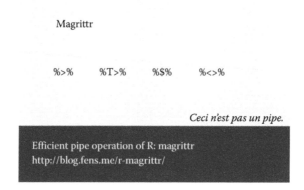

Introduction
It is very convenient to process data with R language, because just a few lines of codes can complete complicated operations. However, for continuous data processing operations, some do not consider codes simplified enough. The codes either are long nested call functions, similar to Lisp in that everything is included in brackets, or assign value to a temporary variable in every operation, which is too tedious. Can it be as elegant as Linux pipeline?

Under this circumstance, the magrittr package is developed. It processes the continuous and complicated data in the way of pipeline, which makes the codes shorter and more readable. In this way, 1 line of codes can achieve as much as the original 10 lines do.

3.3.1 An Introduction to magrittr

The magrittr package is defined as a highly efficient pipe operation toolkit. It passes data or expressions efficiently in the way of pipe connection and it uses operator %>% to directly pass data to the next call function or expression. There are two goals for the magrittr package. The first is to reduce the code development time and improve the readability and maintainability of codes. The second is to make the codes shorter and shorter and shorter.

The magrittr package mainly defines these four pipe operators: %>%, %T>%, %$% and %<>%, where %>% is the most frequently used operator and the rest three, similar to %>%, work better in special occasions. You will love these operators when you have a grasp of them. How excited to rebuild all the codes and chop off all the tedious parts.

magrittr project homepage: https://github.com/smbache/magrittr.

The system environment used here:

- Win10 64bit
- R: 3.2.3 x86_64-w64-mingw32/x64 b4bit

magrittr can be found in the standard library of CRAN. Its installation is very simple, just two commands to go.

```
~ R
> install.packages('magrittr')
> library(magrittr)
```

3.3.2 Basic Usage of magrittr Package

The usage of the magrittr package actually is the usage of these four operators: %>% (forward-pipe operator), %T>% (tee operator), %$% (exposition pipe-operator) and %<>% (compound assignment pipe-operator).

3.3.2.1 %>% (Forward-Pipe Operator)

%>% is the most frequently used operator. It passes the left-side data or expressions to the right-side call function or expressions, which can be continuously operated as a chain.

The practical principle, as shown in Figure 3.3, uses %>% to pass the dataset A of the left-side program to the function B of the right-side program, then pass the result dataset of the function B again to its right-side function C, and finally complete the data calculation.

For example, we are going to do the following things. This is a demand I invent.

1. Take 10,000 random numbers that comply with normal distribution.
2. Find the absolute values of these 10,000 numbers and multiple the values by 50.
3. Form a 100*100 square matrix with the result.
4. Calculate the mean of every row in the square matrix and then round it to the nearest integer.
5. Divide the results by 7 and find the remainders; draw the histogram of the remainders.

We find that the five processes above are continuous, so how do we implement them with normal codes?

```
# Set a random seed
> set.seed(1)

# Start
> n1<-rnorm(10000)              # First step
> n2<-abs(n1)*50                # Second step
> n3<-matrix(n2,ncol = 100)     # Third step
> n4<-round(rowMeans(n3))       # Fourth step
> hist(n4%%7)                   # Fifth step
```

See Figure 3.4 for the output histogram.

In the codes above, every line implements a condition, but there are many temporary variables in the middle. There's another way of coding and everything is included in the brackets.

```
# Set a random seed
> set.seed(1)
> hist(round(rowMeans(matrix(abs(rnorm(10000))*50,ncol=100)))%%7)
```

Forward-pipe operator

A   %>%   B   →   C

Figure 3.3 Forward-pipe operator.

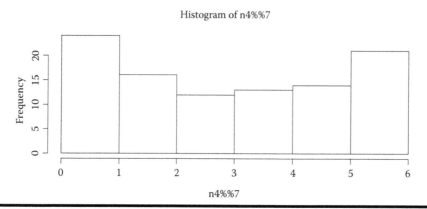

Figure 3.4 Temporary variable coding.

See Figure 3.5 for the output histogram.

Our demand is satisfied with these two common ways of coding. Now let's see what difference can %>% make.

```
# Set a random seed
> set.seed(1)

# Start
> rnorm(10000) %>%
+    abs %>% `*` (50)   %>%
+    matrix(ncol=100)   %>%
+    rowMeans %>% round %>%
+    `%%`(7) %>% hist
```

See Figure 3.6 for the output histogram.

Just one line of structured and readable codes has achieved all. This is the pipe coding, elegant and concise.

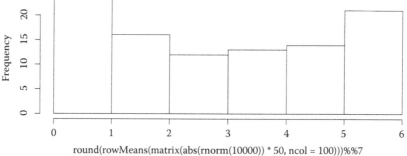

Figure 3.5 Nested function coding.

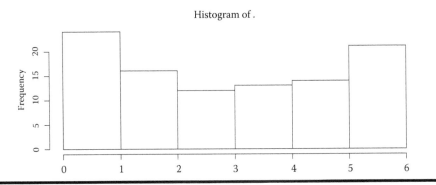

Figure 3.6   Pipe coding.

3.3.2.2 %T>% (Tee Operator)

The function of %T>% (tee operator) is almost the same as %>%, except that it passes on the values of the left side instead of those of the right side. There can be many application scenarios for this. For example, if you need a print output or a picture output in the middle of data processing, the process will be interrupted and the tee operator can help solve this kind of problem.

The practical principle, as shown in Figure 3.7, uses %T>% to pass the dataset A of the left-side program to the function B of the right-side program, then instead of passing the result dataset of function B again to its right-side function C, it passes B's left-side dataset A to its right-side function C and then complete the data calculation.

We can adjust the demand mentioned above, add one more requirement, and then the tee operator will be used.

1. Take 10,000 random numbers that comply with normal distribution.
2. Find the absolute values of these 10,000 numbers and multiple the values by 50.
3. Form a 100*100 square matrix with the result.
4. Calculate the mean of every row in the square matrix and then round it to the nearest integer.
5. Divide the results by 7 and find the remainders; draw the histogram of the remainders.
6. Sum the remainders.

The return value is NULL after outputting the histogram, so if the pipe continues, the null value will be passed to the right side and the final result will error. Here what we need is to pass the remainders to the last step of sum, so %T>% is just in need.

Anomaly will occur if we simply use %>% to pass values to the right.

Tee operator

Figure 3.7   Tee operator.

```
> set.seed(1)
> rnorm(10000) %>%
+   abs %>% `*` (50)   %>%
+   matrix(ncol=100)   %>%
+   rowMeans %>% round %>%
+   `%%`(7) %>% hist %>% sum
Error in sum(.) : invalid 'type' (list) of argument
```

Use %T>% to pass the left-side value to the right side and the result is correct.

```
> rnorm(10000) %>%
+   abs %>% `*` (50)   %>%
+   matrix(ncol=100)   %>%
+   rowMeans %>% round %>%
+   `%%`(7) %T>% hist %>% sum
[1] 328
```

3.3.2.3 %$% (Exposition Pipe-Operator)

%$% allows the function in the right side to use column names for data operations.

The practical principle, as shown in Figure 3.8, uses %$% to pass the dataset A of the left-side program to the function B of the right-side program, meanwhile pass the attribute names of dataset A to the function B as its internal variables, so as to process the dataset A more conveniently and finally complete the data calculation.

Now let's define a data.frame dataset of three columns and ten rows, with its column names as *x*, *y* and *z* and the default values of *x* column greater than 5. Pass the column name *x* directly to the right side with %$%. The symbol "." here represents the complete data object on the left side. The demand is achieved with only one line of codes and no intermediate variable definition is shown here.

```
> set.seed(1)
> data.frame(x=1:10,y=rnorm(10),z=letters[1:10]) %$% .[which(x>5),]
     x           y z
6    6 -0.8204684 f
7    7  0.4874291 g
8    8  0.7383247 h
9    9  0.5757814 i
10  10 -0.3053884 j
```

Without %$%, the common coding for this will be:

```
> set.seed(1)
> df<-data.frame(x=1:10,y=rnorm(10),z=letters[1:10])
> df[which(df$x>5),]
```

Exposition pipe-operator

Figure 3.8　Exposition pipe-operator.

```
     x          y z
6    6 -0.8204684 f
7    7  0.4874291 g
8    8  0.7383247 h
9    9  0.5757814 i
10  10 -0.3053884 j
```

The common coding needs to define a temporary variable df, which is seen to be used for three times in all. This is exactly what %$% improves. It keeps the codes clean.

3.3.2.4 %<>% (Compound Assignment Pipe-Operator)

The function of %<>%, the compound assignment pipe-operator, is almost the same as %>%, except for an extra operation that it writes the result to the left-side object. For example, we need to sort a dataset. It will be convenient to get the sorting result with %<>%.

The practical principle, as shown in Figure 3.9, uses %<>% to pass the dataset A of the left-side program to the function B of the right-side program, then pass the result dataset of function B again to its right-side function C, assign the result dataset of function C to A and complete the data calculation.

Define 100 random numbers that obey normal distribution. Calculate their absolute values and sort them from small to large. Assign the first 10 numbers to *x*.

```
> set.seed(1)
> x<-rnorm(100) %<>% abs %>% sort %>% head(10)
> x
 [1] 0.001105352 0.016190263 0.028002159 0.039240003 0.044933609
0.053805041 0.056128740
 [8] 0.059313397 0.074341324 0.074564983
```

Isn't it simple? A series of operations is implemented by just one line of codes. However, there is a trap here. It should be noted that only when %<>% is used in the object of the first pipe can the assignment operation be achieved; if it is not used in the first place on the left side, the assignment won't work.

```
> set.seed(1)
> x<-rnorm(100)

# The first place on the left, assignment successful
> x %<>% abs %>% sort %>% head(10)
> x
 [1] 0.001105352 0.016190263 0.028002159 0.039240003 0.044933609
0.053805041 0.056128740
 [8] 0.059313397 0.074341324 0.074564983
```

Compound assignment pipe-operator

Figure 3.9 Compound assignment pipe-operator.

```
# The second place in the left, the result is directly printed out and
the value of x doesn't change
> x %>% abs %<>% sort %>% head(10)
 [1] 0.001105352 0.016190263 0.028002159 0.039240003 0.044933609
0.053805041 0.056128740
 [8] 0.059313397 0.074341324 0.074564983
> length(x)
[1] 10

# The third place in the left, the result is directly printed out and the
value of x doesn't change
> x %>% abs %>% sort %<>% head(10)
 [1] 0.001105352 0.016190263 0.028002159 0.039240003 0.044933609
0.053805041 0.056128740
 [8] 0.059313397 0.074341324 0.074564983
> length(x)
[1] 10
```

3.3.3 Extended Functions of magrittr Package

We have already learned the usage of the four operators in the magrittr package. What other functions of the magrittr package are there except for the operators?

- Definitions of sign operators
- %>% for passing to code blocks
- %>% for passing to functions

3.3.3.1 Definitions of Sign Operators

To make the chain transmission more user-friendly, magrittr re-defines the common computing sign operators, so that every operation will adopt a corresponding function and all the codes called in the transmission will be of a unified style. For example, the function add() is equal to +.

The following is the list of correspondences:

```
extract                  `[`
extract2                 `[[`
inset                    `[<-`
inset2                   `[[<-`
use_series               `$`
add                      `+`
subtract                 `-`
multiply_by              `*`
raise_to_power           `^`
multiply_by_matrix       `%*%`
divide_by                `/`
divide_by_int            `%/%`
mod                      `%%`
is_in                    `%in%`
and                      `&`
or                       `|`
equals                   `==`
is_greater_than          `>`
```

```
is_weakly_greater_than      `>=`
is_less_than                `<`
is_weakly_less_than         `<=`
not (`n'est pas`)           `!`
set_colnames                `colnames<-`
set_rownames                `rownames<-`
set_names                   `names<-`
```

Let's check the effects. Multiple a vector which includes 10 random numbers by 5 and then add 5 to it.

```
# Code with signs
> set.seed(1)
> rnorm(10) %>% `*`(5) %>% `+`(5)
 [1]  1.8677309  5.9182166  0.8218569 12.9764040  6.6475389  0.8976581
7.4371453  8.6916235
 [9]  7.8789068  3.4730581


# Code with functions
> set.seed(1)
> rnorm(10) %>% multiply_by(5) %>% add(5)
 [1]  1.8677309  5.9182166  0.8218569 12.9764040  6.6475389  0.8976581
7.4371453  8.6916235
 [9]  7.8789068  3.4730581
```

Replacing the signs with the functions, the result is still the same. Actually, this conversion is very useful when doing the object-oriented encapsulation, e.g. when Hibernate, the DAL framework, encapsulates all the SQL and when XFire encapsulates the WebServices protocol, etc.

3.3.3.2 %>% for Passing to Code Blocks

Sometimes, we may need to adjust a data block into the Whitesmiths style, which is hard to achieve in one statement, so the code block processing is needed here. The dataset is passed to the {} code block, and the incoming dataset is represented with "." A piece of codes, rather than a statement, is needed to complete the operation.

For example, Multiple a vector which includes 10 random numbers by 5 and then add 5 to it. Find the mean and the standard deviation of the result, sort it from small to large and then return to the first five.

```
> set.seed(1)
> rnorm(10)      %>%
+   multiply_by(5) %>%
+   add(5)          %>%
+   {
+     cat("Mean:", mean(.),
+         "Var:", var(.), "\n")
+     sort(.) %>% head
+   }
Mean: 5.661014 Var: 15.23286
[1] 0.8218569 0.8976581 1.8677309 3.4730581 5.9182166 6.6475389
```

It is convenient to deal with these complicated operations with code blocks wrapped in {}.

3.3.3.3 %>% for Passing to Functions

The design of passing to functions, similar to that of passing to code blocks, passes a dataset to an anonymous function for complicated data processing. Here the displayed dataset name will be taken as the parameter of the anonymous function.

For example, process the dataset of iris and keep the first row and the last row as the result.

```
> iris %>%
+     (function(x) {
+         if (nrow(x) > 2)
+             rbind(head(x, 1), tail(x, 1))
+         else x
+     })
    Sepal.Length Sepal.Width Petal.Length Petal.Width   Species
1            5.1         3.5          1.4         0.2    setosa
150          5.9         3.0          5.1         1.8 virginica
```

Here the dataset of iris is the displayed parameter *x*, which is applied in the following data processing.

After the learning of magrittr, we have acquired some special programming skills of R language. The magrittr package coded in R language is different from the coding in the traditional R language. It can make your program simple and efficient.

Programmers who are "lazy" by nature will figure out various ways to make their codes simplified, elegant and, meanwhile, reliable. The less your codes, the closer you are to a master.

3.4 String Processing Package of R: stringr

Question

How do you process strings with R language?

String processing package of R: stringr
http://blog.fens.me/r-stringr/

Introduction

It is troublesome to process strings with R language, for you cannot segment them with vector, you cannot index them with cycle traversal, you cannot remember the family of grep() functions and the paste() function is segmented with spaces by default. Everything is not convenient. There are more and more scenarios needing R language, so it is necessary to process strings. Now I recommend a flexible package for string process, developed by Hadley Wickham, named stringr.

3.4.1 An Introduction to stringr

The stringr package is defined as a consistent and simple string toolkit. All its definitions of functions and parameters are consistent, e.g. the same method can be used in NA processing and 0-length vector processing.

String processing is not the most important function of R language, but a necessary one. It can be used in operations such as data cleansing, visualization, etc. The basic string functions provided by the base package of R language have become inconsistent as time goes. The non-standard naming and parameter definition make it difficult for instant use. The string processing can be easy in other language. R language is lagging in this aspect. The stringr package is the exact solution to this problem. It makes string processing easy and provides with user-friendly string operation interfaces.

stringr project homepage: https://cran.r-project.org/web/packages/stringr/index.html

The system environment used here:

- Win10 64bit
- R: 3.2.3 x86_64-w64-mingw32/x64 b4bit

stringr can be found in the standard library of CRAN. Its installation is very simple, just two commands to go.

```
~ R
> install.packages('stringr')
> library(stringr)
```

3.4.2 An Introduction to API of stringr

The stringr package version 1.0.0 provides with 30 functions for convenient string processing. The names of its common string processing operations begin with str_, for intuitive understanding of the function definitions. The functions can be classified according to our using habits.

String concatenation functions

- str_c: Concatenate strings
- str_join: Join strings, same as str_c
- str_trim: Trim the spaces and TAB() of a string
- str_pad: Pad the string length
- str_dup: Duplicate a string
- str_wrap: Control the output format of a string
- str_sub: Intercept a string
- str_sub<- Intercept a string and assign value to it, same as str_sub

String calculation functions

- str_count: Count strings
- str_length: the length of a string
- str_sort: Sort strings
- str_order: Index sort strings in the same rules as str_sort

String matching functions

- str_split: Segment a string
- str_split_fixed: Segment a string, same as str_split
- str_subset: Return the matching string
- word: Extract words from text
- str_detect: Check the characters matching the string
- str_match: Extract the matching group from a string
- str_match_all: Extract the matching group from a string, same as str_match
- str_replace: Replace a string
- str_replace_all: Replace a string, same as str_replace
- str_replace_na: Replace NA with the NA string
- str_locate: Find the location of the matching string
- str_locate_all: Find the location of the matching string, same as str_locate
- str_extract: Extract the matching character from a string
- str_extract_all: Extract the matching character from a string, same as str_extract

String transformation functions

- str_conv: Convert the character codes
- str_to_upper: Convert the string to uppercase
- str_to_lower: Convert the string to lowercase in the same rules as str_to_upper
- str_to_title: Capitalize the string in the same rules as str_to_upper

the controlling functions of parameters is only suitable for the parameters of constructor and cannot be used alone.

- boundary: Define the boundary
- coll: Define the sorting rules of the string
- fixed: Define the characters for matching, including the escape characters of the regular expression
- regex: Define the regular expression

3.4.2.1 *String Concatenation Functions*

1. Str_c, the string concatenation operation, is totally the same as str_join, but not totally in accordance with the actions of paste().
   Function definition:

   ```
   str_c(..., sep = "", collapse = NULL)
   str_join(..., sep = "", collapse = NULL)
   ```

   Parameter list:
   - ...: Input multiple parameters
   - sep: Concatenate multiple strings into one large string, used in the string delimiter
   - collapse: Concatenate multiple vector parameters into one large string, used in the string delimiter

Concatenate multiple strings into one large string.

```
> str_c('a','b')
[1] "ab"
> str_c('a','b',sep='-')
[1] "a-b"
> str_c(c('a','a1'),c('b','b1'),sep='-')
[1] "a-b"    "a1-b1"
```

Concatenate multiple vector parameters into one large string.

```
> str_c(head(letters), collapse = "")
[1] "abcdef"
> str_c(head(letters), collapse = ", ")
[1] "a, b, c, d, e, f"

# the collapse parameter, invalid for multiple strings
> str_c('a','b',collapse = "-")
[1] "ab"
> str_c(c('a','a1'),c('b','b1'),collapse='-')
[1] "ab-a1b1"
```

NA is still NA when concatenating the string vectors with NA value.

```
> str_c(c("a", NA, "b"), "-d")
[1] "a-d" NA    "b-d"
```

Find the differences between function str_c() and function paste().

```
# When concatenating multiple strings, the default sep parameter is
inconsistent
> str_c('a','b')
[1] "ab"
> paste('a','b')
[1] "a b"

# When concatenating vectors into strings, the collapse parameter is
inconsistent
> str_c(head(letters), collapse = "")
[1] "abcdef"
> paste(head(letters), collapse = "")
[1] "abcdef"

#When concatenating the string vectors with NA value, the NA
processing is inconsistent
> str_c(c("a", NA, "b"), "-d")
[1] "a-d" NA    "b-d"
> paste(c("a", NA, "b"), "-d")
[1] "a -d"   "NA -d" "b -d"
```

2. str_trim: Trim the spaces and TAB() of a string.
   Function definition:

```
str_trim(string, side = c("both", "left", "right"))
```

Parameter list:
– string: The string vector.
– side: Filter modes. Both means filter on both sides, left filter on the left side and right filter on the right side.

Trim the spaces and TAB() of a string

```
#Filter the spaces on the left side
> str_trim("  left space\t\n",side='left')
[1] "left space \n"

#Filter the spaces on the right side
> str_trim("  left space\t\n",side='right')
[1] "  left space"

#Filter the spaces on both sides
> str_trim("  left space\t\n",side='both')
[1] "left space"

#Filter the spaces on both sides
> str_trim("\nno space\n\t")
[1] "no space"
```

3. str_pad: Pad the string length.
   Function definition:

```
str_pad(string, width, side = c("left", "right", "both"), pad = " ")
```

Parameter list:
– string: The string vector.
– width: The length of the string after padding.
– side: Padding direction. both means to pad on both sides, left to pad on the left side and right to pad on the right side.
– pad: The characters used for padding.

Pad the string length.

```
# Pad with spaces on the left side till the string length is 20
> str_pad("conan", 20, "left")
[1] "               conan"

# Pad with spaces on the right side till the string length is 20
> str_pad("conan", 20, "right")
[1] "conan               "

# Pad with spaces on both sides till the string length is 20
> str_pad("conan", 20, "both")
[1] "       conan        "

# Pad with the character x on both sides till the string length is 20
> str_pad("conan", 20, "both",'x')
[1] "xxxxxxxconanxxxxxxxx"
```

4. str_dup: Duplicate a string.
    Function definition:

```
str_dup(string, times)
```

Parameter list:
- string: The string vector.
- times: The duplicate times.

Duplicate a string vector.

```
> val <- c("abca4", 123, "cba2")
```

```
# Duplicate twice
> str_dup(val, 2)
[1] "abca4abca4" "123123"      "cba2cba2"
```

```
# Duplicate by location
> str_dup(val, 1:3)
[1] "abca4"        "123123"        "cba2cba2cba2"
```

5. str_wrap: Control the output format of a string.
    Function definition:

```
str_wrap(string, width = 80, indent = 0, exdent = 0)
```

Parameter list:
- string: The string vector.
- width: Set the width of one line.
- indent: The indentation of the first line of a paragraph.
- exdent: The indentation of the non-first lines of a paragraph.

```
txt<-' R language is a statistical language which used to shine in
small fields. As the big data era comes, it has become a hot tool in
data analysis. With more and more people with engineering background
joining in, the community of R language is expanding rapidly. Now R
language is not only used in statistics field, but also in
education, banks, e-commerce, Internet, etc. '
```

```
# Set the width to 40 characters
> cat(str_wrap(txt, width = 40), "\n")
R language is a statistical language
which used to shine in small fields. As
the big data era comes, it has become a
hot tool in data analysis. With more and
more people with engineering background
joining in, the community of R language
is expanding rapidly. Now R language
is not only used in statistics filed,
but also education, banks, e-commerce,
Internet, etc.
```

```
# Set the width to 60 characters, with 2 characters indentation of
the first line
> cat(str_wrap(txt, width = 60, indent = 2), "\n")
  R language is a statistical language which used to shine
in small fields. As the big data era comes, it has become
a hot tool in data analysis. With more and more people
with engineering background joining in, the community of R
language is expanding rapidly. Now R language is not only
used in statistics filed, but also education, banks, e-
commerce, Internet, etc.


# Set the width to 10 characters, with 4 characters indentation of
the non-first line
> cat(str_wrap(txt, width = 10, exdent = 4), "\n")
R language
    is a
    statistical
    language
    which used
    to shine
    in small
    fields.
    As the big
    data era
    comes, it
    has become
    a hot tool
    in data
    analysis.
    With more
    and more
    people
    with
    engineering
    background
    joining
    in, the
    community
    of R
    language
    is
    expanding
    rapidly.
    Now R
    language
    is not
    only
    used in
    statistics
    filed,
    but also
    education,
    banks, e-
```

```
    commerce,
    Internet,
etc.
```

6. str_sub: Inte rcept a string
    Function definition:

```
str_sub(string, start = 1L, end = -1L)
```

Parameter list:
- string: The string vector.
- start: The starting position
- end: The ending position

Intercept a string.

```
> txt <- "I am Conan."


# Intercept the strings in 1-4 index positions
> str_sub(txt, 1, 4)
[1] "I am"


# Intercept the strings in 1-6 index positions
> str_sub(txt, end=6)
[1] "I am C"


# Intercept the strings in the index positions from 6 to the end
> str_sub(txt, 6)
[1] "Conan."


# Intercept the strings in two segments
> str_sub(txt, c(1, 4), c(6, 8))
[1] "I am C" "m Con"


# Intercept the strings with negative coordinates
> str_sub(txt, -3)
[1] "an."
> str_sub(txt, end = -3)
[1] "I am Cona"
```

Assign values to the intercepted strings.

```
> x <- "AAABBBCCC"


# Assign 1 to the string in 1 position
> str_sub(x, 1, 1) <- 1; x
[1] "1AABBBCCC"


# Assign 2345 to the strings in 2 to -2 positions
> str_sub(x, 2, -2) <- "2345"; x
[1] "12345C"
```

3.4.2.2 String Calculation Functions

1. str_count: Count the strings.
   Function definition:

```
str_count(string, pattern = "")
```

Parameter list:
   - string: The string vector.
   - pattern: The matching character

Count the matching characters in the string.

```
> str_count('aaa444sssddd', "a")
[1] 3
```

Count the matching characters in the string vector.

```
> fruit <- c("apple", "banana", "pear", "pineapple")
> str_count(fruit, "a")
[1] 1 3 1 1
> str_count(fruit, "p")
[1] 2 0 1 3
```

Count the character "." in the string. "." is a matching character in regular expressions, so the direct count result may not be correct.

```
> str_count(c("a.", ".", ".a.",NA), ".")
[1]   2   1   3 NA
```

```
# Match the character with fixed
> str_count(c("a.", ".", ".a.",NA), fixed("."))
[1]   1   1   2 NA
```

```
# Match the character with \\
> str_count(c("a.", ".", ".a.",NA), "\\.")
[1]   1   1   2 NA
```

2. str_length: The string length.
   Function definition:

```
str_length(string)
```

Parameter list:
   - string: The string vector.

Calculate the string length.

```
> str_length(c("I", "am", "张丹", NA))
[1]   1   2   2 NA
```

3. str_sort: Sort the string values, same as the index sorting of str_order.
   Function definition:

```
str_sort(x, decreasing = FALSE, na_last = TRUE, locale = "", ...)
str_order(x, decreasing = FALSE, na_last = TRUE, locale = "", ...)
```

Parameter list:
- *x*: The string vector.
- decreasing: Sort order.
- na_last: The storage place of NA values. There are three values in all: TRUE (in the last place), FALSE (in the first place) and NA (filtered).
- locale: Sort by which language conventions.

Sort the string values.

```
# Sort by ASCII letters
> str_sort(c('a',1,2,'11'), locale = "en")
[1] "1"  "11" "2"  "a"

# Sort in descending order
> str_sort(letters,decreasing=TRUE)
 [1] "z" "y" "x" "w" "v" "u" "t" "s" "r" "q" "p" "o" "n" "m" "l" "k"
"j" "i" "h"
[20] "g" "f" "e" "d" "c" "b" "a"

# Sort by the Chinese phonetic alphabet
> str_sort(c('你','好','粉','丝','日','志'),locale = "zh")
[1] "粉" "好" "你" "日" "丝" "志"
```

Sort the NA value.

```
 #Put NA in the last place
> str_sort(c(NA,'1',NA),na_last=TRUE)
[1] "1" NA  NA

#Put NA in the first place
> str_sort(c(NA,'1',NA),na_last=FALSE)
[1] NA  NA  "1"

#Remove NA
> str_sort(c(NA,'1',NA),na_last=NA)
[1] "1"
```

3.4.2.3 String Matching Functions

1. str_split: Segment a string, same as str_split_fixed.
   Function definition:

```
str_split(string, pattern, n = Inf)
str_split_fixed(string, pattern, n)
```

Parameter list:
- string: The string vector.
- pattern: The matching character
- *n*: Number of segments.

Segment a string.

```
> val <- "abc,123,234,iuuu"

# Segment by ","
> s1<-str_split(val, ",");s1
[[1]]
[1] "abc"  "123"  "234"  "iuuu"

# Segment by "," and save two segments
> s2<-str_split(val, ",",2);s2
[[1]]
[1] "abc"          "123,234,iuuu"

# View the result type list of str_split() function operation
> class(s1)
[1] "list"

# Segment by function str_split_fixed() and the result type is
  matrix
> s3<-str_split_fixed(val, ",",2);s3
     [,1]  [,2]
[1,] "abc" "123,234,iuuu"

> class(s3)
[1] "matrix"
```

2. str_subset: Return the matching string.
   Function definition:

```
str_subset(string, pattern)
```

Parameter list:
- string: The string vector.
- pattern: The matching character

```
   > val <- c("abc", 123, "cba")

# Full match
> str_subset(val, "a")
[1] "abc" "cba"

# Match with the start
> str_subset(val, "^a")
[1] "abc"

# Match with the end
> str_subset(val, "a$")
[1] "cba"
```

3. word: Extract words from text.
   Function definition:

```
word(string, start = 1L, end = start, sep = fixed(" "))
```

Parameter list:
   - string: The string vector.
   - start: The starting position.
   - end: The ending position.
   - sep: The matching character.

```
> val <- c("I am Conan.", "http://fens.me, ok")
```

```
# Segment by spaces in default and take the string at the first
  position
> word(val, 1)
[1] "I"                "http://fens.me,"
> word(val, -1)
[1] "Conan." "ok"
> word(val, 2, -1)
[1] "am Conan." "ok"
```

```
# Segment by "," and take the string at the first position
> val<-'111,222,333,444'
> word(val, 1, sep = fixed(','))
[1] "111"
> word(val, 3, sep = fixed(','))
[1] "333"
```

4. str_detect: The character matching the string.
   Function definition:

```
str_detect(string, pattern)
```

Parameter list:
   - string: The string vector.
   - pattern: The matching character.

```
> val <- c("abca4", 123, "cba2")
```

```
# Check if the string vector includes "a"
> str_detect(val, "a")
[1]  TRUE FALSE  TRUE
```

```
# Check if the string vector starts with "a"
> str_detect(val, "^a")
[1]  TRUE FALSE FALSE
```

```
# Check if the string vector ends with "a"
> str_detect(val, "a$")
[1] FALSE FALSE FALSE
```

5. str_match: Extract the matching group from the string.
   Function definition:

```
str_match(string, pattern)
str_match_all(string, pattern)
```

Parameter list:
- string: The string vector.
- pattern: The matching character.

Extract the matching group from the string

```
> val <- c("abc", 123, "cba")
```

```
# Match character "a" and return to the corresponding character
> str_match(val, "a")
     [,1]
[1,] "a"
[2,] NA
[3,] "a"
```

```
# Match character 0-9, limited to one, and return to the
corresponding character
> str_match(val, "[0-9]")
     [,1]
[1,] NA
[2,] "1"
[3,] NA
```

```
# Match character 0-9, no number limitation, and return to the
corresponding characters
> str_match(val, "[0-9]*")
     [,1]
[1,] ""
[2,] "123"
[3,] ""
```

Extract the matching group from the string and return it in the string format of matrix.

```
> str_match_all(val, "a")
[[1]]
     [,1]
[1,] "a"

[[2]]
     [,1]

[[3]]
     [,1]
[1,] "a"

> str_match_all(val, "[0-9]")
[[1]]
     [,1]
```

```
[[2]]
     [,1]
[1,] "1"
[2,] "2"
[3,] "3"

[[3]]
     [,1]
```

6. str_replace: Replace strings.
   Function definition:

   ```
   str_replace(string, pattern, replacement)
   ```

   Parameter list:
   – string: The string vector.
   – pattern: The matching character.
   – replacement: The character for replacement.

   ```
   > val <- c("abc", 123, "cba")

   # Replace the first "a" or "b" in the target string with "-"
   > str_replace(val, "[ab]", "-")
   [1] "-bc" "123" "c-a"

   # Replace all "a" or "b" in the target string with "-"
   > str_replace_all(val, "[ab]", "-")
   [1] "--c" "123" "c--"

   # Replace all "a" in the target string with the escape characters
   > str_replace_all(val, "[a]", "\1\1")
   [1] "\001\001bc" "123"         "cb\001\001"
   ```

7. str_replace_na: Replace NA with the NA string
   Function definition:

   ```
   str_replace_na(string, replacement = "NA")
   ```

   Parameter list:
   – string: The string vector.
   – replacement: The character for replacement.

   Replace NA with strings.

   ```
   > str_replace_na(c(NA,'NA',"abc"),'x')
   [1] "x"   "NA"  "abc"
   ```

8. str_locate: Find the location of the matching character in the string.
   Function definition:

   ```
   str_locate(string, pattern)
   str_locate_all(string, pattern)
   ```

Parameter list:
- string: The string vector.
- pattern: The matching character.

```
> val <- c("abca", 123, "cba")

# Match the location of a in the string
> str_locate(val, "a")
     start end
[1,]     1   1
[2,]    NA  NA
[3,]     3   3

# Match with the vector
> str_locate(val, c("a", 12, "b"))
     start end
[1,]     1   1
[2,]     1   2
[3,]     2   2

# Return in the string format of matrix
> str_locate_all(val, "a")
[[1]]
     start end
[1,]     1   1
[2,]     4   4

[[2]]
     start end

[[3]]
     start end
[1,]     3   3

# Matching the character "a" or "b" and return in the string format
  of matrix
> str_locate_all(val, "[ab]")
[[1]]
     start end
[1,]     1   1
[2,]     2   2
[3,]     4   4

[[2]]
     start end

[[3]]
     start end
[1,]     2   2
[2,]     3   3
```

9. str_extract: Extract the matching pattern from the string.

Function definition:

```
str_extract(string, pattern)
str_extract_all(string, pattern, simplify = FALSE)
```

Parameter list:
- string: The string vector.
- pattern: The matching character.
- simplify: The return value. If the value is TRUR, matrix will be returned; if it is FALSE, the string vector will be returned.

```
> val <- c("abca4", 123, "cba2")

# Return the matching numbers
> str_extract(val, "\\d")
[1] "4" "1" "2"

# Return the matching characters
> str_extract(val, "[a-z]+")
[1] "abca" NA      "cba"

> val <- c("abca4", 123, "cba2")
> str_extract_all(val, "\\d")
[[1]]
[1] "4"

[[2]]
[1] "1" "2" "3"

[[3]]
[1] "2"

> str_extract_all(val, "[a-z]+")
[[1]]
[1] "abca"

[[2]]
character(0)

[[3]]
[1] "cba"
```

3.4.2.4 String Transform Functions

1. str_conv: Convert the character codes
   Function definition:

   ```
   str_conv(string, encoding)
   ```

   Parameter list:
   - string: The string vector.
   - encoding: The encoding name.

Transcode Chinese.

```
# Convert the Chinese characters into bytes
> x <- charToRaw('你好');x
[1] c4 e3 ba c3

# The default character set in win system is GBK. GB2312 is a GBK
character set and the transcoding is normal
> str_conv(x, "GBK")
[1] "你好"
> str_conv(x, "GB2312")
[1] "你好"

# UTF-8 transcoding failed
> str_conv(x, "UTF-8")
[1] "▨▨▨"
Warning messages:
1: In stri_conv(string, encoding, "UTF-8") :
  input data \xffffffc4 in current source encoding could not be
converted to Unicode
2: In stri_conv(string, encoding, "UTF-8") :
  input data \xffffffe3\xffffffba in current source encoding could
not be converted to Unicode
3: In stri_conv(string, encoding, "UTF-8") :
  input data \xffffffc3 in current source encoding could not be
converted to Unicode
```

Convert unicode to UTF-8.

```
> x1 <- "\u5317\u4eac"
> str_conv(x1, "UTF-8")
[1] "北京"
```

2. str_to_upper: Convert the string to uppercase.
   Function definition:

```
str_to_upper(string, locale = "")
str_to_lower(string, locale = "")
str_to_title(string, locale = "")
```

Parameter list:
   - string: The string.
   - locale: Sort by which language conventions.

Convert the string to uppercase.

```
> val <- "I am conan. Welcome to my blog! http://fens.me"

# All to uppercases
> str_to_upper(val)
[1] "I AM CONAN. WELCOME TO MY BLOG! HTTP://FENS.ME"

# All to lowercases
> str_to_lower(val)
[1] "i am conan. welcome to my blog! http://fens.me"
```

```
# Capitalize the initial letters
> str_to_title(val)
[1] "I Am Conan. Welcome To My Blog! Http://Fens.Me"
```

Strings are frequently used in data processing and they are needed to be segmented, concatenated, converted, etc. stringr introduced in this section is a flexible string processing library that effectively improves the coding efficiency. With this effective tool, it will be convenient and efficient to process strings in R language.

3.5 Chinese Words Segmentation Package of R: jiebaR

Question
How do you segment Chinese words with R language?

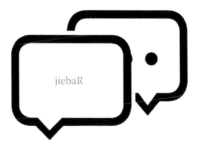

Chinese words segmentation
Package of R : jiebaR
http://blog.fens.me/r-word-jiebar/

Introduction
Text mining is a very important part of data mining. It can be promising in many applications, e.g. with it, we can analyze news events to have knowledge of state affairs; we can also analyze micro-blog information to know the concerns of social opinions. We can find the hidden information of the articles and analyze their structures by text mining and then judge whether they are from the same author; we can also use it, along with bayes algorithm, to analyze emails and find out which are junk mails and which are not.

The first step of text mining is to segment words, which will have a direct impact on the effects of text mining. R language supports word segmentation very well. In the following, I will introduce a good Chinese words segmentation package of R language: jiebaR.

3.5.1 An Introduction to jiebaR Package

jiebaR is a high-efficient Chinese words segmentation package of R language, with C++ as its underlying layer. It will be efficient when called with Rcpp. jiebaR is based on the MIT protocol and it is free and open source. Thanks to the strong support from the Chinese developers, now it is convenient to process Chinese text with R.

The official website of Github: https://github.com/qinwf/jiebaR

The system environment used here:

- Win10 64bit
- R: 3.2.3 x86_64-w64-mingw32/x64 b4bit

The jiebaR package can be found in the standard library of CRAN. Its installation is very simple, just two commands to go.

```
~ R
> install.packages("jiebaR")
> library("jiebaR")
```

You can use the devtools package to install the developer edition. For usage of devtools, please refer to Section 5.2 of *R for Programmers: Advanced Techniques*.

```
> library(devtools)
> install_github("qinwf/jiebaRD")
> install_github("qinwf/jiebaR")
> library("jiebaR")
```

To install the developer edition, the official recommendation will be to use the gcc> = 4.6 compiler of the Linux system. If it is Windows, Rtools should be installed.

3.5.2 Learn jiebaR in Five Minutes

To learn jiebaR in 5 minutes, we can just jump to the first example to segment a paragraph of text.

```
> wk = worker()

> wk["我是《R的极客理想》图书作者"]
[1] "我是"  "R"      "的"     "极客" "理想" "图书" "作者"

> wk["我是R语言的深度用户"]
[1] "我"    "是"     "R"      "语言" "的"     "深度" "用户"
```

The Chinese words segmentation is completed with two lines of codes. Simple.

jiebaR provides three syntaxes of word segmentation. The example above adopted the syntax of [] notation. There are <= notation syntax and segment() function syntax. They are of different forms, but their results of word segmentation are all the same. If the <= notation syntax is adopted, the result will be as follow:

```
> wk<='另一种符合的语法'
[1] "另"    "一种" "符合" "的"     "语法"
```

If the segment() function syntax is adopted, the result will be as follow:

```
> segment( "segment()函数语句的写法" , wk )
[1] "segment" "函数"    "语句"    "的"     "写法"
```

If this amazes you and you want to know how to self-define an operator, you can check the project source file named quick.R.

```
# <= notation definition
`<=.qseg`<-function(qseg, code){
  if(!exists("quick_worker",envir = .GlobalEnv ,inherits = F) ||
      .GlobalEnv$quick_worker$PrivateVarible$timestamp != TIMESTAMP){
```

```
    if(exists("qseg",envir = .GlobalEnv,inherits = FALSE ) )
      rm("qseg",envir = .GlobalEnv)

    modelpath   = file.path(find.package("jiebaR"),"model","model.rda")
    quickparam = readRDS(modelpath)

    if(quickparam$dict == "AUTO") quickparam$dict = DICTPATH
    if(quickparam$hmm == "AUTO") quickparam$hmm = HMMPATH
    if(quickparam$user == "AUTO") quickparam$user = USERPATH
    if(quickparam$stop_word == "AUTO") quickparam$stop_word = STOPPATH
    if(quickparam$idf == "AUTO") quickparam$idf = IDFPATH

    createquickworker(quickparam)
    setactive()
  }

  #..codes omitted
}

# [ notation definition
`[.qseg`<- `<=.qseg`
```

If we want to directly segment the text file, we can create a new text file idea.txt in the current directory.

```
~ notepad idea.txt
```

R的极客理想系列文章，涵盖了R的思想，使用，工具，创新等的一系列要点，以我个人的学习和体验去诠释R的强大。

R语言作为统计学一门语言，一直在小众领域闪耀着光芒。 直到大数据的爆发，R语言变成了一门炙手可热的数据分析的利器。 随着越来越多的工程背景的人的加入，R语言的社区在迅速扩大成长。现在已不仅仅是统计领域，教育，银行，电商，互联网….都在使用R语言。

Of course, the word segmentation program will generate a new file of the segmentation result in the current directory.

```
> wk['./idea.txt']
[1] "./idea.segment.2016-07-20_23_25_34.txt"
```

Open the file idea.segment.2016-07-20_23_25_34.txt, and segment the words of the text with spaces.

```
~ notepad idea.segment.2016-07-20_23_25_34.txt
```

R 的 极客 理想 系列 文章 涵盖 了 R 的 思想 使用 工具 创新 等 的 一系列 要点 以 我个人 的 学习 和 体验 去 诠释 R 的 强大 R 语言 作为 统计学 一门 语言 一直 在 小众 领域 闪耀着 光芒 直到 大 数据 的 爆发 R 语言 变成 了 一门 炙手可热 的 数据分析 的 利器随着 越来越 多 的 工程 背景 的 人 的 加入 R 语言 的 社区 在 迅速 扩大 成长 现在 已不仅仅 是 统计 领域 教育 银行 电商 互联网 都 在 使用 R 语言

Isn't it simple? It takes just 5 minutes to complete the word segmentation task.

3.5.3 *Word Segmentation Engine*

When calling function worker(), we are actually loading the word segmentation engine of jiebaR library. The jiebaR library provides seven-word segmentation engines.

- MixSegment: It is a word segmentation engine with better effect than the other seven and it combines MPSegment and HMMSegment.
- MPSegment: It builds directed acyclic graphs and implements the dynamic programming algorithm according to Trie tree. It is the core of word segmentation algorithm.
- HMMSegment: It segments words according to the HMM model built by the corpus based on People's Daily, etc. Its main algorithm idea is to represent the hidden status of every character with these four statuses: B, E, M and S. The HMM model is provided by dict/hmm_model.utf8. Its word segmentation algorithm is viterbi algorithm.
- QuerySegment: It first segments words by MixSegment and then for those long words, enumerates all the possible words in a sentence and finds the matching words in the library.
- Tag
- Simhash
- Keywods

If engines are not a concern for you, you can just choose the officially recommended MixSegment (the default option).

Check the definition of function worker().

```
worker(type = "mix", dict = DICTPATH, hmm = HMMPATH, user = USERPATH,
  idf = IDFPATH, stop_word = STOPPATH, write = T, qmax = 20, topn = 5,
  encoding = "UTF-8", detect = T, symbol = F, lines = 1e+05,
  output = NULL, bylines = F, user_weight = "max")
```

Parameter list:

- type, the engine type
- dict, the system dictionary
- hmm, the path of HMM model
- user, the user's dictionary
- idf, the IDF dictionary
- stop_word, the library of keywords that are not in use any more.
- write, whether to write the word segmentation results to a file. FALSE by default.
- Qmax, the maximum characters for a word. 20 by default.
- topn, the number of keywords. 5 by default.
- encoding, the encoding of the input file. UTF-8 by default.
- detect, whether to run coding check. TRUE by default.
- symbol, whether to keep the symbols. FALSE by default.
- lines, the maximum lines of every file read, used to control the length of the read file. Big files will be read by several times.
- output, the output path
- bylines, to output by lines
- user_weight, the user weight

When we call worker(), we load the word segmentation engine. We can print it out and check the configurations of the engine.

```
> wk = worker()
> wk
Worker Type: Jieba Segment

Default Method   :   mix       # MixSegment
Detect Encoding  :   TRUE      # Check the encoding
Default Encoding:  UTF-8       # UTF-8
Keep Symbols     :   FALSE     # Do not keep the symbols
Output Path      :             # The file output path
Write File       :   TRUE      # Write the file
By Lines         :   FALSE     # Do not output by lines
Max Word Length  :   20        # The maximum word length
Max Read Lines   :   1e+05     # The maximum lines of a file read

Fixed Model Components:

$dict                    # System dictionary
[1] "D:/tool/R-3.2.3/library/jiebaRD/dict/jieba.dict.utf8"

$user                    # User dictionary
[1] "D:/tool/R-3.2.3/library/jiebaRD/dict/user.dict.utf8"

$hmm                     # HMMSegment
[1] "D:/tool/R-3.2.3/library/jiebaRD/dict/hmm_model.utf8"

$stop_word               # No stop word
NULL

$user_weight             # The weight of user dictionary
[1] "max"

$timestamp               # The time stamp
[1] 1469027302

$default $detect $encoding $symbol $output $write $lines $bylines can be
reset.
```

If we want to change the configurations of the word segmentation engine, we can define the configurations during the call of worker() or we can set the configurations by wk$XX. We need to know the object type of wk, which can be figured out with function otype of the pryr package. For usage of the pryr package, please refer to Section 3.1 The Advanced Toolkit pryr Which Levers the R Kernel of *R for Programmers: Advanced Techniques*.

```
# Load the pryr package
> library(pryr)
> otype(wk)  # Object-oriented type check
[1] "S3"

> class(wk)  # Check the attributes of class
[1] "jiebar"  "segment" "jieba"
```

3.5.4 *Configuration of Dictionary*

Dictionary is the key factor to whether a result of word segmentation is good or bad. There is a default standard dictionary in jiebaR. We'd better use special dictionaries for different industries or different language types. The default standard dictionary can be viewed by the function show_dictpath() of jiebaR. We can make our own dictionary by specifying the configuration items introduced in the previous section. The frequently used daily conversation dictionary can be, for example, Sogo Input Library.

```
# View the path of the default library
> show_dictpath()
[1] "D:/tool/R-3.2.3/library/jiebaRD/dict"

# View the contents
> dir(show_dictpath())
[1] "D:/tool/R-3.2.3/library/jiebaRD/dict"
 [1] "backup.rda"       "hmm_model.utf8"  "hmm_model.zip"
 [4] "idf.utf8"         "idf.zip"         "jieba.dict.utf8"
 [7] "jieba.dict.zip"   "model.rda"       "README.md"
[10] "stop_words.utf8"  "user.dict.utf8"
```

We find many files in the dictionary contents.

- jieba.dict.utf8, a system dictionary file, MPSegment, utf8 encoded
- hmm_model.utf8, a system dictionary file, HMMSegment, utf8 encoded
- user.dict.utf8, a user dictionary file, utf8 encoded
- stop_words.utf8, a stop word file, utf8 encoded
- idf.utf8, an IDF corpus, utf8 encoded
- jieba.dict.zip, the zip of jieba.dict.utf8
- hmm_model.zip, the zip of hmm_model.utf8
- idf.zip, the zip of idf.utf8
- backup.rda, no annotation
- model.rda, no annotation
- README.md, the help file for users

Open the system dictionary file jieba.dict.utf8 and print the first 50 lines.

```
> scan(file="D:/tool/R-3.2.3/library/jiebaRD/dict/jieba.dict.utf8",
+         what=character(),nlines=50,sep='\n',
+         encoding='utf-8',fileEncoding='utf-8')
Read 50 items
 [1] "1号店 3 n"   "1號店 3 n"   "4S店 3 n"    "4s店 3 n"
 [5] "AA制 3 n"    "AB型 3 n"    "AT&T 3 nz"   "A型 3 n"
 [9] "A座 3 n"     "A股 3 n"     "A輪 3 n"     "A轮 3 n"
[13] "BB机 3 n"    "BB機 3 n"    "BP机 3 n"    "BP機 3 n"
[17] "B型 3 n"     "B座 3 n"     "B股 3 n"     "B超 3 n"
[21] "B輪 3 n"     "B轮 3 n"     "C# 3 nz"     "C++ 3 nz"
[25] "CALL机 3 n"  "CALL機 3 n"  "CD机 3 n"    "CD機 3 n"
[29] "CD盒 3 n"    "C座 3 n"     "C盘 3 n"     "C盤 3 n"
[33] "C語言 3 n"   "C语言 3 n"   "D座 3 n"     "D版 3 n"
[37] "D盘 3 n"     "D盤 3 n"     "E化 3 n"     "E座 3 n"
```

```
[41] "E盘 3 n"        "E盤 3 n"        "E通 3 n"        "F座 3 n"
[45] "F盘 3 n"        "F盤 3 n"        "G盘 3 n"        "G盤 3 n"
[49] "H盘 3 n"        "H盤 3 n"
```

We find that there are three columns for every row in the dictionary, segmented by spaces. The first column is the lexical item, the second column is the lexical frequency and the third is the part-of-speech tag.

Open the user dictionary file user.dict.utf8 and print the first 50 lines.

```
> scan(file="D:/tool/R-3.2.3/library/jiebaRD/dict/user.dict.utf8",
+       what=character(),nlines=50,sep='\n',
+       encoding='utf-8',fileEncoding='utf-8')
Read 5 items
[1] "云计算"       "韩玉鉴赏"  "蓝翔 nz"  "CEO"          "江大桥"
```

There are two columns of every row in the user dictionary, with the first column for the lexical item, second column for the part-of-speech tag and no column for the lexical frequency. The default lexical frequency in the user dictionary is the maximum lexical frequency in the system dictionary.

The jiebaR package uses ICTCLAS as its dictionary part-of-speech tagging method. ICTCLAS is the collection of the Chinese part-of-speech tags.

Code	Name	Interpretation That Helps Memorizing
Ag	Adjective morpheme	The adjective morpheme. The code for adjective is *a* and the code for morpheme is *g*. Capitalize *A* for it is the initial letter.
A	Adjective	Take the first letter of "adjective" in English.
Ad	Adverbial adjective	The adjective that can be directly used as an adverbial. *ad* is the combination of the adjective code *a* and the adverb code *d*.
an	Adnoun	The adjective that can function as a noun. *an* is the combination of the adjective code *a* and the noun code *d*.
b	Distinguishing word	*b* is the initial consonant of the Chinese character *bie* (distinguish).
c	Conjunction	*c* is the initial letter of the English word "conjunction."
Dg	Adverb morpheme	The adverb morpheme. The code for adverb is *d* and the code for morpheme is *g*. Capitalize *D* for it is the initial letter.
d	Adverb	*d* is the second letter of "adverb." The initial letter is not taken here because it has been the code of adjective.
e	Exclamation	*e* is the initial letter of the English word "exclamation."
f	Noun of locality	*f* is the initial consonant of the Chinese character *fang* (locality).
g	Morpheme	Most of the morphemes can be the roots of compound words. *g* is the initial consonant of the Chinese character *gen* (root).

(Continued)

(*Continued*)

h	Head constituent	*h* is the initial letter of the English word "head."
i	Idiom	*i* is the initial letter of the English word "idiom."
j	Abbreviation	*j* is the initial consonant of the Chinese character *jian* (abbreviation)
k	Following constituent	
l	Phrase	A phrase is not an idiom. It is a little temporary. *l* is the initial consonant of the Chinese character *lin* (temporary).
m	Numeral	*m* is the third letter of the English word "numeral," for *n* and *u* have been put to other uses.
Ng	N-morpheme	The noun morpheme. The code for noun is *n* and the code for morpheme is *g*. Capitalize *N* for it is the initial letter.
n	Noun	*n* is the initial letter of the English word "noun."
nr	Person name	*nr* is the combination of the noun code *n* and the initial consonant of Chinese character *ren* (person).
ns	Address	*ns* is the combination of the noun code *n* and the space codes.
nt	Organization	*nt* is the combination of the noun code *n* and the initial consonant of Chinese character *tuan* (organization).
nz	Other proper names	*nz* is the combination of the noun code *n* and the initial consonant of Chinese character *zhuan* (proper).
o	onomatopoeia	*o* is the initial letter of the English word "onomatopoeia."
p	Prepositional	*p* is the initial letter of the English word "prepositional."
q	Quantity	*q* is the initial letter of the English word "quantity."
r	Pronoun	*r* is the second letter of the English word "pronoun," for *p* has been put to other use.
s	Space	*s* is the initial letter of the English word "space."
Tg	Time morpheme	The time morpheme. The time code is *t* and the morpheme code is *g*. Capitalize *T* for it is the initial letter.
t	Time	*t* is the initial letter of the English word "time."
u	Auxiliary	*u* is the second letter of the English word "auxiliary," for *a* has been put to other use.
Vg	V-morpheme	The verb morpheme. The code of verb is *v*. the morpheme code is *g*. Capitalize *V* for it is the initial letter.
v	Verb	*v* is the initial letter of the English word "verb."

(*Continued*)

(Continued)

vd	Adverbial verb	The verb that can be directly used as an adverbial. *vd* is the combination of verb and adverb codes.
vn	Nominal verb	The verb that can function as a noun. *vn* is the combination of verb and noun codes.
w	Punctuations	
x	Non-morpheme character	The non-morpheme character is a symbol. The letter *x* is usually used to represent the unknowns and symbols.
y	Modal particle	*y* is the initial consonant of the Chinese character *yu*.
z	Descriptive word	*z* is the initial consonant of the Chinese character *zhuang*.

Next, let's define a user dictionary and see what will happen. Write the dictionary file user. utf8.

```
~ notepad user.utf8
```

```
R语言
R的极客理想
大数据
数据
```

Then segment the text with the user dictionary we defined just now.

```
> wk = worker(user='user.utf8')
> wk['./idea.txt']
[1] "./idea.segment.2016-07-21_11_14_24.txt"
```

Compare these two results of word segmentation: idea.segment.2016-07-20_23_25_34.txt and idea.segment.2016-07-21_11_14_24.txt (Figure 3.10).

In practical use, the default user dictionary of jiebaR, containing only five words, is too simple and cannot be adequate. We can use Sogou dictionary to enrich our user dictionary. In the following, let's configure the Sogou dictionary. First, we need to install a Sogo Input. The detailed installation process will not be introduced here.

What I've installed is the Sogo Five-stroke Input. Go to the installation directory of Sogou Input and find the dictionary file. My Sogou dictionary is installed under the directory:

```
C:\Program Files (x86)\SogouWBInput\2.1.0.1288\scd\17960.scel
```

Copy the 17960.scel file to our own project directory and open the file with text editor. We find that the file is binary. Then we need to convert this binary dictionary file to a text file with tools. The developer of jiebaR has developed a cidian project, which can be used to convert Sogou dictionary, so what we need to do is to install the cidian package.

Install the cidian package.

```
> install.packages("devtools")
> install.packages("stringi")
```

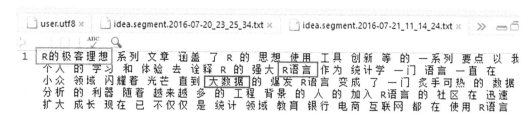

Figure 3.10   Compare these two results of word segmentation.

```
> install.packages("pbapply")
> install.packages("Rcpp")
> install.packages("RcppProgress")
> library(devtools)
> install_github("qinwf/cidian")
> library(cidian)
```

Convert the binary dictionary to a text file.

```
# Conversion
> decode_scel(scel = "./17960.scel",cpp = TRUE)
output file: ./17960.scel_2016-07-21_00_22_11.dict

# View the dictionary file
> scan(file="./17960.scel_2016-07-21_00_22_11.dict",
+      what=character(),nlines=50,sep='\n',
+      encoding='utf-8',fileEncoding='utf-8')
Read 50 items
 [1] "阿坝州 n"          "阿百川 n"          "阿班 n"
 [4] "阿宾 n"            "阿波菲斯 n"        "阿不都热希提 n"
 [7] "阿不都西库尔 n"    "阿不力克木 n"      "阿尔姆格伦 n"
[10] "阿尔沙文 n"        "阿肥星 n"          "阿菲正传 n"
[13] "阿密特 n"          "阿穆 n"            "阿穆隆 n"
[16] "阿帕鲁萨镇 n"      "阿披实 n"          "阿衰 n"
[19] "阿霞 n"            "艾奥瓦 n"          "爱不欻 n"
[22] "爱的错位 n"        "爱得得体 n"        "爱的火焰 n"
[25] "爱的流刑地 n"      "爱得起 n"          "埃夫隆 n"
[28] "爱搞网 n"          "爱国红心 n"        "爱呼 n"
```

```
[31]  "爱就宅一起  n"       "埃克希儿  n"        "爱没有错  n"
[34]  "埃蒙斯  n"          "爱奴新传  n"        "爱起点  n"
[37]  "爱情的牙齿  n"       "爱情海滨  n"        "爱情节  n"
[40]  "爱情美的样子  n"     "爱情无限谱  n"       "爱情占线  n"
[43]  "爱情转移  n"         "爱情左灯右行  n"     "爱上你是一个错  n"
[46]  "矮哨兵  n"          "爱是妥协  n"        "爱似水仙  n"
[49]  "爱太痛  n"          "爱无界  n"
```

Then set the Sogou dictionary to our segmentation lexicon and it is ready for use. Rename the Sogou dictionary from 17960.scel_2016-07-21_00_22_11.dict to user.dict.utf8 and then replace D:/tool/R-3.2.3/library/jiebaRD/dict/user.dict.utf8 with it. Now the default user dictionary is the Sogou dictionary. Pretty cool, right?

3.5.5 Stop Word Filter

A stop word is the word that will not be put into the result during the segmentation. The English words like *a, the, or, and*, etc. and the Chinese words such as 的, 地, 得, 我, 你, 他, etc. can all be categorized as stop words. These words are used in high frequency and there can be great amount of them in a text, which will become the noises for lexical frequency count of the segmentation result, so we usually filter these words.

There are two ways to filter the stop words. One is to filter them by configuring the stop_word file, another is by using the filter_segment() function.

Let's first see how to filter them by configuring the stop_word file. Create a stop_word.txt file.

```
~ notepad stop_word.txt

我
我是
```

Load the word segmentation engine and configure the stop word filtering.

```
> wk = worker(stop_word='stop_word.txt')
> segment<-wk["我是《R的极客理想》图书作者"]
> segment
[1] "R"     "的"     "极客" "理想" "图书" "作者"
```

In the text above, we filtered "我是" by stop words. If we still need to filter the word "作者", we can dynamically call the filter_segment() function.

```
> filter<-c("作者")
> filter_segment(segment,filter)
[1] "R"     "的"     "极客" "理想" "图书"
```

3.5.6 Keyword Abstraction

Keyword abstraction is a very important step of text processing. A classic algorithm can be used here, i.e. the TF-IDF algorithm, of which TF means Term Frequency and IDF means Inverse Document Frequency. If a certain word appears repeatedly and it is not a stop word, it can reflect the character of this article and this is the key word we need. Then we can calculate the weight of every word by IDF. The more frequently a rare word appears, the more weight it has. The calculation formula of TF-IDF is:

```
TF-IDF = TF (Term Frequency) * IDF (Inverse Document Frequency)
```

Calculate the IF-IDF value of every word in the text, sort the result from small to large and then we can get the criticality sorting list of this text.

The jiebaR package also adopts the TF-IDF algorithm to implement the keyword abstraction. The idf.utf8 file under the installation directory is the IDF corpus. View the contents of idf.utf8.

```
> scan(file="D:/tool/R-3.2.3/library/jiebaRD/dict/idf.utf8",
+       what=character(),nlines=50,sep='\n',
+       encoding='utf-8',fileEncoding='utf-8')
Read 50 items
  [1] "劳动防护 13.900677652"      "生化学 13.900677652"
  [3] "奥萨贝尔 13.900677652"      "考察队员 13.900677652"
  [5] "岗上 11.5027823792"        "倒车档 12.2912397395"
  [7] "编译 9.21854642485"        "蝶泳 11.1926274509"
  [9] "外委 11.8212361103"        "故作高深 11.9547675029"
 [11] "尉遂成 13.2075304714"      "心源性 11.1926274509"
 [13] "现役军人 10.642581114"     "杜勃留 13.2075304714"
 [15] "包天笑 13.900677652"       "贾政陪 13.2075304714"
 [17] "托尔湾 13.900677652"       "多瓦 12.5143832909"
 [19] "多瓣 13.900677652"         "巴斯特尔 11.598092559"
 [21] "刘皇帝 12.8020653633"      "亚历山德罗夫 13.2075304714"
 [23] "社会公众 8.90346537821"    "五百份 12.8020653633"
 [25] "两点阈 12.5143832909"      "多瓶 13.900677652"
 [27] "冰天 12.2912397395"        "库布齐 11.598092559"
 [29] "龙川县 12.8020653633"      "银燕 11.9547675029"
 [31] "历史风貌 11.8212361103"    "信仰主义 13.2075304714"
 [33] "好色 10.0088573539"        "款款而行 12.5143832909"
 [35] "凳子 8.36728816325"        "二部 9.93038573842"
 [37] "卢巴 12.1089181827"        "五百五 13.2075304714"
 [39] "畅叙 11.598092559"         "吴栅子 13.2075304714"
 [41] "智力竞赛 13.900677652"     "库邦 13.2075304714"
 [43] "非正义 11.3357282945"      "编订 10.2897597393"
 [45] "悲号 12.8020653633"        "陈庄搭 13.2075304714"
 [47] "二郎 9.62401153296"        "电光石火 11.8212361103"
 [49] "抢球 11.9547675029"        "南澳大利亚 10.9562386728"
```

There are two columns of every row in the idf.utf8 file, with lexical item as the first column and weight the second. Calculate the text TF and multiple it by the IDF value of the corpus; then we get the TF-IDF value and we can extract the keywords from the text.

For example, let's extract the keywords from the text below.

```
> wk = worker()
> segment<-wk["R的极客理想系列文章，涵盖了R的思想，使用，工具，创新等的一系列要点，
以我个人的学习和体验去诠释R的强大。"]

# Calculate the TF
> freq(segment)
    char freq
1   创新    1
```

```
2        了        1
3       文章       1
4       强大       1
5        R         3
6       个人       1
7        的        5
8       诠释       1
9        和        1
10     一系列      1
11      使用       1
12       以        1
13       等        1
14      极客       1
15      理想       1
16      思想       1
17      涵盖       1
18      系列       1
19       去        1
20       我        1
21      工具       1
22      学习       1
23      体验       1
24      要点       1

# Take the first five keyworkds of TF-IDF
> keys = worker("keywords",topn=5)

# Calculate the keywords
> vector_keywords(segment,keys)
11.7392  8.97342  8.23425   8.2137  7.43298
 "极客"    "诠释"    "要点"    "涵盖"    "体验"
```

It is simple to process word segmentation with the jiebaR package. Various algorithm operations of segmentation can be implemented by only a few lines of codes. With this tool, we can find the language rules in texts and run text mining as well. When there is time, I am going to text mine the announcements of listed companies. Maybe I can discover some market rules and make money from them.

The usage of the jiebaR package introduced in this section is just a beginning. For more detailed operations, please refer to the official introduction from the package developer (https://jiebar.qinwf.com/). Thank the jiebaR developer @qinwenfeng again for the contribution of this great R language package on the Chinese words segmentation.

Chapter 4

High Performance
Computing of R

The performance of R language is always a problem of everyone's concern. There are many solutions to improve the performance of R, but due to its single-threaded kernel design, it is difficult to come out with leaps in its performance improvement. In this chapter, three external technologies will be introduced to help the performance of R language meet the requirements of production environment.

4.1 OpenBlas Speeds Up Matrix Calculation of R

Question
How can you improve the matrix calculation efficiency of R language?

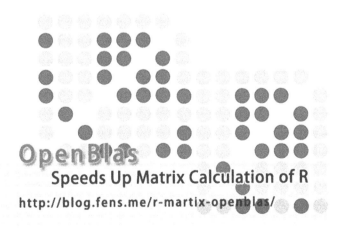

OpenBlas
Speeds Up Matrix Calculation of R
http://blog.fens.me/r-martix-openblas/

Introduction

I went to the sharing themed OpenBlas again in the 2015 IBM Open Source Community Conference. This time I must try it by myself. I learned various advantages of OpenBlas for the first time in the R Conference 2 years ago, but I did not attach much importance to it at that time. After 2 years, we come back to it. My team is stable and the project has been widely accepted and is progressing well. Now it is time to try OpenBlas.

In IAIS2016 (2016 China Optics International A.I. Industries Industry Forum), I saw the theme of OpenBlas again and the author Zhang Xianyi has got his PhD degree and found his own company PerfXLab. OpenBlas has become a standard high-performance computing library.

4.1.1 An Introduction to OpenBlas

OpenBlas is an open source project initiated by the Laboratory of Parallel Software and Computational Science Institute of Software Chinese Academy of Sciences. It is a high-performance implementation of the open source BLAS (Basic Linear Algebra Subprograms) library based on the GotoBLAS2 1.13 BSD (Berkeley Software Distribution) version.

BLAS is an application program interface (API) standard, used to regulate the database where the basic linear algebra operations (e.g. the vector or matrix multiplication) are published. The subprograms were first released in 1979, used to build larger numerical packages (e.g. LAPACK (Linear Algebra Package)). BLAS is widely used in the area of high performance computing. For example, the computing performance of LINPACK (Linear System Package) depends largely on the performance of DGEMM (Double-Precision General Matrix Multiply), a subprogram of BLAS. To improve the performance, many software vendors and hardware vendors are trying to highly optimize the BLAS interface implementation of their products.

Project homepage: http://www.openblas.net/.

4.1.2 Installation of R and OpenBlas

OpenBlas can improve the performance of matrix computing for every language substratum. Let's try to combine R and OpenBlas.

The system environment used here:

■ Linux Ubuntu 14.01.1
■ CPU dual-core Intel(R) Xeon(R) CPU E5-2650 v2 @ 2.60GHz
■ Memory 4G
■ R: version 3.2.2 x86_64-pc-linux-gnu (64-bit)

Check the system parameters by commands.

```
# Operation system
~ cat /etc/issue
Ubuntu 14.04.1 LTS \n \l

# CPU
~ cat /proc/cpuinfo
processor    : 0
vendor_id    : GenuineIntel
cpu family   : 6
```

```
model       : 62
model name  : Intel(R) Xeon(R) CPU E5-2650 v2 @ 2.60GHz
stepping    : 4
microcode   : 0x428
cpu MHz     : 2600.048
cache size  : 20480 KB
physical id : 0
siblings    : 2
core id     : 0
cpu cores   : 2
apicid      : 0
initial apicid  : 0
fpu     : yes
fpu_exception   : yes
cpuid level : 13
wp      : yes
flags       : fpu vme de pse tsc msr pae mce cx8 apic sep mtrr pge mca
              cmov pat clflush mmx fxsr sse sse2 ht syscall nx rdtscp lm
              constant_tsc rep_good nopl pni ssse3 cx16 sse4_1 sse4_2
              popcnt aes hypervisor lahf_lm
bogomips    : 5200.09
clflush size    : 64
cache_alignment : 64
address sizes   : 46 bits physical, 48 bits virtual
power management:

processor   : 1
vendor_id   : GenuineIntel
cpu family  : 6
model       : 62
model name  : Intel(R) Xeon(R) CPU E5-2650 v2 @ 2.60GHz
stepping    : 4
microcode   : 0x428
cpu MHz     : 2600.048
cache size  : 20480 KB
physical id : 0
siblings    : 2
core id     : 1
cpu cores   : 2
apicid      : 2
initial apicid  : 2
fpu     : yes
fpu_exception   : yes
cpuid level : 13
wp      : yes
flags       : fpu vme de pse tsc msr pae mce cx8 apic sep mtrr pge mca
              cmov pat clflush mmx fxsr sse sse2 ht syscall nx rdtscp lm
              constant_tsc rep_good nopl pni ssse3 cx16 sse4_1 sse4_2
              popcnt aes hypervisor lahf_lm
bogomips    : 5200.09
clflush size    : 64
cache_alignment : 64
address sizes   : 46 bits physical, 48 bits virtual
power management:
```

```
# Memory
~ cat /proc/meminfo
MemTotal: 4046820 kB
MemFree: 1572372 kB
Buffers: 40588 kB
Cached: 709684 kB
SwapCached: 0 kB
Active: 1953940 kB
Inactive: 418084 kB
Active(anon):    1621840 kB
Inactive(anon):     5732 kB
Active(file):     332100 kB
Inactive(file):   412352 kB
Unevictable: 0 kB
Mlocked: 0 kB
SwapTotal: 0 kB
SwapFree: 0 kB
Dirty: 24 kB
Writeback: 0 kB
AnonPages: 1623792 kB
Mapped: 34936 kB
Shmem: 5828 kB
Slab: 58024 kB
SReclaimable: 45252 kB
SUnreclaim: 12772 kB
KernelStack: 1512 kB
PageTables: 8980 kB
NFS_Unstable: 0 kB
Bounce: 0 kB
WritebackTmp: 0 kB
CommitLimit: 2023408 kB
Committed_AS: 2556460 kB
VmallocTotal: 34359738367 kB
VmallocUsed: 9664 kB
VmallocChunk: 34359723308 kB
HardwareCorrupted: 0 kB
AnonHugePages: 1562624 kB
HugePages_Total: 0
HugePages_Free: 0
HugePages_Rsvd: 0
HugePages_Surp: 0
Hugepagesize: 2048 kB
DirectMap4k: 28672 kB
DirectMap2M: 4296704 kB
```

First, we can use just one command in Linux Ubuntu to install the operation environment of R language.

```
# Install R language
~ sudo apt-get install r-base

#View the version of R language
~ R --version
R version 3.2.2 (2015-08-14) -- "Fire Safety"
```

```
Copyright (C) 2015 The R Foundation for Statistical Computing
Platform: x86_64-pc-linux-gnu (64-bit)
R is free software and comes with ABSOLUTELY NO WARRANTY.
You are welcome to redistribute it under the terms of the
GNU General Public License versions 2 or 3.
For more information about these matters see
http://www.gnu.org/licenses/.
```

The latest version of R language we have installed is 3.2.2. Now let's run a matrix computing, e.g. multiple two square matrices of 3,000 rows and 3,000 columns.

```
# Launch R
~ R

# Multiple two square matrices
> x <- matrix(1:(3000 * 3000), 3000, 3000)

# Calculate the consumption of time
> system.time(tmp <- x %*% x)
   user  system elapsed
 33.329   0.332  33.788
```

Next, we will install OpenBlas to improve the computing performance. It will take only one command to install OpenBlas in Ubuntu. Pretty simple.

```
~ sudo apt-get install libopenblas-base
```

Switch the computing engine of blas from libblas to openblas.

```
~ sudo update-alternatives --config libblas.so.3
There are 2 choices for the alternative libblas.so.3 (providing /usr/lib/
libblas.so.3).
```

Selection	Path	Priority	Status
* 0	/usr/lib/openblas-base/libblas.so.3	40	auto mode
1	/usr/lib/libblas/libblas.so.3	10	manual mode
2	/usr/lib/openblas-base/libblas.so.3	40	manual mode

```
Press enter to keep the current choice[*], or type selection number: 0
```

Choose 0 and use the openblas-base engine.

Let's re-launch the operation environment of R and execute the matrices multiplication again.

```
~ R
> x <- matrix(1:(3000 * 3000), 3000, 3000)

# Calculate the consumption of time
> system.time(tmp <- x %*% x)
   user  system elapsed
  7.391   0.127   3.869
```

Something miraculous has happened. The calculation has been speeded up by more than four times. OpenBlas can accelerate the calculation of matrices. Let's try it on all the matrix operations.

4.1.3 Speed Up R Language

I have found two scripts for the R language performance tests on the Internet. We can test them in our environment. Download the scripts from the release page of the benchmarks.

I found that Revolution Analytics Company has tested this script, so I compared the Revolution's enterprise edition with the official release of R.

Download the scripts.

```
~ wget http://brettklamer.com/assets/files/statistical/
  faster-blas-in-r/R-benchmark-25.R
--2015-09-24 12:06:05-- http://brettklamer.com/assets/files/statistical/
faster-blas-in-r/R-benchmark-25.R
Resolving brettklamer.com (brettklamer.com)... 199.96.156.242
Connecting to brettklamer.com (brettklamer.com)|199.96.156.242|:80...
connected.
HTTP request sent, awaiting response... 200 OK
Length: 13666 (13K)
Saving to: 'R-benchmark-25.R'

100%[====================================================================
=======================>] 13,666       --.-K/s   in 0s

2015-09-24 12:06:06 (203 MB/s) - 'R-benchmark-25.R' saved [13666/13666]
```

Run the scripts.

```
~ R

# Run the script
> source("R-benchmark-25.R")
Loading required package: Matrix
Loading required package: SuppDists


   R Benchmark 2.5
   ===============
Number of times each test is run_____ : 3

   I. Matrix calculation
   ---------------------
Creation, transp., deformation of a 2500x2500 matrix (sec):  1.103
2400x2400 normal distributed random matrix ^1000____ (sec):
0.812333333333333
Sorting of 7,000,000 random values_____ (sec):
0.962666666666667
2800x2800 cross-product matrix (b=a' * a)_____ (sec):  1.547
Linear regr. over a 3000x3000 matrix (c=a \ b')___ (sec):
0.828000000000001
                   ------------------------------------------
                Trimmed geom. mean (2 extremes eliminated):
0.957989159036612

   II. Matrix functions
   --------------------
```

```
FFT over 2,400,000 random values_____ (sec):
0.365333333333335
Eigenvalues of a 640x640 random matrix_____ (sec):
1.43466666666667
Determinant of a 2500x2500 random matrix_____ (sec):
0.895999999999998
Cholesky decomposition of a 3000x3000 matrix_____ (sec):
0.832000000000003
Inverse of a 1600x1600 random matrix_____ (sec):
0.724333333333334
                 -----------------------------------------
             Trimmed geom. mean (2 extremes eliminated):
0.814310314522547


   III. Programmation
   ------------------
3,500,000 Fibonacci numbers calculation (vector calc)(sec):
0.776666666666661
Creation of a 3000x3000 Hilbert matrix (matrix calc) (sec):
0.269666666666671
Grand common divisors of 400,000 pairs (recursion)__ (sec):
0.570666666666663
Creation of a 500x500 Toeplitz matrix (loops)_____ (sec):
0.506666666666665
Escoufier's method on a 45x45 matrix (mixed)_____ (sec):
0.533000000000001
                 -----------------------------------------
             Trimmed geom. mean (2 extremes eliminated):
0.536138937440438


Total time for all 15 tests_____ (sec):   12.162
Overall mean (sum of I, II and III trimmed means/3)_ (sec):
0.747841037469598
                      --- End of test ---
```

Then we switch to the default blas engine of R language and run the scripts again.

```
~ sudo update-alternatives --config libblas.so.3
There are 2 choices for the alternative libblas.so.3 (providing /usr/lib/
libblas.so.3).


  Selection    Path                                    Priority   Status
  --------------------------------------------------------------------
* 0            /usr/lib/openblas-base/libblas.so.3      40        auto mode
  1            /usr/lib/libblas/libblas.so.3            10        manual mode
  2            /usr/lib/openblas-base/libblas.so.3      40        manual mode


Press enter to keep the current choice[*], or type selection number: 1
update-alternatives: using /usr/lib/libblas/libblas.so.3 to provide /usr/
lib/libblas.so.3 (libblas.so.3) in manual mode
```

Choose 1 and switch to the libblas engine. Re-launch the environment of R language and run the scripts.

```
~ R
> source("R-benchmark-25.R")
Loading required package: Matrix
Loading required package: SuppDists

    R Benchmark 2.5
    ===============
Number of times each test is run_____: 3

    I. Matrix calculation
    ---------------------
Creation, transp., deformation of a 2500x2500 matrix (sec):
1.09366666666667
2400x2400 normal distributed random matrix ^1000____ (sec):
0.817333333333333
Sorting of 7,000,000 random values_____ (sec):
0.954333333333333
2800x2800 cross-product matrix (b=a' * a)_____ (sec):
15.3033333333333
Linear regr. over a 3000x3000 matrix (c=a \ b')___ (sec):  7.155
                  --------------------------------------------
                  Trimmed geom. mean (2 extremes eliminated):
1.95463154033118

    II. Matrix functions
    --------------------
FFT over 2,400,000 random values_____ (sec):
0.363666666666669
Eigenvalues of a 640x640 random matrix_____ (sec):  1.131
Determinant of a 2500x2500 random matrix_____ (sec):  5.061
Cholesky decomposition of a 3000x3000 matrix_____ (sec):  5.634
Inverse of a 1600x1600 random matrix_____ (sec):  4.142
                  --------------------------------------------
                  Trimmed geom. mean (2 extremes eliminated):
2.87278425762591

    III. Programmation
    ------------------
3,500,000 Fibonacci numbers calculation (vector calc)(sec):
0.775000000000006
Creation of a 3000x3000 Hilbert matrix (matrix calc) (sec):
0.259666666666665
Grand common divisors of 400,000 pairs (recursion)__ (sec):
0.633333333333345
Creation of a 500x500 Toeplitz matrix (loops)_____ (sec):
0.533666666666666
Escoufier's method on a 45x45 matrix (mixed)_____ (sec):
0.647999999999996

                  --------------------------------------------
                  Trimmed geom. mean (2 extremes eliminated):
0.602780428790226


Total time for all 15 tests_____ (sec):  44.505
```

```
Overall mean (sum of I, II and III trimmed means/3)_ (sec):
1.5014435867612
                          --- End of test ---
```

We can see in the result that the performance advantage of OpenBlas in matrix computing is very significant. It takes OpenBlas just 12 seconds to finish the 15 tests, while the default Blas 44 seconds. It costs just a switching of an underlying algorithm library to significantly improve the computing performance. Fellow users of R, let's start using it.

4.2 R Calls C++ across Boundary

Question
How do you call C ++ programs with R language?

R Calls C++ Across Boundary
http://blog.fens.me/r-cpp-rcpp

Introduction
I know many R packages, for I have been using them for years. However, maybe because of my long-term use of Java, I did not get to know those packages of C++. As the development of many languages and the popularity of cross-language applications, methods of crossing the language boundaries have matured. It is now simple to communicate between R and C++.

Let's catch up with the inter-language communication of R and C++ and solve the performance problem of R with C++.

4.2.1 A Brief Introduction to Rcpp

The Rcpp (Seamless R and C++ integration) package is a communication package that communicates between R language and C++ language and it provides function calls between R and C++. The data types of R and C++ are completely mapped with the Rcpp package.

Rcpp official website: https://cran.r-project.org/web/packages/Rcpp/index.html.

This section is just a getting started tutorial with a simple introduction to the communication between R and C++. We will not do in-depth discussion here. There are similar implementations of the communications between R and other languages. For calls between R and JAVA, please refer to Section 4.3 High-Speed Channel between R and rJava of *R for Programmers: Mastering the Tools*; for calls between R and Nodejs, please refer to Section 4.4 Cross-Platform Communication between Node.js and R of *R for Programmers: Mastering the Tools*.

4.2.2 Learn Rcpp in Five Minutes

As a 5-minute tutorial, we will simply jump to an example and will not discuss API.

The system environment used here:

- Win10 64bit
- R: 3.2.3 x86_64-w64-mingw32/x64 b4bit

It will need Rtools support under the environment of Windows system, so we need to manually download the Rtools package of a corresponding version. My version of R language is 3.2.3, so I need to install Rtools33.exe. We do not need to talk much about the EXE program installation. Just double clicks will do (Figure 4.1).

It takes only one line of codes to download the Rcpp package and install it.

```
> install.packages("Rcpp")
trying URL 'https://mirrors.tuna.tsinghua.edu.cn/CRAN/bin/windows/
contrib/3.2/Rcpp_0.12.6.zip'
Content type 'application/zip' length 3221864 bytes (3.1 MB)
downloaded 3.1 MB

package 'Rcpp' successfully unpacked and MD5 sums checked
Warning in install.packages :
  cannot remove prior installation of package 'Rcpp'

The downloaded binary packages are in
    C:\Users inkpad\AppData\Local\Temp\RtmpKkg8zo\downloaded_packages
```

4.2.2.1 Begin with "hello world"

Let's begin with a simple program "hello world." R language can call the hello() function of C++. Create two files in the same directory.

- demo.cpp, the source file of C++ program.
- demo.r, the source file of R program.

Download	R compatibility	Frozen?
Rtools34.exe	R 3.3.x and later	No
Rtools33.exe	R 3.2.x to 3.3.x	Yes
Rtools32.exe	R 3.1.x to 3.2.x	Yes
Rtools31.exe	R 3.0.x to 3.1.x	Yes
Rtools30.exe	R >2.15.1 to R 3.0.x	Yes
Rtools215.exe	R >2.14.1 to R 2.15.1	Yes
Rtools214.exe	R 2.13.x or R 2.14.x	Yes
Rtools213.exe	R 2.13.x	Yes
Rtools212.exe	R 2.12.x	Yes
Rtools211.exe	R 2.10.x or R 2.11.x	Yes
Rtools210.exe	R 2.9.x or 2.10.x	Yes
Rtools29.exe	R 2.8.x or R 2.9.x	Yes
Rtools28.exe	R 2.7.x or R 2.8.x	Yes
Rtools27.exe	R 2.6.x or R 2.7.x	Yes
Rtools26.exe	R 2.6.x, R 2.5.x or (untested) earlier	Yes

Figure 4.1   Rtools release note.

First, edit demo.cpp and define the hello() function.

```
~ notepad demo.cpp


#include <Rcpp.h>
#include <string>

using namespace std;
using namespace Rcpp;

//[[Rcpp::export]]
string hello(string name) {
  cout << "hello " << name << endl;
  return name;
}

/*** R
hello('world')
hello('Conan')
*/
```

We can divide the code above into three parts.

- ■ #Include and using part: The statement of the package references and its name space. <Rcpp.h> and namespace must be loaded by C++ program. The parameters and the return values are of string type, so the class library <string> should be introduced and the name space is stated with namespace std.
- ■ Performance function part: It defines a hello(string name) function with a parameter and the return value of string type. What needs to be noted is that the functions open to R must be added with the annotation statement of //[[Rcpp::export]].
- ■ Code execution: Codes that begins with /*** R and ends with */ is the R language codes and will be executed by default.

Edit demo.r and call the hello() function of demo.cpp.

```
~ notepad demo.r

library(Rcpp)

sourceCpp(file='demo.cpp')
hello('R')
```

Execute the R language codes.

```
# Load the Rcpp package
> library(Rcpp)

# Encode and load the demo.cpp file
> sourceCpp(file='demo.cpp')
```

```
# Execute the R codes encapsulated in demo.cpp
> hello('world')
hello world

[1] "world"
> hello('Conan')
hello Conan
[1] "Conan"

# Execute the hello function
> hello('R')
hello [1]R
  "R"
```

A simple helloworld program has completed.

4.2.2.2 Mix Codes of R and Rcpp

The sourceCpp() function is really powerful. It completes R's call of C++ program with just two lines of codes. It also provides a method of mix coding, i.e. to embed C++ codes in R codes.

```
sourceCpp(code='
  #include >Rcpp.h<
  #include >string<

  using namespace std;
  using namespace Rcpp;

  //[[Rcpp::export]]
  string hello(string name) {
    cout << "hello " << name << endl;
    return name;
  }
')
hello('R2')
```

Execute the codes.

```
> sourceCpp(code='
+   #include >Rcpp.h<
+   #include >string<
+
+   using namespace std;
+   using namespace Rcpp;
+
+   //[[Rcpp::export]]
+   string hello(string name) {
+     cout << "hello " << name << endl;
+     return name;
+   }
+ ')
```

```
> hello('R2')
hello R2
[1] "R2"
```

We do not recommend the syntax of mix coding with multiple languages, but if there are just a few lines, it can be very convenient.

4.2.2.3 Generate cpp File with RStudio IDE

It will be convenient if you develop with RStudio IDE. You can use it to create a new C++ Program and generate a standard code template (Figure 4.2).

The code template generated is as follow:

```cpp
#include <Rcpp.h>
using namespace Rcpp;

// This is a simple example of exporting a C++ function to R. You can
// source this function into an R session using the Rcpp::sourceCpp
// function (or via the Source button on the editor toolbar). Learn
// more about Rcpp at:
//
//   http://www.rcpp.org/
//   http://adv-r.had.co.nz/Rcpp.html
//   http://gallery.rcpp.org/
//

// [[Rcpp::export]]
NumericVector timesTwo(NumericVector x) {
  return x * 2;
}
// You can include R code blocks in C++ files processed with sourceCpp
```

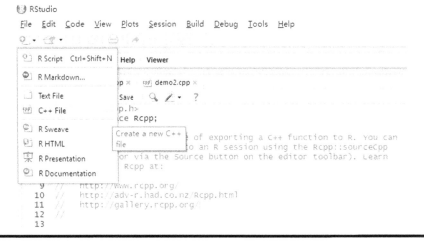

Figure 4.2   Generate a CPP file with RStudio.

```
// (useful for testing and development). The R code will be automatically
// run after the compilation.
//

/*** R
timesTwo(42)
*/
```

A standard code template can be generated rapidly with RStudio, and the template can be in immediate use after a few adjustments.

4.2.3 Data Type Conversion

We have tested the call of string type in the above example. There are many data types in R language. We will test them all in the following.

4.2.3.1 Primitive Type

The primitive type is the default mapping relation of C++ in R language. The codes of C++ are as below:

```
// [[Rcpp::export]]
char char_type(char x){
  return x;
}

// [[Rcpp::export]]
int int_type(int x){
  return x;
}

// [[Rcpp::export]]
double double_type(double x){
  return x;
}

// [[Rcpp::export]]
bool bool_type(bool x){
  return x;
}

// [[Rcpp::export]]
void void_return_type(){
  Rprintf( "return void" );
}
```

Execute the R calls.

```
# char type
> a1<-char_type('a')
> a1;class(a1)          # Correspond to the character type of R by default
```

```
[1] "a"
[1] "character"
> char_type('bbii')      # Only process the first byte of the string
[1] "b"

# int type
> a2<-int_type(111)
> a2;class(a2)           # Correspond to the integer type of R by default
[1] 111
[1] "integer"
> int_type(111.1)        # Remove the decimal places directly
[1] 111

# double type
> a3<-double_type(111.1)
> a3;class(a3)           # Correspond to the numeric type of R by default
[1] 111.1
[1] "numeric"
> double_type(111)
[1] 111

# boolean type
> a4<-bool_type(TRUE)
> a4;class(a4)           # Correspond to the logical type of R by default
[1] TRUE
[1] "logical"
> bool_type(0)           # 0 means FALSE
[1] FALSE
> bool_type(1)           # Non 0 means TRUE
[1] TRUE

# Function with no parameter and no return value
> a5<-void_return_type()
return void
> a5;class(a5)           # The return value is NULL
NULL
[1] "NULL"
```

4.2.3.2 Vector Type

The vector type is the default mapping relation of C++ in R language. The codes of C++ are as below:

```
// [[Rcpp::export]]
CharacterVector CharacterVector_type(CharacterVector x){
  return x;
}

// [[Rcpp::export]]
StringVector StringVector_type(StringVector x){
  return x;
}

// [[Rcpp::export]]
```

```
NumericVector NumericVector_type(NumericVector x) {
  return x;
}

// [[Rcpp::export]]
IntegerVector IntegerVector_type(IntegerVector x) {
  return x;
}

// [[Rcpp::export]]
DoubleVector DoubleVector_type(DoubleVector x){
  return x;
}

// [[Rcpp::export]]
LogicalVector LogicalVector_type(LogicalVector x){
  return x;
}

// [[Rcpp::export]]
DateVector DateVector_type(DateVector x){
  return x;
}

// [[Rcpp::export]]
DatetimeVector DatetimeVector_type(DatetimeVector x){
  return x;
}
```

Execute the R calls.

```
# Character vector
> a6<-CharacterVector_type(c('abc','12345'))
> a6;class(a6)                                    # Correspond to the
  character type of R by default
[1] "abc"    "12345"
[1] "character"
> CharacterVector_type(c('abc',123.5, NA, TRUE))  # NA means no
  processing
[1] "abc"    "123.5" NA       "TRUE"

# String vector, totally the same as the Character vector
> a7<-StringVector_type(c('abc','12345'))
> a7;class(a7)                                    # Correspond to the
  character type of R by default
[1] "abc"    "12345"
[1] "character"
> StringVector_type(c('abc',123.5, NA, TRUE))
[1] "abc"    "123.5" NA       "TRUE"

# Numeric vector
> a8<-NumericVector_type(rnorm(5))
> a8;class(a8)                                    # Correspond to the
  numeric type of R by default
[1] -0.2813472 -0.2235722 -0.6958443 -1.5322172  0.5004307
```

```
[1] "numeric"
> NumericVector_type(c(rnorm(5),NA,TRUE))          # NA means no
  processing, TRUE is 1
[1]   0.1700925   0.5169612 -0.3622637   1.0763204 -0.5729958
[6]          NA  1.0000000

# Integer vector
> a9<-IntegerVector_type(c(11,9.9,1.2))            # Remove the decimal
  places directly
> a9;class(a9)                                     # Correspond to the
  integer type of R by default
[1] 11  9  1
[1] "integer"
> IntegerVector_type(c(11,9.9,1.2,NA,TRUE))        # NA means no
  processing, TRUE is 1
[1] 11  9  1 NA  1

# Double vector,same as the Numeric vector
> a10<-DoubleVector_type(rnorm(5))
> a10;class(a10)                                   # Correspond to the
  numeric type of R by default
[1]   0.9400947 -0.8976913   0.2744319 -1.5278219   1.2010569
[1] "numeric"
> DoubleVector_type(c(rnorm(5),NA,TRUE))           # NA means no
  processing, TRUE is 1
[1]   2.0657148   0.2810003   2.1080900 -1.2783693   0.2198551
[6]          NA  1.0000000

# Logical vector
> a11<-LogicalVector_type(c(TRUE,FALSE))
> a11;class(a11)                                   # Correspond to the
  logical type of R by default
[1]   TRUE FALSE
[1] "logical"
> LogicalVector_type(c(TRUE,FALSE,TRUE,0,-1,NA))   # NA means no
  processing, 0 means FALSE, non 0 means TRUE
[1]   TRUE FALSE   TRUE FALSE   TRUE      NA

# Date vector
> a12<-DateVector_type(c(Sys.Date(),as.Date('2016-10-10')))
> a12;class(a12)                                   # Correspond to the
  Date type of R by default
[1] "2016-08-01" "2016-10-10"
[1] "Date"
> DateVector_type(c(Sys.Date(),as.Date('2016-10-10'),NA,TRUE,FALSE))   #
  NA means no processing, TRUE means 1970-01-02, FALSE means 1970-01-01
[1] "2016-08-01" "2016-10-10" NA           "1970-01-02"
[5] "1970-01-01"

# Datetime vector
> a13<-DatetimeVector_type(c(Sys.time(),as.POSIXct('2016-10-10')))
> a13;class(a13)                                   # Correspond to the
  POSIXct type of R by default
[1] "2016-08-01 20:05:25 CST" "2016-10-10 00:00:00 CST"
[1] "POSIXct" "POSIXt"
```

```
> DatetimeVector_type(c(Sys.time(),as.POSIXct('2016-10-
  10'),NA,TRUE,FALSE))  # NA means no processing
[1] "2016-08-01 20:05:25 CST" "2016-10-10 00:00:00 CST"
[3] NA                        "1970-01-01 08:00:01 CST"
[5] "1970-01-01 08:00:00 CST"
```

4.2.3.3 Matrix Type

The matrix type is the default mapping relation of C++ in R language. The codes of C++ are as below:

```
// [[Rcpp::export]]
CharacterMatrix CharacterMatrix_type(CharacterMatrix x){
  return x;
}

// [[Rcpp::export]]
StringMatrix StringMatrix_type(StringMatrix x){
  return x;
}

// [[Rcpp::export]]
NumericMatrix NumericMatrix_type(NumericMatrix x){
  return x;
}

// [[Rcpp::export]]
IntegerMatrix IntegerMatrix_type(IntegerMatrix x){
  return x;
}

// [[Rcpp::export]]
LogicalMatrix LogicalMatrix_type(LogicalMatrix x){
  return x;
}

// [[Rcpp::export]]
ListMatrix ListMatrix_type(ListMatrix x){
  return x;
}
```

Execute the R calls.

```
# Character matrix
> a14<-CharacterMatrix_type(matrix(LETTERS[1:20],ncol=4))
> a14;class(a14)
     [,1] [,2] [,3] [,4]
[1,]  "A"  "F"  "K"  "P"
[2,]  "B"  "G"  "L"  "Q"
[3,]  "C"  "H"  "M"  "R"
[4,]  "D"  "I"  "N"  "S"
[5,]  "E"  "J"  "O"  "T"
[1] "matrix"
```

```
# String matrix, same as the Character matrix
> a15<-StringMatrix_type(matrix(LETTERS[1:20],ncol=4))
> a15;class(a15)
     [,1] [,2] [,3] [,4]
[1,]  "A"  "F"  "K"  "P"
[2,]  "B"  "G"  "L"  "Q"
[3,]  "C"  "H"  "M"  "R"
[4,]  "D"  "I"  "N"  "S"
[5,]  "E"  "J"  "O"  "T"
[1] "matrix"

# Numeric matrix
> a16<-NumericMatrix_type(matrix(rnorm(20),ncol=4))
> a16;class(a16)
            [,1]        [,2]       [,3]       [,4]
[1,]   1.2315498  2.3234269  0.5974143  0.9072356
[2,]   0.3484811  0.3814024 -0.2018324  0.8717205
[3,]  -0.2025285  2.1076947 -0.3433948  1.1523710
[4,]  -1.4948252 -0.7724951 -0.7681800 -0.5406494
[5,]   0.4815904  1.4930873 -1.1444258  0.2537099
[1] "matrix"

# Integer matrix
> a17<-IntegerMatrix_type(matrix(seq(1,10,length.out = 20),ncol=4))
> a17;class(a17)
     [,1] [,2] [,3] [,4]
[1,]    1    3    5    8
[2,]    1    3    6    8
[3,]    1    4    6    9
[4,]    2    4    7    9
[5,]    2    5    7   10
[1] "matrix"

# Logical matrix
> a18<-LogicalMatrix_type(matrix(c(rep(TRUE,5),rep(FALSE,5),rnorm(10)),
  ncol=4))
> a18;class(a18)
      [,1]  [,2] [,3] [,4]
[1,]  TRUE FALSE TRUE TRUE
[2,]  TRUE FALSE TRUE TRUE
[3,]  TRUE FALSE TRUE TRUE
[4,]  TRUE FALSE TRUE TRUE
[5,]  TRUE FALSE TRUE TRUE
[1] "matrix"

# List matrix, supports multiple matrix types
> a19<-ListMatrix_type(matrix(rep(list(a=1,b='2',c=NA,d=TRUE),10),ncol=5))
> a19;class(a19)
     [,1] [,2] [,3] [,4] [,5]
[1,]    1    1    1    1    1
[2,]  "2"  "2"  "2"  "2"  "2"
[3,]   NA   NA   NA   NA   NA
[4,] TRUE TRUE TRUE TRUE TRUE
[5,]    1    1    1    1    1
```

```
[6,]   "2"   "2"   "2"   "2"   "2"
[7,]    NA    NA    NA    NA    NA
[8,]  TRUE  TRUE  TRUE  TRUE  TRUE
[1]  "matrix"
```

4.2.3.4 Other Data Types

Other data types include data.frame, environment, S3, S4, RC, etc. where data.frame is a special data type of R language and S3, S4, RC, etc. are object types.

```
// [[Rcpp::export]]
Date Date_type(Date x){
  return x;
}

// [[Rcpp::export]]
Datetime Datetime_type(Datetime x){
  return x;
}

// [[Rcpp::export]]
S4 S4_type(S4 x){
  return x;
}

// [[Rcpp::export]]
RObject RObject_type(RObject x){
  return x;
}

// [[Rcpp::export]]
SEXP SEXP_type(SEXP x){
  return x;
}

// [[Rcpp::export]]
Environment Environment_type(Environment x){
  return x;
}
```

Execute the R calls.

```
# data.frame type
> a19<-DataFrame_type(data.frame(a=rnorm(3),b=1:3))
> a19;class(a19)
           a b
1 -1.8844994 1
2  0.6053935 2
3 -0.7693985 3
[1]  "data.frame"

# list type
> a20<-List_type(list(a=1,b='2',c=NA,d=TRUE))
> a20;class(a20)
```

```
$a
[1] 1
$b
[1] "2"
$c
[1] NA
$d
[1] TRUE
[1] "list"

# Date type
> a21<-Date_type(Sys.Date())
> a21;class(a21)
[1] "2016-08-01"
[1] "Date"
> Date_type(Sys.time())                # Cannot correctly process data of
  POSIXct type
[1] "4026842-05-26"

# POSIXct type
> a22<-Datetime_type(Sys.time())
> a22;class(a22)
[1] "2016-08-01 20:27:37 CST"
[1] "POSIXct" "POSIXt"
> Datetime_type(Sys.Date())            # Cannot correctly process data of
  Date type
[1] "1970-01-01 12:43:34 CST"

# S3 object-oriented type, and its corresponding S4 type definition
> setClass("Person",slots=list(name="character",age="numeric"))
> s4<-new("Person",name="F",age=44)
> a23<-S4_type(s4)
> a23;class(a23)
An object of class "Person"
Slot "name":
[1] "F"
Slot "age":
[1] 44
[1] "Person"
attr(,"package")
[1] ".GlobalEnv"

# S3 object-oriented type, no corresponding type, passing values by
  RObject
> s3<-structure(2, class="foo")
> a24<-RObject_type(s3)
> a24;class(a24)
[1] 2
attr(,"class")
[1] "foo"
[1] "foo"

# RObject can process S4 objects as well
> a25<-RObject_type(s4)
> a25;class(a25)
```

```
An object of class "Person"
Slot "name":
[1] "F"
Slot "age":
[1] 44
[1] "Person"
attr(,"package")
[1] ".GlobalEnv"


# RObject can process RC objects as well
> User<-setRefClass("User",fields=list(name="character"))
> rc<-User$new(name="u1")
> a26<-RObject_type(rc)
> a26;class(a26)
Reference class object of class "User"
Field "name":
[1] "u1"
[1] "User"
attr(,"package")
[1] ".GlobalEnv"


# RObject can process the function type as well
> a27<-RObject_type(function(x) x+2)
> a27;class(a27)
function(x) x+2
[1] "function"


# environment type
> a28<-Environment_type(new.env())
> a28;class(a28)
<environment: 0x0000000015350a80>
[1] "environment"


# SEXP means any type and its type will be defined when it is called
> SEXP_type('fdafdaa')
[1] "fdafdaa"


> SEXP_type(rc)
Reference class object of class "User"
Field "name":
[1] "u1"


> SEXP_type(data.frame(a=rnorm(3),b=1:3))
          a b
1 -0.5396140 1
2  0.1694799 2
3 -1.8818596 3


> SEXP_type(function(x) x+2)
function(x) x+2
```

In the end, we can conclude the corresponding relations of R and Rcpp as below.

C++ Type	R Type
char	character
int	integer
double	numeric
bool	logical
Rcpp::Date	Date
Rcpp::Datetime	POSIXct
Rcpp::CharacterVector	character
Rcpp::StringVector	character
Rcpp::NumericVector	numeric
Rcpp::IntegerVector	integer
Rcpp::DoubleVector	numeric
Rcpp::LogicalVector	logical
Rcpp::DateVector	Date
Rcpp::DatetimeVector	POSIXct
Rcpp::CharacterMatrix	matrix
Rcpp::StringMatrix	matrix
Rcpp::NumericMatrix	matrix
Rcpp::IntegerMatrix	matrix
Rcpp::LogicalMatrix	matrix
Rcpp::ListMatrix	matrix
Rcpp::DataFrame	data.frame
Rcpp::List	list
Rcpp::S4	S4
Rcpp::Environment	environment
Rcpp::RObject	Any Type
Rcpp::SEXP	Any Type

In this section, we have briefly introduced a method of calling C++ programs by using the Rcpp package of R language. The key of the calls is the matching of data types, which guarantees the data transmission between R and C++. According to the tests above, all data types of R language can be mapped to the programs of C++ with the Rcpp package. Then we can, according to our needs, implement some programs that require better performance in C++ to improve the computing efficiency.

4.3 When R Meets Docker

Question
How do you achieve application-level parallelization with R language?

Introduction
R language, as a tool for data analysis, has been widely accepted and used. However, it is rather difficult to engineer the R language project, to deploy it in the production environment and to put it in use for users online. The main reason for this is that R itself is single-threaded without any support to the parallel processing.

So what will happen when R meets Docker? A detailed explanation will be given in this section.

4.3.1 When R Meets Docker

As mentioned above, when R runs, its environment is single-threaded and it does not support the parallel processing, so it is hard for us to directly apply R in the production environment. However, when R meets Docker, a solution to the problem appears.

The R application can be Dockerized by the containerization of Docker. Every time a user makes a request, the program can automatically initiate a Dockerized container online to load R's tasks, deployment, operations and calculation and then return the result (Figure 4.3).

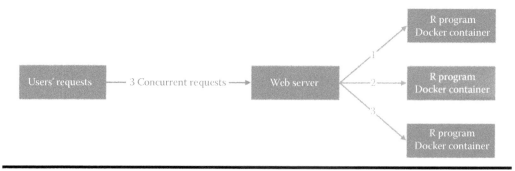

Figure 4.3 Architecture of three concurrences.

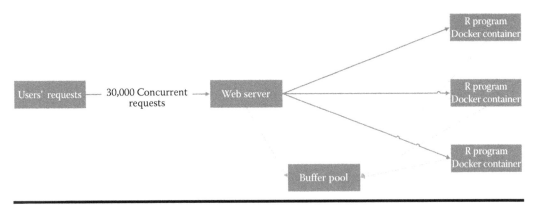

Figure 4.4   30,000 concurrences architecture.

Approached from the extremes, if we are faced with one million concurrent requests, we need to initiate one million Docker containers and every container executes its own task. We should try to avoid this, because R itself is good at processing data tasks, not web requests. If the large amount of user requests can be converted to a small amount of data calculation, it will perfectly solve the problem of R's engineering in concurrence (Figure 4.4).

For example, for the repeated calculations of users' requests in a large amount, we can save the calculation result of R to a buffer pool.

4.3.2 Manage R's Programs with Docker

Since the design plan has been settled, now we are going to implement it.

The operation process can be divided into four steps:

1. Set up an environment for Docker.
2. Find a mature Docker image of R language from a third party.
3. Install R programs in it.
4. Pack, run and upload the programs.

4.3.2.1 Set Up an Environment for Docker

We will not introduce how to install the Docker environment here. For more, please refer to the Appendix of this book.

The system environment used here:

■ Linux Ubuntu 14.04.4
■ R: version 3.3.1 x86_64-pc-linux-gnu (64-bit)

4.3.2.2 Find a Mature Docker Image of R Language from a Third Party

Search the key word "r" in Docker Hub and there are 535 results in all. We can just use the first one, r-base, as the base of the Docker container (Figure 4.5).

Download the r-base image from the repository.

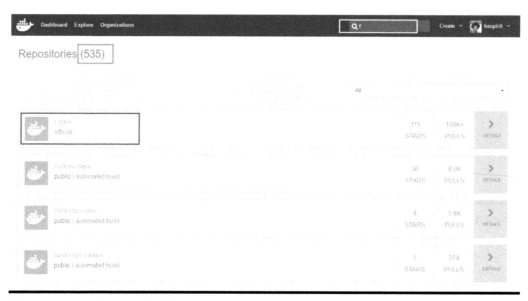

Figure 4.5 Docker repositories.

```
# Download the r-base image, about 300mb. It will take a while
~ sudo docker pull r-base
Using default tag: latest
latest: Pulling from library/r-base
9cd73496e13f: Pull complete
f10af350cd29: Pull complete
eea7b33eea97: Pull complete
c91475e50472: Pull complete
1e5e5f6785b4: Pull complete
8c4091261ff6: Pull complete
Digest: sha256:5f06e5a89cc64cbc513d02a8c650ea8bcbf0499795add57d1879306979
        5c6f8d
Status: Downloaded newer image for r-base:latest

# Check the local image list
~ sudo docker images
REPOSITORY          TAG         IMAGE ID        CREATED         SIZE
bsspirit/fensme     latest      8496b10e857a    2 hours ago     182.8 MB
ubuntu              latest      f8d79ba03c00    2 weeks ago     126.4 MB
r-base              latest      e2abe45e47d7    3 weeks ago     959.9 MB
```

4.3.2.3 Install R Programs in It

Before we install any R program, we need to check what it is like in the r-base container with the interactive commands.

Run the r-base container and a command window of R will pop up.

```
~ sudo docker run -ti --rm r-base

R version 3.3.1 (2016-06-21) -- "Bug in Your Hair"
Copyright (C) 2016 The R Foundation for Statistical Computing
```

```
Platform: x86_64-pc-linux-gnu (64-bit)

R is free software and comes with ABSOLUTELY NO WARRANTY.
You are welcome to redistribute it under certain conditions.
Type 'license()' or 'licence()' for distribution details.

  Natural language support but running in an English locale

R is a collaborative project with many contributors.
Type 'contributors()' for more information and
'citation()' on how to cite R or R packages in publications.

Type 'demo()' for some demos, 'help()' for on-line help, or
'help.start()' for an HTML browser interface to help.
Type 'q()' to quit R.

>
```

Let's execute an R program to get some information of the Docker environment.

```
# The boot path of R program
> getwd()
[1] "/"

# The contents in the current path
> dir()
 [1] "bin"    "boot"  "dev"   "etc"    "home"   "lib"    "lib64" "media"
 "mnt"
[10] "opt"    "proc"  "root"  "run"    "sbin"   "srv"    "sys"   "tmp"
"usr"
[19] "var"

# User ID
> system('whoami')
root

# System information
> sessionInfo()
R version 3.3.1 (2016-06-21)
Platform: x86_64-pc-linux-gnu (64-bit)
Running under: Debian GNU/Linux stretch/sid

locale:
 [1] LC_CTYPE=en_US.UTF-8       LC_NUMERIC=C
 [3] LC_TIME=en_US.UTF-8        LC_COLLATE=en_US.UTF-8
 [5] LC_MONETARY=en_US.UTF-8    LC_MESSAGES=en_US.UTF-8
 [7] LC_PAPER=en_US.UTF-8       LC_NAME=C
 [9] LC_ADDRESS=C               LC_TELEPHONE=C
[11] LC_MEASUREMENT=en_US.UTF-8 LC_IDENTIFICATION=C

attached base packages:
[1] stats      graphics  grDevices utils     datasets  methods   base

# The version of R language
> R.Version()
```

```
$platform
[1] "x86_64-pc-linux-gnu"

$arch
[1] "x86_64"

$os
[1] "linux-gnu"

$system
[1] "x86_64, linux-gnu"

$status
[1] ""

$major
[1] "3"

$minor
[1] "3.1"

$year
[1] "2016"

$month
[1] "06"

$day
[1] "21"

$`svn rev`
[1] "70800"

$language
[1] "R"

$version.string
[1] "R version 3.3.1 (2016-06-21)"

$nickname
[1] "Bug in Your Hair"
```

With the commands above, the system environment of the r-base container has been successfully installed. Next we will write an algorithm of R, run the program in the r-base container and then exit the container.

Create a new project directory.

```
~ mkdir ret && cd ret
~ pwd
/home/conan/ret
```

Let's write a program with a.r to calculate the stock return of VANKE (China Vanke Co., Ltd.) (000002.SZ). The data is collected from Yahoo Finance. R language will be used for the return calculation and the result will be printed out in the console, as shown in Figure 4.6.

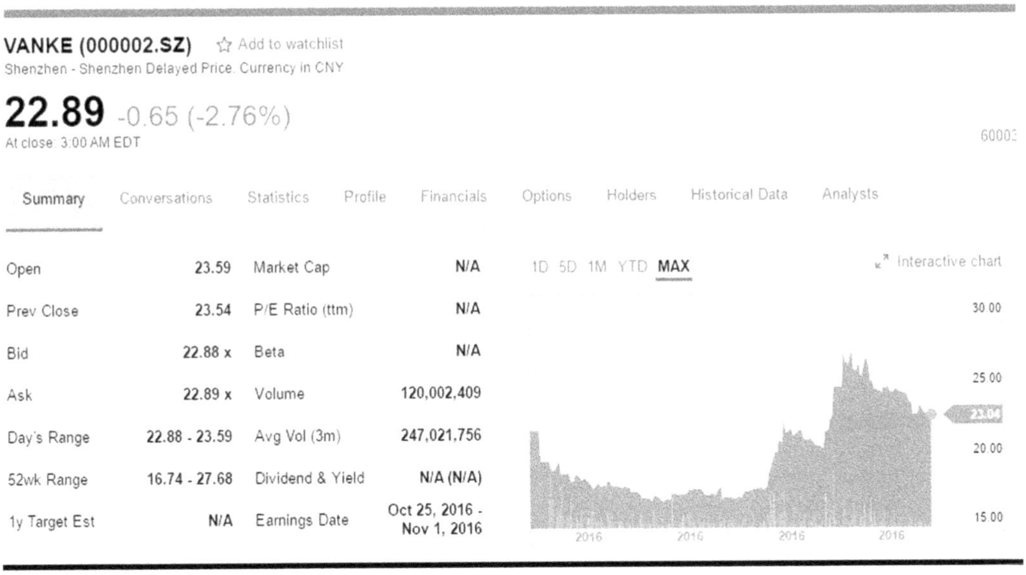

VANKE (000002.SZ)   ☆ Add to watchlist
Shenzhen - Shenzhen Delayed Price. Currency in CNY

22.89 -0.65 (-2.76%)
At close: 3:00 AM EDT   6000:

| Summary | Conversations | Statistics | Profile | Financials | Options | Holders | Historical Data | Analysts |

Open	23.59	Market Cap	N/A	1D 5D 1M YTD **MAX** ↗ Interactive chart
Prev Close	23.54	P/E Ratio (ttm)	N/A	30.00
Bid	22.88 x	Beta	N/A	
Ask	22.89 x	Volume	120,002,409	25.00
Day's Range	22.88 - 23.59	Avg Vol (3m)	247,021,756	23.0◄
52wk Range	16.74 - 27.68	Dividend & Yield	N/A (N/A)	20.00
1y Target Est	N/A	Earnings Date	Oct 25, 2016 - Nov 1, 2016	15.00
				2016 2016 2016 2016

Figure 4.6   Stock of VANKE.

Create an R language algorithm file a.r.

```
~ vi a.r

install.packages(c('quantmod','PerformanceAnalytics'))
library(quantmod)
library(PerformanceAnalytics)
VANKE<-getSymbols("000002.SZ",auto.assign=FALSE, from='2010-10-10')
close<-VANKE$'000002.SZ.Close'
ret<-CalculateReturns(close, method="discrete")
cumret<-cumprod((ret+1)[-1])-1
VANKE_ret<-merge(close,ret,cumret)
names(VANKE_ret)<-c('close','ret','cumret')
print(tail(VANKE_ret))
```

First, run the code on our local machine.

```
# Install the class library
> install.packages(c('quantmod','PerformanceAnalytics'))
# Load the class library
> library(quantmod)
> library(PerformanceAnalytics)

# Obtain the candlestick chart of VANKE
> VANKE<-getSymbols("000002.SZ",auto.assign=FALSE, from='2010-10-10')

# Closing price
> close<-VANKE$'000002.SZ.Close'

# Daily return=(T day closing price - (T-1day closing price))/T-1day
  closing price
> ret<-CalculateReturns(close, method="discrete")
```

```
# Accumulative daily return=(T day return+1) * (T+1day return+1)* ... *
(T+N day return+1)-1
> cumret<-cumprod((ret+1)[-1])-1

# Merge the datasets
> VANKE_ret<-merge(close,ret,cumret)
> names(VANKE_ret)<-c('close','ret','cumret')

# View the recent daily returns of VANKE
> print(tail(VANKE_ret))
           close         ret    cumret
2016-08-18 25.58 -0.010444874 1.893665
2016-08-19 24.59 -0.038702111 1.781674
2016-08-22 24.70  0.004473363 1.794118
2016-08-23 24.70  0.000000000 1.794118
2016-08-24 23.99 -0.028744939 1.713801
2016-08-25 23.54 -0.018757816 1.662896
```

Next, program the method of Dockerfile loading external files.

```
~ vi Dockerfile

FROM r-base
COPY . /usr/local/src/myscripts
WORKDIR /usr/local/src/myscripts
CMD ["Rscript", "a.r"]
```

4.3.2.4 Pack, Run and Upload the Programs

Let's pack the programs and generate the image file of Docker a.r.

```
~ sudo docker build -t a.r .
[sudo] password for conan:
Sending build context to Docker daemon 3.072 kB
Step 1 : FROM r-base
 ---> e2abe45e47d7
Step 2 : COPY . /usr/local/src/myscripts
 ---> e6ef215d3683
Removing intermediate container aaabfdfe92ab
Step 3 : WORKDIR /usr/local/src/myscripts
 ---> Running in e3f2c65b947a
 ---> c667baee06bf
Removing intermediate container e3f2c65b947a
Step 4 : CMD Rscript a.r
 ---> Running in dc040bbdd3b9
 ---> 9a48d6dc02fe
Removing intermediate container dc040bbdd3b9
Successfully built 9a48d6dc02fe
```

Initiate the r-base container and run the script of a.r.

```
~  sudo docker run a.r
```

After piles of logs run through your screen, there is the result of the VANKE returns (Figure 4.7).

```
In library(package, lib.loc = lib.loc, character.only = TRUE, logical.return = TRUE,  :
  there is no package called 'PerformanceAnalytics'
Loading required package: xts
Loading required package: zoo

Attaching package: 'zoo'

The following objects are masked from 'package:base':

    as.Date, as.Date.numeric

Loading required package: TTR
Loading required package: methods
Version 0.4-0 included new data defaults. See ?getSymbols.

Attaching package: 'PerformanceAnalytics'

The following object is masked from 'package:graphics':

    legend

    As of 0.4-0, 'getSymbols' uses env=parent.frame() and
auto.assign=TRUE by default.

This behavior will be phased out in 0.5-0 when the call will
default to use auto.assign=FALSE. getOption("getSymbols.env") and
getOptions("getSymbols.auto.assign") are now checked for alternate defaults

This message is shown once per session and may be disabled by setting
options("getSymbols.warning4.0"=FALSE). See ?getSymbols for more details.
              close        ret   cumret
2016-08-18 25.58 -0.010444874 1.893665
2016-08-19 24.59 -0.038702111 1.781674
2016-08-22 24.70  0.004473363 1.794118
2016-08-23 24.70  0.000000000 1.794118
2016-08-24 23.99 -0.028744939 1.713801
2016-08-25 23.54 -0.018757816 1.662896
conan@ubuntu:~/ret$ sudo docker run a.r
```

Figure 4.7   Command line log.

And don't forget the last step, to upload the result to docker hub. Here is the repository website: https://hub.docker.com/r/bsspirit/ret/

The operating commands of uploading the image:

```
~ sudo docker tag 9a48d6dc02fe bsspirit/ret
~ sudo docker push bsspirit/ret
```

If you have already installed the environment of Docker, you can directly use the command below to download and run the container.

```
~ sudo docker run bsspirit/ret
```

The meeting of R and Docker enables R to implement the parallel computing. This meeting also gives Docker a chance to enter the data processing field, a wider application prospect. Thank both R and Docker for bringing in new opportunities in the programmer's world.

FINANCIAL STRATEGY PRACTICE

Chapter 5

Bonds and Repurchase

There are not only stocks in the great financial market, but also bonds, which take a large part of the market. In this chapter, we will focus on how to use R language to analyze bonds and make bond investments and arbitrages. The low-risk bond investment can be a better choice of investment and financing.

5.1 Treasury Bonds

Question
How do you trade treasury bonds?

Treasury Bonds
http://blog.fens.me/finance-treasury-bond/

Introduction
Treasury bonds are based on the national credit and it sets the benchmark interest rate for various market economy pricings. We need to have knowledge of treasury bonds to understand the policies and the market.

Recently, I have read a book titled *China is a History of Finance* (《中国是部金融史》). It approaches the falls and rises of more than 2,000 years' feudal dynasty in China from the perspective of finance. In the history of China, wars were started again and again, which cost a lot. At that time, the treasury bond of western finance had not been introduced yet. The feudal government accumulated wealth by producing new coins or increasing the issuance of old coins, which caused inflations of 200%–2,000%. People were robbed of their wealth. If the feudal dynasty could be a little merciful, issuing treasury bonds, paying their people back in due time, and giving some hope to them, the history may be completely different.

5.1.1 A Basic Introduction to Treasury Bonds

Treasury bond is a debtor-creditor relationship formed when a country, based on its credit, raises funds from the public according to the general principle of debts. A treasury bond is a government debt issued by the central government to raise financial funds. It is a certificate of credit rights and indebtedness provided by the central government to the investors, where the government promises to pay the interests at stated periods and pay the principal at its maturity. For its issuer is a country, it is of the highest credit level and it is recognized as the safest investment instrument.

Treasury bond is the main form of national credit. The central government issues treasury bonds usually with the purpose of making up the national fiscal deficit, or raising funds for some costly construction projects, some special economic policies or even for wars. A treasury bond is secured by the tax income of the central government, so it is of low risk and high liquidity, but it bears a lower interest rate than other bonds.

The interest rate of treasury bond is the benchmark interest rate in the capital market, while the bank savings deposit cannot function as this credit instrument. Issued by the central government, a treasury bond is a credit certificate only next to currency and it nearly functions as quasi-money. Due to its best cashability and its most convenience, the interest rate of treasury bond is the lowest. Therefore, it objectively functions as the benchmark. In the capital asset pricing model (CAPM), the riskless return is converted by the interest rate if short-term treasury bond.

The regulated market operations in capital market should keep the position of the treasury bond interest rate as the benchmark. Those market credit relations that harm the benchmark position of treasury bond should be nonstandard. And the bank saving deposit does not have this credit function of treasury bond. On the premise of the treasury bond's existence, if the credit relationship of the financial market is not twisted, the interest rate of bank saving deposit should not and is not allowed to become the benchmark interest rate. In reality, it will be uncommon if the interest rate of treasury bond is higher than that of the bank saving deposit.

Next we will draw a plot of a 1-year bond and a 1-year fixed term deposit, for an intuitive comparison. Export the data of 2002–2015 from Wind. The data format is as follow:

```
2002-01-04,2.5850,
2002-01-07,2.6009,
2002-01-08,1.9156,
2002-01-09,1.9040,
2002-01-10,1.8987,
2002-01-11,1.8757,
2002-01-14,1.6794,
2002-01-15,1.7265,
2002-01-16,1.7403,
2002-01-17,1.7165,
```

There are three columns of the data:

- Index name.
- ChinaBond Treasury Bond Yield Upon Maturity: 1 year.
- Term Deposit Rate: 1 year (Lump-sum Time Deposit). It is null for most time and there are values only during an interest rate adjustment.

The system environment used here:

- Win10 64bit
- R: 3.2.3 x86_64-w64-mingw32/x64 b4bit

Let's draw a plot with R language.

```
# Load the class library
> library(xts)
> library(ggplot2)

# Read data from bondSaving.csv
> bs<-read.csv("bondSaving.csv",header=TRUE)

# Re-define the column names: bond for treasury bond and saving for
  deposit
> names(bs)<-c("date","bond","saving")

# Converted to time series format
> bsxts<-xts(bs[-1],order.by = as.Date(bs$date))

# The null value is filled backward
> bsxts<-na.locf(bsxts,fromLast=FALSE)

# Draw a plot as shown in Figure 5-1
> g<-ggplot(aes(x=Index, y=Value, colour=Series),data=fortify(bsxts,melt=
  TRUE))
> g<-g+geom_line()
> g
```

In Figure 5.1, the *x*-axis means time and *y*-axis the value of interest rate. Let's compare the yields of the 1-year bond upon maturity and the 1-year saving from 2002 to 2015. The steep curve (bond) is the ChinaBond Treasury Bond Yield Upon Maturity, while the flat curve (saving) is the 1-year fixed term deposit issued by People's Bank of China. For most of the time, the interest rate of treasury bond is higher than the deposit. Why is so? To find the answer, we need to get a more profound understanding of China's economy.

5.1.2 Significance of Treasury Bonds

5.1.2.1 The Necessity to Issue Treasury Bonds

5.1.2.1.1 To cover military spending

During war times, the military spending is massive, so treasury bonds will be issued to raise funds for war if there is no other way. Treasury bond for military spending is a common way for every government to raise funds during war, which is also the origin of treasury bonds.

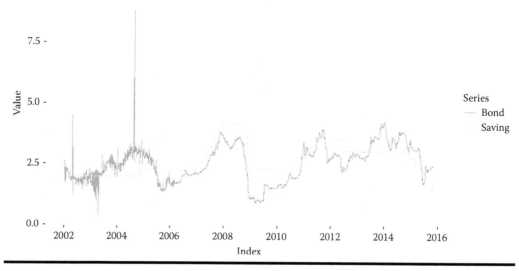

Figure 5.1 1-Year bond and 1-year fixed term deposit.

In ancient China, to raise funds for military spending, the government did not issue bonds, but currency.

5.1.2.1.2 To balance fiscal revenue and expenditure

Usually, fiscal revenue and expenditure can be balanced by increasing tax revenues, increasing currency issuance or issuing treasury bonds.

Increasing tax revenues: Taxes are collected from people and spent on people, which is surely a good way. However, if the taxes surpass certain limits and they go beyond what enterprises and individuals can bear, it will harm the production development and thus the future tax income.

Increasing currency issuance: This is the simplest way, but also the worst way, because it will lead to serious inflation, which will severely impact the economy.

When it is difficult to increase taxes and we do not adopt currency issuance increasing, it can be a possible measure to issue treasury bonds to make up the fiscal deficit. Governments can collect idle money from organizations and individuals by issuing bonds, so as to help their countries through the tough financially difficult times. However, the treasury bonds for deficit covering must be issued in a moderate volume; otherwise, it may result in severe deflation.

5.1.2.1.3 To raise funds for constructions

When the government needs large amounts of mid-term and long-term funds to construct the infrastructure and the public facilities, it will issue treasury bonds of mid-term or long-term, which will help convert part of the short-term funds to the mid-term or long-term funds for the large-scale construction projects of the country, so as to stimulate its economy development.

5.1.2.1.4 To pay the treasury bonds at maturity

When there is a treasury bond at maturity, the government may issue a new treasury bond to raise fund and pay the debt for the previous bond. This can reduce and diffuse the debt burden of the country at the peak of debt payment.

5.1.2.2 Classification of Treasury Bonds

By the forms of bonds, the treasury bonds issued in China can be classified into three types: certificate treasury bonds, physical treasury bonds and book-entry treasury bonds.

- Certificate treasury bonds: Cannot be listed in circulation and the interest shall be accrued from the date of purchase. During the holding period, the bond holder can redeem it in the purchase outlets in advance if he/she needs to withdraw cash under special circumstance. When redeeming it in advance, besides the principal, the interest will be paid by the corresponding interest rate bracket and the actual holding days. 1‰ of the redeemed principal will be paid to the management institution as commission.
- Physical treasury bonds: Saving bonds (also called electronic treasury bonds) are non-negotiable registered treasury bonds issued by governments to individual investors, with the purpose of collecting the individual savings and satisfying the need of long-term savings investment. The electronic saving bond is a saving bond recording the credit rights in electronic ways.
- Book-entry treasury bonds: It records the credit rights in book keeping way. It is issued by the Ministry of Finance, and it is issued and traded in the trading systems of the securities exchanges. It can be registered, reported loss and listed for transfer.

5.1.3 Book-Entry Treasury Bonds

To trade book entry securities, an investor must open an account in the securities exchange. Due to the paperless issuance and trading of the book-entry treasury bonds, the trading is of high efficiency, low cost and transaction safety.

Here is the list of book-entry treasury bonds issued by the exchanges (Figure 5.2):

We can query the treasury bonds by ourselves according to the securities codes.

5.1.4 Treasury Bond 101308

In the following, a treasury bond coded 101308 will be taken for illustration. 2013 Book-entry Coupon Bearing Treasury Bonds (Issue No.8) (hereinafter referred to as "the treasury bond") have finished the issuance. According to the notice of the Ministry of Finance, the treasury bond would be listed in the exchange for trading on April 24, 2013.

The treasury bond is a fixed-rate coupon bearing treasury bond of 7 years, with 101308 as its code and T-bond 1308 as its short name. Its coupon rate is 3.29%, the standard board lot is 10 and on the maturity date of April 18, 2020, the principal and the final interest will be paid.

All Treasury Bonds

Update Time: 2017-02-17

Code	Short Name	Full Price	APR (%)	Term of Maturity (year)	Remaining Maturity	Net Price	Accrued Days	Accrued Interest	Int Payment	YTM (%)	Modified Duration	Convexity	Quotations
130779	15贵州20	100.50	3.23	10	8.85	100.00	56	0.50	年付	3.23%	NaN	NaN	📊
101917	国债917	108.36	4.26	20	4.45	106.00	202	2.36	年付	2.80%	NaN	NaN	📊
101617	国债1617	100.74	2.74	10	9.47	100.00	198	0.74	半年付	2.74%	NaN	NaN	📊
101614	国债1614	102.20	2.95	7	6.33	100.20	247	2.00	年付	2.91%	NaN	NaN	📊
101613	国债1613	98.37	3.70	50	49.29	97.00	271	1.37	半年付	3.84%	NaN	NaN	📊
101612	国债1612	101.56	2.51	2	1.25	99.67	275	1.89	年付	2.78%	NaN	NaN	📊
101611	国债1611	102.20	2.30	1	0.21	99.90	285	2.30	固定单利	0.46%	NaN	NaN	📊
101608	国债1608	101.88	3.52	30	29.20	99.00	299	2.88	年付	3.58%	NaN	NaN	📊
101606	国债1606	102.83	2.75	7	6.08	100.28	338	2.55	年付	2.70%	NaN	NaN	📊
101605	国债1605	102.24	2.22	1	0.00	100.02	362	2.22	固定单利	-7.25%	NaN	NaN	📊
101603	国债1603	99.43	2.55	3	1.95	99.28	21	0.15	年付	2.94%	NaN	NaN	📊
101602	国债1602	100.05	2.53	5	3.91	99.81	35	0.24	年付	2.58%	NaN	NaN	📊
101528	国债1528	115.91	3.89	50	48.80	115.45	87	0.46	半年付	3.26%	NaN	NaN	📊
101522	国债1522	103.18	2.92	3	1.60	102.00	147	1.18	年付	1.64%	NaN	NaN	📊
101516	国债1516	99.60	3.51	10	8.41	98.56	217	1.04	半年付	3.71%	NaN	NaN	📊

Figure 5.2   Book-entry treasury bonds.

Basic Information

Field	*Explanation*
Bond name	2013 Book-entry Coupon Bearing Treasury Bonds (Issue No.18)
Short name of the bond	T-bond 1308
Bond code	101308
T-bond type	Coupon bearing bond
Issue date	2013-04-17
Issue volume (*yuan*)	30000000000.00
Issue price (*yuan*)	100
Listing date	2013-04-24
Exchange	Shenzhen Stock Exchange
Board lot	Round block

(*Continued*)

(*Continued*)

Par value	100.00
Term of maturity (year)	7
Method of calculating interest	Fixed interest rate
Annual interest rate (%)	3.29
Basic spread	—
Int payment	Paid on an annual basis
Maturity date	2020-04-18
Delisting date	—

Screenshot of transaction: Time sharing picture of recent 10 days. Screenshot time is 11:29 of December 30, 2013 (Figure 5.3).

Relevant Explanations

- Very inactive transactions and poor liquidity.
- In the recent 10 days, there are notable transactions on December 25, resulted from the low price of 92 *yuan* per bond, nearly the bottom line.
- There are four transactions today, all traded in the highest bid price, but only 2 round blocks (20 bonds).
- The accrued interest today is 2.32. It will grow a little bit every day. Its maximum value is 3.29 and it will return zero after the payment every year.
- Full settlement price = Trading price + Accrued interest = 92.4 + 2.32 = 94.72
- Yield upon maturity (4.81) = (Sum of principal and interest at maturity − Bid price of the bond)/(Bid price of the bond * Remaining maturity) * 100%

When the treasury bonds we hold are at maturity, the yield upon maturity is still 4.81, unless the country goes broke. Basically, it is risk-free, but with a comparatively low return.

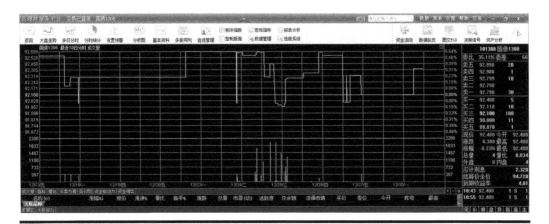

Figure 5.3　Screenshot of T-bond 1308.

5.1.5 Historical Performance of Treasury Bonds

In the following, I will respectively intercept data of 2002–2017 ChinaBond treasury bond yield upon maturity of 1 year term, 3 years term and 10 years term for comparison.

Draw the yield curves with R language.

```
# Read data from bonds.csv
> bonds<-readcsv("bonds.csv",header=TRUE)

# Rename the columns: ytm1 for one year term, ytm3 for three years term
and ytm10 for ten years term
> names(bonds)<-c("date","ytm1","ytm3","ytm10")

# Convert to time series format
> bondsxts<-xts(bonds[-1],order.by = as.Date(bonds$date))
> bondsxts<-na.locf(bondsxts,fromLast=FALSE)

# Draw a plot as shown in Figure 5-4
> g<-ggplot(aes(x=Index, y=Value, colour=Series),data=fortify(bondsxts,
melt=TRUE))
> g<-g+geom_line()
> g
```

In Figure 5.4, the blue curve is the 10-year T-bond yield upon maturity, always on the top; the green one is the 3-year T-bond yield upon maturity, in the middle; and the red curve is the 1-year T-bond yield upon maturity, at the bottom.

From the curve comparison, we find that:

■ The 10-year T-bond yield upon maturity is more stable than those of 1 and 3 years. The curve of the 1-year-term T-bond fluctuates the most.

■ The 10-year T-bond yield upon maturity is basically above 3%, mostly fluctuating around 4% and it is significantly impacted by the interest rate regulation.

Figure 5.4   Comparison of T-bond yields in different terms.

■ The 1-year T-bond yield upon maturity fluctuates around 1%–4% and is impacted notably by the short-term liquidity of the monetary market.

Next, let's quantify the data, calculate the index in the way of data analysis and verify the intuitive judgment.

For example, calculate the standard deviations of these three yield curves and measure the volatilities with them.

```
# Calculate the standard deviations
> apply(bonds[,-1],2,sd)
     ytm1        ytm3       ytm10
0.7291878 0.6523377 0.5888107
```

According to the calculation result, the standard deviation of the yield upon maturity for 1-year term is 0.729, the largest, while that for 10 years term is 0.588, the smallest.

Then let's calculate the standard deviations of these three yield curves by years.

```
# Group the data by years and calculate the standard deviations
> sdy<-apply.yearly(bondsxts,function(cols) apply(cols,2,sd));sdy
                   ytm1        ytm3       ytm10
2002-12-31 0.24506549 0.19079566 0.33013846
2003-12-31 0.41372346 0.31208564 0.31722444
2004-12-31 0.47182531 0.43472113 0.55582216
2005-12-31 0.54615193 0.56419153 0.65071361
2006-12-31 0.19039665 0.17720776 0.11174525
2007-12-29 0.56432016 0.54567296 0.55400302
2008-12-31 0.67966696 0.75470889 0.56060732
2009-12-31 0.27320510 0.39522147 0.25356530
2010-12-31 0.42360918 0.32081918 0.22876109
2011-12-31 0.38883385 0.26012716 0.17028720
2012-12-31 0.25408047 0.22746152 0.09955422
2013-12-31 0.49976235 0.49377643 0.38454713
2014-12-31 0.26920789 0.25424162 0.31266992
2015-12-31 0.43946060 0.19512158 0.21551456
2016-12-31 0.16860824 0.14493450 0.12929623
2017-02-17 0.08070615 0.05674466 0.10786616
```

Find the standard deviation with the maximum volatility. Which year is it?

```
# Maximum value
> apply(sdy,2,max)
     ytm1        ytm3       ytm10
0.6796670 0.7547089 0.6507136
```

```
# Time when the maximum value appeared
> sdy[unique(apply(sdy,2,which.max)),]
                ytm1      ytm3      ytm10
2005-12-31 0.5461519 0.5641915 0.6507136
2008-12-31 0.6796670 0.7547089 0.5606073
```

The maximum volatilities of the 1-year T-bond and the 3-year T-bond, reaching their maximum in 2008, both surpassed that of the 10-year T-bond. The 10-year T-bond volatility was at its peak in 2005.

What happened in 2005? And what happened in 2008? Recall the state economic situations at those times and maybe we would know.

Let's switch to another perspective and check the yield of the 10-year T-bond by years.

```
# Load the class library
> library(lubridate)
> library(reshape2)

# Organize data
> ytm<-bondsxts$ytm10['2010/']                    # Take data from 2012
> years<-unique(year(index(ytm)))
> df<-data.frame(matrix(NA,ncol=length(years),nrow=366))
> names(df)<-years

# Group data by years
> apply.yearly(ytm,function(x){
+     ycol<-format(index(x),'%Y')[1]
+     df[yday(x),ycol]<<-x
+ })

# Organize data
> df<-cbind(id=1:366,df)
> dt<-na.omit(melt(df,id.vars="id"))
>
> # Draw a plot
> g<-ggplot(aes(x=id,y=value, colour=variable),data=dt)
> g<-g+geom_line(size=1)
> g<-g+xlab('Day of Year')
> g
```

In Figure 5.5, *x*-axis stands for the 365 days of a year, while *y*-axis represents the yield value. It is obvious that the yield started to decline since the second half of 2015, which was greatly related

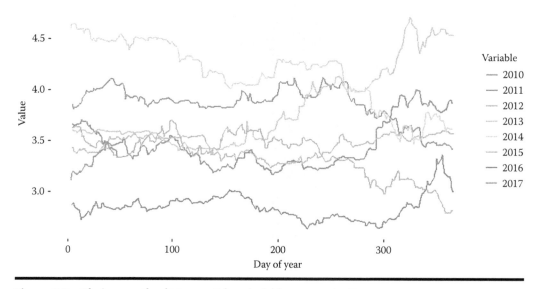

Figure 5.5   Chain growth of 10-year T-bond yield upon maturity.

to the interest reduction of 2015. The yield started to increase since 2017, which indicates that the central bank may start raising interest rates.

The treasury bond may be a little bit far away from us, but it is useful to know. There are ways to profit from treasury bonds. In 2013, the treasury bond futures launched by China Financial Futures Exchange gave a new way of trading treasury bonds. Treasury bonds and treasury bond futures are financial instruments that share the same objects in different markets. Due to their different prices in different markets, there will be opportunities for arbitrage. For arbitrage ways, please refer to Section 6.3, "Build Quantitative Model of Pairs Trading with R." Besides, the inter-bank market of treasury bonds is much bigger than the exchange market. News reports once said that the inter-bank market might be open. Once the market is open, there will be many opportunities for us to discover.

When we are learning the treasury bond markets, we need to, meanwhile, learn the politics, economy, macro factors, international relations, etc. of the country, so that we can seize the opportunities.

5.2 Enterprise Bonds and the Arbitrage

Question

How do you invest enterprise bonds?

Enterprise Bonds and the Arbitrage
http://blog.fens.me/finance-enterprise-bond/

Introduction

Debt is an important part of the financial system. The economic crises in the recent 40 years were caused by debts, such as the 1980 American debt crisis, the 1997 financial crisis in southeast Asia, the 2007 subprime crisis and the 2009 European debt crisis. There are serious debt problems in China right now. Local government bonds and enterprise bonds are all faced with default risks.

What is an enterprise bond and how to find transaction opportunities for enterprise bonds in the financial market will be the emphases of this section.

5.2.1 What Is an Enterprise Bond?

An enterprise bond is a bond issued by an affiliation of the central government departments, a state-owned enterprise or a state holding enterprise. It is issued by the enterprise engaged in economic activities like production, trading, transportation, etc. In western countries, only joint-stock companies are permitted to issue enterprise bonds, so for them, an enterprise bond is a corporate bond. In China, enterprise bonds refer to the bonds issued by enterprises of all kinds of ownerships.

5.2.1.1 Classification of Enterprise Bonds

1. By terms: Enterprise bonds can be divided into short-term enterprise bonds, mid-term enterprise bonds and long-term enterprise bonds. In China, short-term enterprise bonds refer to the bonds with terms of less than 1 year, mid-term enterprise bonds those with terms of more than 1 year but less than 5 years and long-term enterprise bonds those with terms of above 5 years.
2. By whether early redemption is allowed: Enterprise bonds can be classified into callable bonds and uncallable bonds. If the enterprise has rights to buy back all or part of the bonds at regular intervals or at any time before the maturity date, those bonds are called the callable bonds; otherwise, the bonds are the uncallable bonds.
3. By whether the coupon rate changes: Enterprise bonds can be classified into fixed rate bonds, floating rate bonds and step-up bonds Fixed rate bonds refer to the bonds with fixed interest rates during the packback period. Floating rate bonds are those bonds whose coupon rates change periodically as the market interest rate varies. Step-up bonds mean the bonds whose interest rate will be accumulated as the increase of bond term.
4. By whether the investors are given options: Enterprise bonds can be categorized into enterprise bonds with attached options and enterprise bonds without attached options. Enterprise bonds with attached options mean that the bond issuer grants the holders certain rights. Convertible bonds, bonds with attached warrant, putable bonds, etc. are examples. The holder of convertible bonds can convert the bonds into the stocks issued by the company at a stated price in a certain period of time. The holder of bonds with attached warrant can purchase the stocks of the company according to the terms of the warrant. Putable bonds can be sold back within a specified time period. If the bond holder does not have the options mentioned above, the bond he/she holds is an enterprise bond without attached options.
5. By issue ways: Enterprise bonds can be divided into public offering bonds and private placement bonds. Public offering bonds refer to the bonds that are, according to legal procedures, issued openly to the public investors with the approval of securities authorities. While private placement bonds are bonds that are issued specially for a minority of investors, with simple issue procedures and cannot be listed for public trading.

5.2.1.2 Issue Terms

■ The company should profit for three consecutive years before the bond issuance and the use of the funds raised should accord with the state industrial policies.
■ The accumulated bond balance should not exceed 40% of the company's net assets. The enterprise bond issuance in recent 3 years should be directionally loosing. The annual average distributable profit of the security housing bond contrarian should be sufficient to pay the interest of the corporate bonds for 1 year.
■ If the bond is issued for technological upgrading projects, the offer amount should not exceed 30% of its aggregate investment; if it is for infrastructure projects, the offer amount should not exceed 20% of its aggregate investment.

- The bond issuance proposal should be approved by the board of the company or the municipal SASAC.

The term of bonds is usually above 5 years, mostly 10 and 20 years and the int payment is once a year. The issue scale should not exceed 40% of its net assets, should not exceed 30% of its aggregate investment and should not be less than 1 billion *yuan*. Its issuing target should all the domestic institutions, except for those prohibited from purchasing by national laws and regulations.

Here is the list the enterprise bonds we found in the market, as shown in Figure 5.6.

Basically, the names of the bonds indicate their investment purposes, mostly for the major local construction projects. These bonds are vouched by the government credit and their lowest credit level is A+. The yields upon maturity vary from 3.5% to 8.3%, which are more attractive than the treasury bonds and the bank saving deposits mentioned in the previous section.

5.2.2 *What Is a Corporate Bond?*

Corporate bonds are issued by limited companies or limited liability companies, of which Securities Law has made specific provisions. Therefore, those unincorporated entities are not allowed to issue corporate bonds.

Code	Name	Remaining Maturity	Latest Net Price	Trading Volume (10,000)	Latest YTM	Percentage	Rating	Time
代码	名称	剩余年限 ▲	最新净价	成交量(万)	最新YTM	涨跌幅	主体评级	时间
1080031	10红谷城投债	0.05	100.1573	1420.00	3.7770	-0.99%	AA	02/17
1080047	10南昌城投债	0.19	100.3694	530.00	4.0434	-0.06%	AA+	02/17
1080087	10渝交通债	0.45	99.9309	14000.00	5.1887	0.00%	AA-	02/17
078046	07永煤债	0.54	100.3866	19000.00	4.5000	-1.72%	AA+	02/17
1080111	10渝江北嘴债	0.61	100.4516	4000.00	4.0650	-0.03%	AA+	02/17
1080162	10东城发债	0.82	101.6339	8000.00	4.7051	-0.45%	AA	02/17
1180004	11晋中公投债	0.88	101.9165	3309.00	3.7563	0.00%	AA	02/17
1380027	13金桥盐化债	0.93	101.1969	10000.00	4.7386	-0.07%	AA	02/17
1280024	12东胜债	1.02	50.0000	2760.00	8.3795	0.00%	A+	02/17
1180097	11三门峡债	1.18	102.7000	16000.00	4.4714	0.54%	AA	02/17
1180123	11泰矿债	1.34	99.4000	10000.00	7.1919	0.08%	AA	02/17
1280201	12赣城债	1.39	50.8054	6000.00	4.4499	-0.05%	AA+	02/17
1480507	14抚顺城投小微	1.58	100.8407	3000.00	5.7000	-1.06%	AA	02/17
1280350	12兴化债	1.67	51.1145	8000.00	5.1922	-1.10%	AA-	02/17
1280032	12淮安水利债	2.05	62.5000	22000.00	4.0645	-0.20%	AA+	02/17
1280047	12郑新债	2.06	106.4433	1000.00	4.7499	-1.68%	AA	02/17
1280071	12渝李渡债	2.09	103.7753	8000.00	5.0000	0.44%	AA	02/17
1280077	12芜湖建投债02	2.10	101.7402	4000.00	4.9561	0.38%	AA	02/17
1280093	12融强债	2.11	92.2202	5000.00	6.5584	0.75%	AA	02/17
1280178	12仪征债	2.32	62.1797	4000.00	4.8091	-1.19%	AA-	02/17
1280197	12北京建工债	2.37	60.9293	2000.00	4.7247	-0.48%	AA	02/17
1280242	12毕节信泰债	2.50	61.8742	5000.00	4.8923	0.45%	AA	02/17
1680367	16兴泰小微债01	2.55	99.6390	10000.00	3.6440	0.00%	AA+	02/17
1280329	12港航债	2.66	102.1513	5000.00	5.4514	-0.03%	AA	02/17
1280345	12昆明产投债	2.67	61.4795	6000.00	4.8601	0.06%	AA+	02/17

Figure 5.6 **Enterprise bonds in the market (screenshot of wind).**

Corporate bonds are securities issued by listed companies in accordance with legal procedures and it is agreed that the principal and the interest should be paid within a term of more than 1 year. A corporate bond is a direct financing variety of mid or long term regulated by the China Securities Regulatory Commission. Corporate bonds promise to pay their creditors the par values at maturity without preconditions and the interest at fixed intervals according to the agreed interest rate. Corporate bond issuance is an important external financing means of a company, an important financing source of a corporate, as well as one of the important financial instruments in the financial market.

It is issued according to the specific needs of the company operations, mainly used for fixed-asset investment, technological upgrading, fund source structure improvement, asset structure adjustment, financial cost reduction, merger and acquisition support, asset restructuring, etc.

As for the debtor-creditor relationship of corporate bonds, the bond holders are the creditors of the company, not the owners, which is the biggest difference from the stock holders; the bond holders have the right to get interest from the company according to the agreed terms and get the principal back at maturity. And the bond holder's interest is given priority over the dividends to the shareholders. Upon bankruptcy liquidation of the company, the bond holders have priority over the shareholders to take back their principal. However, the bond holders cannot take part in the company activities like the operation, the management, etc.

5.2.2.1 Issue Terms

The bond issue terms refer to the factors that the issuers must take into consideration when issuing bonds, mainly the bond issue volume, the issue price, the repayment terms, the coupon rate, the int. payment, whether there is guarantee, etc. The issue terms decide the profitability, liquidity and security of the bond and directly impact the cost of the financing and the investor's income of the investment. Usually, corporate bonds are classified by the issue terms, so the bond types will be determined if the issue terms are decided.

5.2.2.2 Risk Factors

The risk factors of corporate bonds mainly include:

1. Interest rate risk. Interest rate is one of the important factors influencing the bond price. When the interest rate rises, the bond price will go down. This is where the risk lies. The longer the remaining maturity of a bond, the higher the interest rate risk.
2. Liquidity risk. An investor cannot sell a bond with poor liquidity at a reasonable price within a short time and thus will suffer a loss or lose a new investment opportunity from it.
3. Credit risk. It means the loss an investor may suffer when the bond issuer is not able to pay the interest on time or even pay back the principal.
4. Reinvestment risk. If an investor buys a short-term bond, rather than a long-term bond, there will be reinvestment risk. For example, the interest rate of a long-term bond is 10%, and that of a short-term is 8%; an investor purchases the short-term bond to reduce the interest rate risk. The principal is paid back when the bond is mature. However, at this time the interest rate is reduced to 5% and it is not easy to find another investment opportunity

with an interest rate higher than 5%. Therefore, it may be better to invest the long-term bond at first. If so, the investor can still receive a return of 10%. The reinvestment risk is still an issue concerning the interest rate risk.

5. Call risk. It is for the bonds with call provisions, whose issuers have the right to buy them back before the maturity date. It may happen when the interest rate of the market declines and the investors expect to receive interests by the coupon rate. At the time, the investors will suffer a loss and receive lower income than the expected return, which is called the call risk.

6. Inflation risk. It refers to the risk of the purchasing power decrease due to the inflation. During inflation, the actual interest rate of the investors should be the coupon rate deducted by the inflation rate. If the coupon rate of the bond is 7% and the inflation rate is 3%, the actual interest rate will be 4%. The purchasing power risk is the most common risk of bond investment.

Here is the list of the corporate bonds we found in the market, as shown in Figure 5.7.

Mostly, we can know its issuer by the naming of a corporate bond. For example, Suning Bond (112196) is a corporate bond issued by Suning Commerce Group (002024). The credit rating of corporate bonds is lower than that of enterprise bonds, and the lowest rating can even be BBB−. Of course, higher risk comes along with higher profit. Some corporate bonds can achieve yields upon maturity of more than 10% and mostly around 8%.

Code	Name	Remaining Maturity	Latest Net Price	Trading Volume (10,000)	Latest YTM	Percentage	Rating	Time
代码	名称	剩余年限 ▲	最新净价	成交量(万)	最新YTM	涨跌幅	主体评级	时间
112220	14福星01	2.52	108.6180	20.90	5.4134	0.01%	AA	02/17
112235	15福星01	2.92	105.0990	4.30	5.5476	-0.10%	AA	02/17
122071	11海航02	4.26	105.0000	179.70	4.8611	-0.94%	AAA	02/17
122329	14伊泰01	2.64	104.8500	2039.00	4.9616	0.14%	AA+	02/17
112048	11凯迪债	1.75	104.6500	413.50	5.6101	0.00%	AA	02/17
112204	14好想债	2.18	104.5000	447.90	6.1997	0.00%	AA	02/17
112142	12格林债	3.84	104.4000	376.20	5.3406	0.01%	AA	02/17
112196	13苏宁债	2.73	103.9000	1.60	4.3905	-0.02%	AAA	02/17
122328	12开滦02	3.60	103.1000	7.00	5.3151	-0.39%	AAA	02/17
112231	14金贵债	2.70	103.1000	5.00	5.7573	0.00%	AA	02/17
112215	14万马01	2.42	103.0000	40.00	5.8079	-0.87%	AA	02/17
122143	12亿利01	3.17	102.6800	3.10	6.3239	-0.07%	AA	02/17
122310	13苏新城	2.42	102.5600	2384.30	7.6556	0.05%	AA+	02/17
122080	11康美债	1.33	102.3500	2101.00	4.1268	0.05%	AA+	02/17
112107	12云内债	1.52	102.3000	1300.50	4.4214	-0.14%	AA	02/17
122046	10中铁G2	2.94	102.0500	5000.00	4.1191	-1.40%	AAA	02/17
136769	16欣捷01	2.66	102.0000	0.60	6.1366	2.05%	AA	02/17
112243	15东旭债	3.24	102.0000	170.41	5.2982	0.00%	AA+	02/17
122028	09华发债	0.65	102.0000	2.00	4.3397	-0.27%	AA+	02/17
112441	16凯迪01	2.55	101.9000	0.10	5.2617	1.77%	AA	02/17
112030	11鲁西债	1.38	101.8800	542.50	4.7034	0.08%	AA	02/17
122267	13永泰债	1.46	101.8300	934.50	5.9073	-0.01%	AA+	02/17
122337	13魏桥02	2.72	101.8000	39.90	4.7626	0.00%	AA	02/17
122049	10营口港	1.03	101.7900	529.40	4.0794	-0.21%	AA+	02/17
136351	16永泰01	2.11	101.7900	3848.30	6.5421	-0.01%	AA+	02/17

Figure 5.7   Corporate bonds in the market (screenshot of wind).

5.2.3 Differences between Enterprise Bonds and Corporate Bonds

The use of funds raised by enterprise bonds is limited mainly in the fixed-asset investment and the technological upgrading, and it is directly related to the projects approved by the government departments.

In the environment of market economy, the credit of a corporate bond comes from the issuer's asset quality, operating conditions, profitability, sustained profitability, etc. The credit of an enterprise bond in China comes from the endorsement of the state-own enterprise by the government credit and the guarantee mechanism is implemented by the administrative enforcement. Therefore, its actual credit rating is almost the same as the government bonds.

According to the regulations of Securities Law, the minimum bond issue amounts of limited companies and limited liability companies are 12 million *yuan* and 24 million *yuan* respectively. However, in accordance with the internal control indicators of enterprise bonds, the issue amount of every enterprise bond should not be under 1 billion *yuan*. Therefore, only a few large-scale enterprises are able to issue enterprise bonds.

In the market economy, the regulator of corporate bonds usually requires strict bond credit rating and strict disclosure of the issuer's information, and it pays special attention to the market supervision after the bond issuance. In China, the issuance of enterprise bonds should be examined and approved by the National Development and Reform Commission and the State Council and should be guaranteed by a bank. After the issuance, the examination and approval department would not disclose the issuer's information or supervise the market behavior. The enterprise bond issuance is strictly controlled by the administrative mechanism. Its issue amount for 1 year is much lower than that of treasury bonds, central bank bills, financial bonds or the stocks.

The differences between enterprise bonds and corporate bonds mainly lie in the issuing bodies, the uses of funds, the credit bases, the regulatory procedures, the market functions, the legal bases, the issue terms, the issue procedures, etc.

5.2.4 Statistical Analyses of Enterprise Bonds

In the following, we will do some statistical analyses for the enterprise bond data to see if there is any rule for the trading of enterprise bonds.

The system environment used here:

- Win10 64bit
- R: 3.2.3 x86_64-w64-mingw32/x64 b4bit

Organize the data of enterprise bonds and save it to a CSV file for R language reading. Save the data of the enterprise bonds issued in Shanghai Stock Exchange as sh_bond.csv and save the data of the enterprise bonds issued in Shenzhen Stock Exchange as sz_bond.csv. The data collecting time is December 30, 2013.

The data format is as follow:

```
2013 Wuzhong Urban And Rural Construction Investment And Developoment
Co.,Ltd Corporate Bond,124369,13 Wuchengtou,1380303,13 Wuzhongchengtou
Bond,2020-10-12
2011 Yichang Construction Investment&Development Co.,Ltd Municipal
Project Construction Bond,122766,11 Yijiantou,1180162,11 Yichangchengtou
Bond,2019-11-17
```

2013 Chongqing Shuangfu Construction Development Co.,Ltd Corporate
Bond,124400,13 Yushuangfu,1380285,13 Yushuangfu Bond,2020-10-23
2013 Pingliang Investment Development Co.,Ltd Corporate Bond,124352,13
Pingliang Bond,1380279,13 Pingliang Bond,2020-09-17
2012 Ying Kou Port Group CORP. Corporate Bond,124053,12
Yingkouport,1280391,12 Yingkouport,2020-11-13
2013 Tianjin Infrastructure Investment Group Corporate Bond,124437,13
Jinjinghai,1380368,13 Jinjinghai Bond,2020-11-26

The data includes six columns:

- Name
- Code
- Abbr.
- Inter-bank Code
- Inter-bank Abbr.
- Maturity Date

Read data with R language and do some data analyses.

```
# Load the class library
> library(ggplot2)
> library(scales)

# Enterprise bonds of SSE
>
sh<-read.table(file="sh_bond.csv",header=FALSE,sep=",",colClasses="charac
ter", fileEncoding="utf-8", encoding = "utf-8")
>
names(sh)<-c("name","code","abbr","bank_code","bank_abbr","maturity_
date")
> sh$maturity_date<-as.Date(sh$maturity_date,format="%Y-%m-%d")

# Enterprise bonds of SZSE
>
sz<-read.table(file="sz_bond.csv",header=FALSE,sep=",",colClasses="charac
ter", fileEncoding="utf-8", encoding = "utf-8")
>
names(sz)<-c("name","code","abbr","bank_code","bank_abbr","maturity_
date")
> sz$maturity_date<-as.Date(sz$maturity_date,format="%Y-%m-%d")

# Merge all the undue enterprise bonds
> ss<-rbind(sh,sz)

# Print data
> head(ss)
                      Name   Code   Abbr. Inter-bank Code Inter-bank
Abbr. Maturity Date
1 2013 Wuzhong Urban And Rural Construction Investment And Developoment
Co., Ltd Corporate Bond 124369 13 Wuchengtou     1380303 13
Wuzhongchengtou Bond 2020-10-12
```

```
2 2011 Yichang Construction Investment&Development Co.,Ltd Municipal
Project Construction Bond 122766 11 Yijiantou     1180162 11
Yichangchengtou Bond
3 2013 Chongqing Shuangfu Construction Development Co.,Ltd Corporate Bond
124400 13 Yushuangfu     1380285    13 Yushuangfu Bond 2020-10-23
4 2013 Pingliang Investment Development Co.,Ltd Corporate Bond 124352 13
Pingliang Bond     1380279     13 Pingliang Bond 2020-09-17
5 2012 Ying Kou Port Group CORP. Corporate Bond 124053 12 Yingkouport
1280391     12 Yingkouport 2020-11-13
6 2013 Tianjin Infrastructure Investment Group Corporate Bond 124437 13
Jinjinghai     1380368    13 Jinjinghai Bond 2020-11-26
```

Next, we can use the basic bond data to do some researches and analyses of the bond market.

First of all, calculate the first 10 bonds which would be due recently.

```
> tmp<-ss[order(ss$maturity_date),]
> head(tmp[,c(2,3,6)],10)
      code      abbr maturity_date
959 111022 04  ShoulvBond  2014-02-19
678 122985 09  ZhenengBond    2014-02-23
689 120483 04 Sinopec    2014-02-24
110 122979  09 Jintou 2    2014-03-25
690 120482 04 GenertecBond    2014-03-31
82  122970 09 ThreeGorges 01    2014-04-08
91  122960 09 PolyGroup    2014-05-07
691 122998 Changzhaizanting    2014-05-26
104 122947 09 Hejiantou    2014-07-08
285 122943 09 Waigaoqiao    2014-09-04
```

For the data acquisition date is December 30, 2013, the first bond would mature on February 19, 2014, still 2 months to go.

Count the bonds by their maturity dates and output a histogram to illustrate the distribution of the dates.

```
> g<-ggplot(ss, aes(x=maturity_date))
> g<-g+geom_histogram(binwidth=50,position="identity")
> g<-g+scale_x_date(breaks = date_breaks(width="1 year"),labels = date_
format("%Y"), limits = c(as.Date("2014-01-01"),as.Date("2024-01-01")))
> g<-g+xlab("Maturity dates")+ylab("Bond numbers")
> g
```

In Figure 5.8, there are not many bonds mature in 2014, but many will be at maturity from 2018 to 2020. In that period of time, the enterprise will be faced with huge debts and the bank will also bear high risks. Once there is default, the consequence will be disastrous.

Then let's calculate the first 30 bonds that need to pay interests recently.

```
>
ss$interest_date<-as.Date(paste(2014,format(ss$matur
ity_date,format='-%m-%d'),sep=""))
> tmp<-ss[order(ss$interest_date),]
> head(tmp[,c(2,3,6,7)],20)
      code      abbr maturity_date interest_date
40  122841 11 YujinBond    2018-01-06    2014-01-06
257 122842 11 HechengBond    2018-01-06     2014-01-06
```

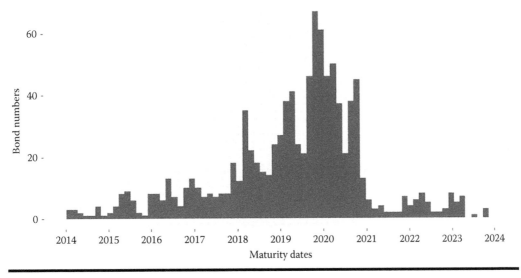

Figure 5.8 Distribution of the enterprise bond maturity dates.

```
509 122756 12 Gannongken      2019-01-06      2014-01-06
975 111047 08 ChangxingBond   2016-01-06       2014-01-06
766 123007 11 WeikuangBond    2021-01-07       2014-01-07
866 124110 13 Ningxinkai      2020-01-08       2014-01-08
541 124103 13 TongchuangBond   2020-01-09       2014-01-09
542 124139 13 Tonggangzha     2020-01-09       2014-01-09
547 124112 13 Yushengrun      2019-01-10       2014-01-10
548 124157 13 Pangaoxin       2020-01-10       2014-01-10
549 124131 13 Guoanzi      2020-01-10       2014-01-10
27  124135 13 Diangongtou     2019-01-11       2014-01-11
338 122849 11 XinyuBond       2018-01-11       2014-01-11
451 124134 13 Quanchengtou    2020-01-11       2014-01-11
555 124136 13 Taichengtou     2020-01-11       2014-01-11
770 122749 12 CNPC 02      2022-01-11       2014-01-11
771 122748 12 CNPC 01      2019-01-11       2014-01-11
337 122844 11 Zhuchengtou     2018-01-12       2014-01-12
439 122855 11 Yuqingfang      2018-01-12       2014-01-12
720 122617 12 XiangtouBond    2019-01-12       2014-01-12
```

Count the bonds by their interest payment dates and output a histogram to illustrate the distribution of the dates.

```
# Interest payment dates distribution
> g<-ggplot(ss, aes(x=interest_date))
> g<-g+geom_histogram(binwidth=1,position="identity")
> g<-g+scale_x_date(breaks=date_breaks("1 month"),labels = date_
format("%Y%m"), limits = c(as.Date("2014-01-01"),as.Date("2014-12-31")))
> g<-g+xlab("Interest payment date ")+ylab("Bond numbers")
> g
```

According to the analysis, the interest payment of every month was steady and it paid slightly more interests in November of 2014 than other months (Figure 5.9).

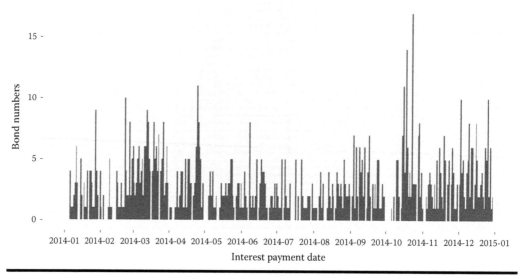

Figure 5.9 **Distribution of the enterprise bonds interest payment dates.**

5.2.5 An Example for Enterprise Bond

Let's take an enterprise bond for example to see what indicators of enterprise bonds we should know. We will take 11 Sinovel 01(122115) as an example. Check the disclosure information of the enterprise.

Field	Explanation
Bond name	Sinovel Wind Group Co., Ltd. 2001 Corporate Bond (Issue No.1)
Short name of the bond	11 Sinovel 01
Bond code	122115
Bond type	Registered book-entry bond
Credit rating	AA+
Par value (year)	100.00
Term of maturity (year)	5
Interest rate (%)	6
Method of calculating interest	Fixed interest rate
Int payment	Paid on an annual basis
Offer amount	950000000.00

(Continued)

(Continued)

Issue price (*yuan*)	100
Issue date	12/27/2011
Listing date	1/18/2012
Redemption date	December 27 of every year (in case of holidays, it shall be postponed to the first following working day)
Delisting date	12/27/2016
First-level trusteeship	China Securities Depository and Clearing Corporation Limited (CSDC) – Shanghai Branch
Second-level trusteeship	UBS Securities Co. Limited, Essence Securities Co., Ltd. and Qilu Securities Co., Ltd.
Way of underwriting	Standby commitment
Lead underwriter	UBS Securities Co. Limited, Essence Securities Co., Ltd. and Qilu Securities Co., Ltd.
Listing sponsor	UBS Securities Co. Limited
Exchange	Shanghai Stock Exchange

The latest interest payment of this bond was on December 27, 2013, that is, a few days ago. The coupon rate is 6.0%, the date of registration is December 26, 2013 and the interest payment date is December 27, 2013.

The interest will be taxed by 20% as the interest tax on the interest payment date, so many main bond holders will sell their bonds before the date of registration and then buy them back after the interest payment date, so as to evade tax. Usually, the bond trading volume is not very active; however, due to the tax avoidance, active trading may occur before the interest payment date. Let's see if there is any significant change of the bond trading volume before the interest payment date.

According to Figure 5.10, the trading volume of 11 Sinovel 01(122115) began to rise before December 27, 2013, the interest payment date, and it reached its maximum on the interest payment date, which means the bond holders bought in the bonds after evading the tax.

5.2.6 Trading Operations of Enterprise Bonds

There are usually two types of transactions for bonds: one is to earn interest at its maturity and the other is to buy it at a low price and sell it at a high price before its maturity, similar to the stock trading.

First, let's give an example of earning interest at its maturity and selling it after receiving the interest.

1. Assume that A bought 100 bonds of 11 Sinovel 01 (122115) at a price of 90 *yuan* (full settlement price) each, so the transaction amount is $90 * 100 = 9000$ *yuan*.
2. A held these bonds till the interest payment date of December 27, 2013, and A received the interest $= 100 * 100 * 6\% = 600$ *yuan* on the date, so the net income = interest income − interest tax $= 600 * (1 - 20\%) = 480$ *yuan*.

Figure 5.10   Candlestick chart of bond 11 Sinovel 01.

3. On December 28, 2013, A sold the bonds at a price of 91 *yuan* each, and the net income = (settlement price − cost price) * amount = (91 − 90) * 100 = 100 *yuan*.
4. The final net income = 480 + 100 = 580 *yuan*.

For the other transaction, people can buy the bond when its price is low and sell it when the price rises.

1. Assume that B bought 100 bonds of 11 Sinovel 01 (122115) at a price of 90 *yuan* (full settlement price) each, so the transaction amount is 90 * 100 = 9000 *yuan*.
2. On December 27, 2013, B sold the bonds at a price of 90 *yuan* each plus the accrued interest of 6% on the interest payment date, so the net income = (settlement price − cost price) * amount = ((trading price + accrued interest) − cost price) * amount = (90 + 6 − 90) * 100 = 600 *yuan*.
3. The final net income = 600 *yuan*.

From the above two transactions, we can learn that it profits more to buy the bond at a low price and sell it at a high price than to earn interest at its maturity.

Why do we need to buy it at a low price and sell it at high? Because if the issue price of the bond is 100 *yuan*, its price at maturity will return to 100 *yuan*. Therefore, when a bond is listed, we can buy it if its price is low, e.g. lower than 80 *yuan*, and then wait for the price rising (Figure 5.11).

Bond 11 Sinovel 01 fell to less than 80 *yuan* in 2013 and rose back to 100 *yuan* on the settlement date of 2016. If we seized the chance, bought it at the price of 80 *yuan* and held it till the settlement date of 2016, we would receive a rather good profit of low risk, with a coupon rate of 6% every year plus a price earning of 25%.

There are many enterprise bonds at maturity or at interest payment every month. If we can handle the time window well, we can earn a decent return from the enterprise bonds. Make good use of the funds and earn more with lower risks.

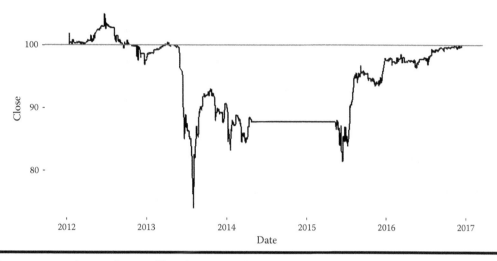

Figure 5.11   Price tendency of bond 11 Sinovel 01.

5.3 Convertible Bonds Arbitrage Practice

Question
How do you arbitrage from convertible bonds?

Article series using IT technology to play with finance

Convertible Bond Arbitrage Practice

http://blog.fens.me/finance-convertible-bond/

Introduction
Convertible bond is an often overlooked financial product. It is a hybrid security with both debt- and equity-like features. Its debt features will be significant if its price is around 100 *yuan*, while the equity features will be significant when the price is over 110 *yuan*. Comparing to the price of its underlying stock, the option value of the stock shows a nonlinear wave. Therefore, an arbitrage portfolio of convertible bonds can be built based on the feature.

In China's market, a convertible bond trading is a T+0 trading without any up or down limit, so convertible bonds can be operated in more ways than stocks. When the price of the underlying stock rises above the price of the convertible bond, we can convert the bond into stock and then sell the stock for arbitrage. In those mature markets overseas, the strategy of convertible bond arbitrage has been widely adopted.

5.3.1 An Introduction to Convertible Bonds

A convertible bond is a bond which can be converted to the stock shares of the issuing company and it has a comparatively lower coupon rate. Essentially, a convertible bond is a corporate bond with attached options and its holders are allowed to convert the bond into the underlying stock within a period of time.

A convertible bond embodies both the debt- and equity-like features and it has three main characteristics:

- Debt-like feature: A convertible bond, like other bonds, has a fixed rate and a fixed term, so the investors can choose to hold the bond till its maturity and then get back the principle and earn the interest.
- Equity-like feature: A convertible bond is a pure bond before it is converted into stock shares; however, after it is converted to stock, the bond holder becomes the shareholder of the company who has right to take part in the operation decision and the dividend distribution of the company, which will, to some degree, influence the company's capital structure.
- Convertibility: It is a significant symbol of convertible bonds. The bond holder can convert the bond into stock shares according to the agreed terms. The convertibility is an option attached to all the convertible bonds, but not the common bonds. It is agreed at the issuance that the holders can convert the convertible bonds to the common stocks of the company at an agreed price. If the holders are not willing to convert the bonds, they can continue to hold the bonds till the maturity date, and then get back the principal and receive the interest, or they can sell the bonds for cash in the circulation market.

Let's view the list of the current trading convertible bonds.

```
Code    Name   Remaining years Latest net price   Trading volumn(10,000)
Latest YTM  Up/Down Credit rating      Time
1  128013.SZ Hongtao Decoration Convertible Bond   5.4356 106.9062
354.25  1.0563  0.0011      AA 17/02/20
2  128012.SZ huifeng joint-stock Convertible Bond   5.1644 104.7008
330.20  0.4997  0.0002      AA 17/02/20
3  128011.SZ TQM Convertible Bond   5.0274 119.6913      29.80 -1.5675
0.0015    AA- 17/02/20
4  128010.SZ AUCKSUN Convertible Bond   4.9205 120.7805      843.31
-1.3974  0.0063      AA 17/02/20
5  128009.SZ Goertek Convertible Bond   3.8082 125.1955      2558.56
-2.9275  0.0031      AA+ 17/02/20
6  127003.SZ HIGHSUN Convertible Bond   5.2959 110.2466      104.57
0.8402  0.0018      AA- 17/02/20
7  123001.SZ BlueFocus Convertible Bond   4.8247 107.5723      480.79
1.0215  0.0010      AA 17/02/20
8  113010.SH Jiangnan Water Convertible Bond   5.0712 113.3905      400.00
-0.1011  0.0033      AA 17/02/20
```

```
9  113009.SH GAC GROUP Convertible Bond   4.9205 118.8289    28761.60
-1.4984  0.0159      AAA 17/02/20
10 113008.SH SHANGHAI ELECTRIC Convertible Bond   3.9507   0.0000
0.00  0.0000  0.0000      AAA 17/02/20
11 110035.SH GBIAC Convertible Bond   4.0164 125.8927    1007.00 -3.6463
0.0128      AAA 17/02/20
12 110034.SH Jointown Convertible Bond   4.9014 123.7595    1513.30
-2.1554  0.0009      AA+ 17/02/20
13 110033.SH ITG Convertible Bond   4.8740 115.3756    1455.00 -0.5405
0.0094      AA+ 17/02/20
14 110032.SH SANY Convertible Bond   4.8712 113.7342    4422.10 -0.5943
0.0073      AA 17/02/20
15 110031.SH AISINO CORP Convertible Bond   4.3068 106.7421    935.60
0.9607  0.0008      AA+ 17/02/20
16 110030.SH Gree Real Estate Convertible Bond   2.8438 112.3411
439.10 -1.2792 -0.0009      AA 17/02/20
```

Bond elements

- Maturity and conversion period: The maturity of a convertible bond is mostly 5 years, usually not more than 6 years. Convertible bonds have specified maturity and conversion period. For example, the listing date of Gree Real Estate Convertible Bond (110030) is January 13, 2015 and its delisting date is December 25, 2019. The conversion period is from June 30, 2015 to December 24, 2019. The bond cannot be converted in the half year after its issuance.
- Stock interest rate or dividend yield: The coupon rate usually rises year by year, but is still lower than the corporate bonds of the same level and the same term. The coupon rate of Gree Real Estate Convertible Bond (110030): 0.60% for the first year, 0.80% for the second year, 1.00% for the third year, 1.50% for the fourth year and 2.00% for the fifth year. The interest is paid once a year and the next interest payment date is December 25, 2017. Now it is the third year, the remaining maturity is 2.84 years and the current coupon rate is 1.0%.
- The pure debt value of convertible bonds: Similar to the enterprise bonds, it is the current value of the sum of principal and coupon income by a certain discount rate.
- Pure debt premium rate: Refers to the ratio of the convertible bond market price and its pure debt value. The lower the pure debt premium rate of a convertible bond, the closer its market price to the pure debt value.
- The conversion value of convertible bonds: Refers to the market value of the company stock shares that the convertible bond is converted to. The conversion ratio can be calculated by 100/conversion price, where the conversion price is the consideration of converting the convertible bond to one share of stock.
- The conversion premium rate: Refers to the ratio of the market price of a convertible bond and its conversion value. The lower the conversion premium rate, the closer its market price to the conversion value and the more approximate the convertible bond to the underlying stock. When the conversion premium rate is a minus, the convertible bond holders can convert the bond to the stock and then sell the stock for arbitrage. The conversion premium rate is an indicator we should pay special attention to.
- Conversion price adjustment: Due to the market changes, in order to prevent large-scale arbitrages, the conversion price will be adjusted under two circumstances. One is when the company issues bonus shares, increases share capital from accumulation fund, issues seasoned equity offerings, makes allotment, distributes cash dividends, etc. The other circumstance

is the adjustment clause. When the closing price of the underlying stock is lower than the current conversion price by a certain percentage in multiple consecutive trading days, the board of the company has rights to propose an amendment to lower the conversion price.

■ Call provision: It means that in a certain period after the convertible bond issuance, the company is well-run and will not need bond anymore, so it can redeem the issued convertible bonds prior to maturity. Call provisions can be triggered when the price of the underlying stock is higher than the conversion price by a certain percentage in a consecutive period of time, and then the company can buy back the issued bonds that have not been converted to stock at an agreed price before maturity.

■ Put provision: It means that if the stock of the company is consecutively lower than the conversion price by a certain percentage in a period of time, the investors are allowed to sell the convertible bonds back to the issuer at an agreed price before maturity, so as to protect the rights of the investors.

■ Investment value: The convertible bond allows its investor to gain the minimum benefit. The current yield of a convertible bond is higher than the common stock dividend and it has prior claims to the stock.

5.3.2 Convertible Bond Operations

In the ups and downs of the market, the rising or falling of convertible bonds is not always in accord with the underlying stocks, which makes it possible that the conversion value of a convertible bond is higher than its market price and, therefore, there can be opportunities for arbitrage. There are two investment strategies for convertible bonds. One is to partake in the issuance of new convertible bonds, i.e. to buy new bonds; another is to do investment analysis of the listing convertible bonds, including the arbitrage analysis of the premium rate, the event arbitrage analysis when the situation is close to the put provision or the conversion price adjustment clause, etc.

Trading rules of convertible bonds:

■ Convertible bond: The transaction is T+0 and it can be day trading, i.e. the bond can be bought in and then sold out on the same day.

■ Board lot: Round block (1 round block = 1000 *yuan* par value).

■ Price unit: The bond price for every 100 *yuan* par value.

■ Tick size: 0.01 *yuan*.

■ Each declaration limit: The minimum declaration is one round block and the maximum is 10000 round blocks.

■ The declaration price limit will be exercised in accordance with the provisions of the trading rules.

■ Conversion rules: The convertible bond can be converted on the day when it is bought. If you apply for the conversion of the convertible bond on T day, you can get stock shares before the opening on T+1 day and thus you can do intraday trading of the stocks.

■ Commission: No commission will be charged for conversion. The transaction cost lies mainly on the trading commission of the convertible bonds and the stocks.

There are two significant phenomena in the convertible bond market:

■ When the underlying stock price of the convertible bond is in the doldrums, almost meeting the put provision, a mysterious force will pull up the price to a safe zone from the put provision.

- After an amendment of the conversion price for the convertible bond, its underlying stock price will usually rebound with the stock market. Therefore, when the situation is close to put provision or when there is an amendment of conversion price, there can be a transitory investment opportunity.

We can testify this rule with data in the market.

5.3.3 Arbitrage Strategy of Negative Premium Rate

In the following, we will do a quantitative research for the convertible bonds, the operation progress of which will be from the basic statistical rules to the arbitrage practice.

The system environment used here:

- Win10 64bit
- R: 3.2.3 x86_64-w64-mingw32/x64 b4bit

5.3.3.1 Data Acquisition

First, let's capture some data from the Internet. Go to the convertible bond page of xueqiu.com and we find that the data is transferred in the way of Ajax and saved in the format of JSON, so we can directly crawl this JSON data (Figure 5.12).

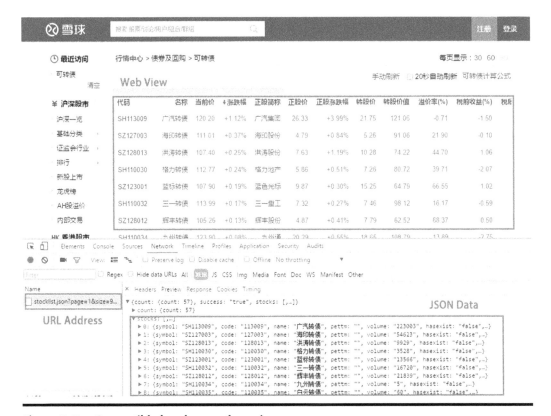

Figure 5.12   Convertible bond page of xueqiu.com.

The URL of the data is https://xueqiu.com/stock/cata/stocklist.json?page=1&size=90&order=desc&orderby=percent&exchange=CN&industry=可转债. For it needs to verify the user name and password, we can manually download it. How to crawl data from the web page that needs verification will be introduced separately in the future.

The format of the downloaded data is JSON and it is saved to file cb.json. The data is as follow:

```
[
  {
    "benefit_after_tax" : "-0.0188",
    "benefit_before_tax" : "-0.015",
    "change" : "1.86",
    "code" : "113009",
    "convert_bond_ratio" : "0.03",
    "convertrate" : "0.78",
    "current" : "118.87",
    "hasexist" : "false",
    "high" : "119.06",
    "high52w" : "128.4",
    "interestrtmemo" : "0.2% for the first year, 0.5% for the second
year, 1.0% for the third year, 1.5% for the fourth year, 1.5% for the
fifth year and 1.6% for the sixth year",
    "kzz_convert_price" : "21.75",
    "kzz_convert_time" : "2016.07.22-2022.01.21",
    "kzz_covert_value" : "116.41",
    "kzz_cpr" : "0.0211",

    // Omitted...
]
```

Next, let's process the JSON data with R language and turn it into the data format of data.frame in R language, which we are already familiar with. And the rjson package can convert JSON into data.frame. If you are not familiar with the rjson package, please refer to Section 1.6 of *R for Programmers: Mastering the Tools*.

```
> library(rjson)

# Read the json data
> json <- fromJSON(paste(readLines("cb.json"), collapse=""))

# Convert JSON into data.frame
> df <- data.frame(matrix(unlist(json), nrow=length(json), byrow=TRUE),
stringsAsFactors = FALSE)

# Rename the columns
> names(df)<-names(json[[1]])

# View data of the first two rows
> head(df,2)
    symbol   code   name pettm volume hasexist marketcapital current
percent change high
1 SH113009 113009 GAC GROUP Convertible Bond   287616    false
0.0  118.87    1.59   1.86 119.06
```

```
2 SH110035 110035 GBIAC Convertible Bond    10070    false           0.0
126.09    1.28    1.59   126.1

     low high52w low52w    open kzz_stock_symbol kzz_stock_name
kzz_stock_current
1 117.0    128.4   112.8 117.01           SH601238        GAC GROUP
25.32
2 124.5    135.0   121.3   124.5          SH600004        GBIAC
14.73

  kzz_convert_price kzz_covert_value kzz_cpr kzz_putback_price
kzz_convert_time
1           21.75            116.41  0.0211            100.0
2016.07.22-2022.01.21
2           12.56            117.28  0.0751            103.0
2016.09.05-2021.02.25

kzz_redempt_price kzz_straight_price kzz_stock_percent pb net_assets
benefit_before_tax
1        106.0             89.87           3.77 3.79     6.6775
-0.015
2        106.0             92.16           2.79 1.59     9.2473
-0.0364

benefit_after_tax convert_bond_ratio totalissuescale outstandingamt
maturitydate remain_year
1    -0.0188            0.03        4.10558E9         407021.9
20220122      4.923
2    -0.0403            0.21          3.5E9           349919.0
20210226      4.019

convertrate
interestrtmemo
1 0.78 0.2% for the first year, 0.5% for the second year, 1.0% for the
third year, 1.5% for the fourth year, 1.5% for the fifth year, 1.6% for
the sixth year
2 0.82    The convertible bond coupon rate: 0.2% for the first year, 0.4%
for the second year, 1.0% for the third year, 1.2% for the fourth year
and 1.5% for the fifth year
```

There are 37 columns in this dataset and the column names are:

- symbol, the symbol of the convertible bond
- code, the code of the convertible bond
- name, the name of the convertible bond
- pettm, the price/earning ratio in trailing 12 months
- volume, the volume of the convertible bond
- hasexist, whether to operate, which is a self-defined function by xueqiu.com
- current, the current price
- percent, the percentage of rising or falling
- high, the highest price
- low, the lowest price
- high52w, the highest price in 52 weeks

- low52w, the lowest price in 52 weeks
- open, the opening price
- kzz_stock_symbol, the code of the underlying stock
- kzz_stock_name, the short name of the underlying stock
- kzz_stock_current, the current price of the underlying stock
- kzz_stock_percent, the percentage of rising or falling of the underlying stock
- kzz_convert_price, the conversion price
- kzz_covert_value, the conversion value, which is a derivative indicator. The conversion value = the current price of the underlying stock/the conversion price * 100
- kzz_cpr, the premium rate (%), which is a derivative indicator. The premium rate = (the current price of the convertible bond − the conversion value)/the conversion value
- kzz_putback_price, the price which triggers the put provision. It is agreed in the prospectus that if the bond price is lower than the conversion price by X% in N consecutive days, the bond can be put back before maturity. The trigger price = the current conversion price * X%
- kzz_convert_time, the conversion period
- kzz_straight_price, the pure debt value
- net_assets, the net asset value of the underlying stock
- benefit_before_tax, the income before tax (%)
- benefit_after_tax, the income after tax (%)
- pb, the P/B ratio of the underlying stock
- convert_bond_ratio, the convertible bond ratio (%)
- kzz_redempt_price, the redemption price at maturity
- totalissuescale, the scale of the convertible bond (100 billion)
- outstandingamt, the remaining scale (100 billion)
- maturitydate, the maturity date
- remain_year, the remaining years
- convertrate, the conversion rate of the standardized bond, which is a derivative indicator. The conversion rate of the standardized bond = the conversion rate published by CSDC/the current price of the convertible bond
- interestrtmemo, the specific interest rate

5.3.3.2 Data Cleansing and Filtering

In the following, we will process the data. The saved data is of string type in the JSON format. We need to convert the data of string type into numeric type for the following numerical value calculation.

```
# View the data type. All string type
> str(df)
'data.frame':   57 obs. of  37 variables:
$ symbol           : chr  "SH113009" "SH110035" "SH110033" "SH110032"
...
 $ code             : chr  "113009" "110035" "110033" "110032" ...
 $ name             : chr  " GAC GROUP Convertible Bond" " GBIAC
Convertible Bond" " ITG Convertible Bond " " SANY Convertible Bond " ...
 $ pettm            : chr  "" "" "" "" ...
 $ volume           : chr  "287616" "10070" "14550" "44221" ...
 $ hasexist         : chr  "false" "false" "false" "false" ...
 $ marketcapital    : chr  "0.0" "0.0" "0.0" "0.0" ...
```

```
 $ current            : chr  "118.87" "126.09" "115.44" "113.8" ...
 $ percent            : chr  "1.59" "1.28" "0.94" "0.73" ...
 $ change             : chr  "1.86" "1.59" "1.08" "0.83" ...
 $ high               : chr  "119.06" "126.1" "115.6" "114.0" ...
 $ low                : chr  "117.0" "124.5" "114.35" "112.96" ...
 $ high52w            : chr  "128.4" "135.0" "131.5" "119.2" ...
 $ low52w             : chr  "112.8" "121.3" "109.9" "105.93" ...
 $ open               : chr  "117.01" "124.5" "114.35" "112.96" ...
 # Omitted...

# Convert string type into numeric type

> df$volume<-as.numeric(df$volume)
> df$current<-as.numeric(df$current)
> df$change<-as.numeric(df$change)
> df$percent<-as.numeric(df$percent)
> df$high<-as.numeric(df$high)
> df$low<-as.numeric(df$low)
> df$high52w<-as.numeric(df$high52w)
> df$low52w<-as.numeric(df$low52w)
> df$open<-as.numeric(df$open)
> df$kzz_stock_current<-as.numeric(df$kzz_stock_current)
> df$kzz_convert_price<-as.numeric(df$kzz_convert_price)
> df$kzz_covert_value<-as.numeric(df$kzz_covert_value)
> df$kzz_cpr<-as.numeric(df$kzz_cpr)
> df$kzz_putback_price<-as.numeric(df$kzz_putback_price)
> df$kzz_redempt_price<-as.numeric(df$kzz_redempt_price)
> df$kzz_straight_price<-as.numeric(df$kzz_straight_price)
> df$kzz_stock_percent<-as.numeric(df$kzz_stock_percent)
> df$pb<-as.numeric(df$pb)
> df$net_assets<-as.numeric(df$net_assets)
> df$benefit_before_tax<-as.numeric(df$benefit_before_tax)
> df$benefit_after_tax<-as.numeric(df$benefit_after_tax)
> df$convert_bond_ratio<-as.numeric(df$convert_bond_ratio)
> df$totalissuescale<-as.numeric(df$totalissuescale)
> df$outstandingamt<-as.numeric(df$outstandingamt)
> df$remain_year<-as.numeric(df$remain_year)
> df$convertrate<-as.numeric(df$convertrate)

# View the data type again. str type has been turned into num type.
> str(df)
'data.frame':   57 obs. of  37 variables:
 $ symbol             : chr  "SH113009" "SH110035" "SH110033" "SH110032"
...
 $ code               : chr  "113009" "110035" "110033" "110032" ...
 $ name               : chr  " GAC GROUP Convertible Bond " " GBIAC
Convertible Bond " " ITG Convertible Bond " " SANY Convertible Bond "
...
 $ pettm              : chr  "" "" "" "" ...
 $ volume             : num  287616 10070 14550 44221 84331 ...
 $ hasexist           : chr  "false" "false" "false" "false" ...
 $ marketcapital      : chr  "0.0" "0.0" "0.0" "0.0" ...
 $ current            : num  119 126 115 114 121 ...
 $ percent            : num  1.59 1.28 0.94 0.73 0.63 0.33 0.31 0.18 0.15
0.11 ...
```

```
$ change          : num  1.86 1.59 1.08 0.83 0.755 0.37 0.39 0.2 0.179
0.114 ...
$ high            : num  119 126 116 114 121 ...
$ low             : num  117 124 114 113 120 ...
$ high52w         : num  128 135 132 119 159 ...
$ low52w          : num  113 121 110 106 114 ...
# Omitted...
```

5.3.3.3 Negative Premium Rate Arbitrage Strategy

If we are to make a negative premium rate arbitrage strategy, first we need to find the convertible bonds at conversion stage. The codes below will mainly use two data processing packages: stringr and magrittr. If you are not familiar with their usage, please refer to Section 3.4 String Processing Package of R: stringr and Section 3.3 Efficient Pipe Operation of R: Magrittr in this book.

```
# Load the packages
> library(stringr)
> library(magrittr)

# Continuous data conversion
> df2<-df[,c('kzz_convert_time')]
> df2 %<>% str_split(pattern='-') %>%
+   unlist %>% matrix(ncol=2, byrow=TRUE) %>%
+   data.frame(stringsAsFactors=FALSE) %>%
+   cbind(df$symbol,df$name)

# Set the date of today
> today<-Sys.Date();today
[1] "2017-02-21"

# Find the convertible bonds at conversion stage
> df2 %<>% {
+   df2$X1<-as.Date(df2$X1,format='%Y.%m.%d')
+   df2$X2<-as.Date(df2$X2,format='%Y.%m.%d')
+   df2[intersect(which(df2$X1 < today),which(df2$X2 > today)),]
+ }
            X1          X2 df$symbol   df$name
1  2016-07-22 2022-01-21   SH113009  GAC GROUP Convertible Bond
2  2016-09-05 2021-02-25   SH110035  GBIAC Convertible Bond
3  2016-07-05 2022-01-04   SH110033  ITG Convertible Bond
4  2016-07-04 2022-01-03   SH110032  SANY Convertible Bond
5  2016-07-29 2022-01-21   SZ128010  AUCKSUN Convertible Bond
6  2016-09-26 2022-03-17   SH113010  Jiangnan Water Convertible Bond
7  2015-06-19 2020-12-11   SZ128009  Goertek Convertible Bond
8  2016-12-16 2022-06-07   SZ127003  HIGHSUN Convertible Bond
9  2016-09-09 2022-03-02   SZ128011  TQM Convertible Bond
10 2017-02-06 2022-07-28   SZ128013  Hongtao Decoration Convertible Bond
11 2016-06-27 2021-12-17   SZ123001  BlueFocus Convertible Bond
12 2016-07-21 2022-01-14   SH110034  Jointown Convertible Bond
13 2015-12-14 2021-06-11   SH110031  AISINO CORP Convertible Bond
14 2016-10-28 2022-04-21   SZ128012  huifeng joint-stock Convertible Bond
39 2015-08-03 2021-02-01   SH113008  SHANGHAI ELECTRIC Convertible Bond
57 2015-06-30 2019-12-24   SH110030  Gree Real Estate Convertible Bond
```

For most of the time, we may have our unique perspective when observing the market and this personal data demand is important. However, the tools in the market can barely fulfill our demands, so we have to implement them by our own programming. For example, Figure 5.13 Form of Conversion Premium Rate Arbitrage is my market observation method.

By realtime market scanning, I can immediately discover the arbitrage chances. When the expected return is above 3% and the loss probability is below 10%, I will decidedly make a bet on it. Of course, you can make more precise quantization standards to totally remove the speculation.

In the following, a self-defined table will be created to observe the data of convertible bonds.

```
# Create a self-defined table
> df3<-df[which(df$symbol %in% df2$`df$symbol`),]
> df3$name2<-str_c(df3$name,'(',df3$code,')')
> df3$time<-format(Sys.time(),'%H:%M:%S')
> df3$current2<-str_c(df3$current,'(',df3$percent,')')
> f3$current3<-str_c(df3$kzz_stock_current,'(',df3$kzz_stock_percent,')')

> cols<-c('name2','time','current2','current3','kzz_cpr')
> df4<-df3[order(df3$kzz_cpr),cols]
> names(df4)<-c('Convertible bond name','Update time','Convertible bond
price (Percentage)','Underlying stock price (Percentage)','Premium rate')
> df4
            Convertible bond name Update time   Convertible bond price
(Percentage)  Underlying stock price (Percentage) Premium rate
1   GAC GROUP Convertible Bond (113009) 13:30:31    118.87(1.59)
25.32(3.77) 0.0211
9   TQM Convertible Bond (128011) 13:30:31   120.179(0.15)      6.46(1.89)
0.0734
2   GBIAC Convertible Bond (110035) 13:30:31    126.09(1.28)     14.73(2.79)
0.0751
7   Goertek Convertible Bond (128009) 13:30:31    125.39(0.31)
29.95(1.91) 0.0982
12 Jointown Convertible Bond (110034) 13:30:31     123.8(0.09)
20.18(0) 0.1441
5   AUCKSUN Convertible Bond (128010) 13:30:31    120.838(0.63)
9.88(2.7) 0.1497
```

Conversion Premium Rate Arbitrage

Convertible Bond Name	Update Time	Price of Ask One	Volume of Ask One	Price of Underlying Stock	Conversion Premium Rate (%)	Expected Return (%)	Volatility in 20 Days (%)	Loss Probability (%)	Annual Sharpe Ratio
中鼎转债 (125887)	10.36.43	117.5	18	8.06	1.17	-1.16	2.52	67.64	-7.09
燕京转债 (126729)	10.36.55	112.4	15.4	8.1	1.3	-1.28	1.63	78.43	-12.19
工行转债 (113002)	10.38.02	101.46	61	3.56	0.61	-0.6	0.66	81.94	-14.15
重工转债 (113003)	10.38.02	116.25	20	5.53	2.38	-2.32	1.75	90.74	-20.52
海直转债 (127001)	10.36.01	122.94	3	8.43	2.81	-2.74	1.96	91.83	-21.59
川投转债 (110016)	10.37.28	125.97	2	11.17	2.51	-2.45	1.07	98.88	-35.4
同仁转债 (110022)	10.37.28	124.33	1	20.6	5.44	-5.16	1.71	99.87	-46.68
泰尔转债 (128001)	10.37.01	103	18	7.67	14.01	-12.29	3.66	99.93	-49.31
国电转债 (110018)	10.38.02	103.69	70	2.31	7.73	-7.18	1.24	100	-89.79
中行转债 (113001)	10.37.37	97.2	97	2.59	5.83	-5.51	0.91	100	-94.22
海运转债 (110012)	10.34.51	104.99	31	3.5	35.29	-26.08	3.13	100	-129.28
石化转债 (110015)	08.55.00	96.68	30000	4.4	12.72	-11.28	1.08	100	-161.73
深机转债 (125089)	10.35.54	94.28	202.7	4.3	22.12	-18.12	1.44	100	-195.19

Figure 5.13   Form of conversion premium rate arbitrage.

```
3  ITG Convertible Bond (110033) 13:30:31   115.44(0.94)        8.9(2.3)
0.1583
4  SANY Convertible Bond (110032) 13:30:31    113.8(0.73)       7.3(0.14)
0.1629
8  HIGHSUN Convertible Bond (127003) 13:30:31    110.6(0.18)
4.75(1.06) 0.2247
6  Jiangnan Water Convertible Bond (113010) 13:30:31   113.67(0.33)
8.1(3.32) 0.3205
57 Gree Real Estate Convertible Bond (110030) 13:30:31   112.5(-0.09)
5.83(0.69) 0.4009
39 SHANGHAI ELECTRIC Convertible Bond (113008) 13:30:31       114.16(0)
8.42(0) 0.4439
10 Hongtao Decoration Convertible Bond (128013) 13:30:31  107.133(0.11)
7.54(0.53) 0.4606
11 BlueFocus Convertible Bond (123001) 13:30:31    107.697(0.1)
9.84(1.13) 0.6691
14 huifeng joint-stock Convertible Bond (128012) 13:30:31   105.12(0.02)
4.85(0.41) 0.6884
13 AISINO CORP Convertible Bond (110031) 13:30:31   107.09(0.08)
20.52(0.15) 1.2467
```

We can monitor the changes of the convertible bonds and the underlying stocks. When the price of the underlying stock rises, and the convertible bond and the underlying stock are not totally synchronized, there can be chances for arbitrage.

Besides the basic indicators based on the raw data, we still need some self-defined derivative indicators, such as expected return, loss probability, Sharpe ratio, etc. These indicators are more intuitive for us and we won't need to translate the numbers in our mind, which relaxes our head and gives it more time for rational decisions.

We should add four derivative indicators, i.e. expected return, volatility in 20 days, loss probability and annual Sharpe ratio.

The calculation formulas of the four derivative indicators are as follows:

```
Expected return = (Conversion value - Convertible bond price)/Convertible
bond price * 100%
Volatility in 20 days = Standard deviation (Moving average return in
20 days) * Square root(250)
Loss probability  = Normalized expected return, with 0 as mean and
volatility in 20 days as variance
Annual Sharpe ratio = Expected return / Volatility in 20 days * Square
root(250)
```

Let's take Gree Real Estate Convertible Bond (110030) as an example to calculate its four derivative indicators. The data above is not enough for the indicator calculation, so we need to add more, mainly the price data of Gree Real Estate Convertible Bond (110030) and Gree Real Estate (600185).

Save the Gree Real Estate Convertible Bond (110030) price data to file 110030.csv. The dataset, including 13 columns, is as follow.

```
Date, Closing price, Ups and downs, Percentage, Trading volume (round
block), Trading amount (10,000), Pure debt YTM, BP of Ups and downs,
Conversion value, Pure debt value, conversion premium rate, Pure debt
premium rate, Parity premium rate
```

```
2017-02-20,112.5,-0.1,-0.000888099,4391,494.4315,-
1.2792,2.74,80.303003,96.7976,0.400944,0.162219,-0.170402954
2017-02-17,112.6,-0.21,-0.001861537,7046,794.8019,-
1.3066,6.39,79.752039,96.627,0.411877,0.165306,-0.174640225
2017-02-16,112.81,-0.25,-0.002211215,6555,740.1899,-
1.3705,7.59,80.027521,96.58,0.40964,0.168047,-0.171386198
2017-02-15,113.06,-0.35,-0.003086148,3002,340.3308,-
1.4464,10.62,79.88978,96.5201,0.415199,0.171362,-0.172299034
```

Save the raw data of the Gree Real Estate (600185) price to file 600185.csv. The dataset, including 7 columns, is as follow.

```
Date, Opening price, Highest price, Lowest price, Closing price, Trading
amount (10,000), Trading volume (round block)

2017-02-10,5.83,5.87,5.82,5.84,32.67,5587300
2017-02-13,5.84,5.88,5.83,5.87,32.54,5555300
2017-02-14,5.87,5.88,5.84,5.86,23.72,4051000
2017-02-15,5.84,5.86,5.79,5.8,34.97,6005000
2017-02-16,5.79,5.83,5.78,5.81,29.6,5096700
```

Extract data from Gree Real Estate Convertible Bond and Gree Real Estate and calculate the indicators.

```
> library(PerformanceAnalytics)

# Read data of Gree Real Estate Convertible Bond
> df110030<-read.csv(file="110030.csv",header=TRUE)
> x110030<-xts(df110030[,c('Closing price','Conversion value')],order.by
= as.Date(as.character(df110030$Date),format='%Y-%m-%d'))
> x110030<-na.omit(x110030)
> names(x110030)<-c('close','value')   # Reserve closing price and
conversion value

# Read data of Gree Real Estate
> df600185<-read.csv(file="600185-2.csv",header=TRUE)
> x600185<-xts(df600185$close,order.by = as.Date(as.character(df600185$da
te),format='%Y-%m-%d'))
> names(x600185)<-'stock'              # Reserve the closing price

# Merge the data of Gree Real Estate Convertible Bond and Gree Real
Estate by the conversion date
> x110030<-merge(x110030,x600185,all = FALSE)

# View the dataset
> head(x110030)
             close      value stock
2014-12-25 100.0016   93.44519 19.53
2014-12-26 100.0033   98.56482 20.60
2014-12-29 100.0082 105.02416 21.95
2014-12-30 100.0099 102.15335 21.35
2014-12-31 100.0115 105.26340 22.00
2015-01-05 100.0197 105.59833 22.07
```

The dataset we've built includes three columns, i.e. the closing price of Gree Real Estate Convertible Bond, the conversion value of Gree Real Estate and the closing price of Gree Real Estate. For the conversion price, which can be adjusted according to the announcements released by the listing companies, has a direct influence on the premium rate, we need to set a field for the conversion price in the data.

```
# The conversion price changes for three times
> x110030$convetPrice<-NA
> x110030$convetPrice['2014-12-25']<-20.9
> x110030$convetPrice['2016-05-26']<-7.39
> x110030$convetPrice['2016-08-25']<-7.26

# Fill NA value with its previous value
> x110030<-na.locf(x110030)
```

Next, let's calculate the derivative indicators.

```
# The premium rate of the convertible bond
> x110030$Premium<-(x110030$close-x110030$value)/x110030$value

# Expected return
> x110030$expRet<-(x110030$value-x110030$close)/x110030$close

# The accumulated return of the convertible bond price
> x110030$ret<-Return.calculate(x110030$close,method="log")
> x110030$ret[which(is.na(x110030$ret))]<-0
> x110030$ret<-cumprod(1 + x110030$ret)-1

# Volatility in 20 days
> x110030$vol <-volatility(x110030$close,n=20)

# Loss probability
> x110030$loss<- pnorm(-1 * x110030$expRet, 0, x110030$vol)

# Annual Sharpe ratio
> x110030$sharpe<-x110030$expRet/x110030$vol*sqrt(250)
```

Multiple all the ratios with 100%.

```
# Ratio*100%
> x110030$Premium<-x110030$Premium*100
> x110030$expRet<-x110030$expRet*100
> x110030$ret<-x110030$ret*100
> x110030$vol<-x110030$vol*100
> x110030$loss<-x110030$loss*100
```

View the calculation result of the self-defined data. There are 10 columns in all, including close (the Gree Real Estate Convertible Bond price), value (the conversion value), stock (the Gree Real Estate price), convetPrice (the conversion price), Premium (the premium rate %), expRet (the expected return %), ret (the return of convertible bond %), vol (the volatility in 20 days %), loss (the loss probability %) and sharpe (the annual Sharpe ratio %).

```
# View the calculation result
> tail(x110030)
            close   value  stock  convetPrice  Premium    expRet      ret
2017-02-13 113.45 80.85397  5.87         7.26 40.31470 -28.73163 -12.24956
2017-02-14 113.41 80.71623  5.86         7.26 40.50459 -28.82795 -12.28050
2017-02-15 113.06 79.88978  5.80         7.26 41.51998 -29.33860 -12.55164
2017-02-16 112.81 80.02752  5.81         7.26 40.96401 -29.05991 -12.74522
2017-02-17 112.60 79.75204  5.79         7.26 41.18761 -29.17226 -12.90780
2017-02-20 112.50 80.30300  5.83         7.26 40.09439 -28.61955 -12.98518
                 vol      loss     sharpe
2017-02-13  7.579765  99.99248  -59.93418
2017-02-14  7.431161  99.99476  -61.33764
2017-02-15  7.425625  99.99611  -62.47070
2017-02-16  7.428001  99.99543  -61.85748
2017-02-17  3.914910 100.00000 -117.81978
2017-02-20  3.919023 100.00000 -115.46624
```

According to the printed result of the recent 6 days, the premium rate of Gree Real Estate Convertible Bond is up to 40% and the convertible bond value is much higher than the real estate value, so it does not fit for conversion arbitrage at all. Let's have a try and see if we can find an interval where the return is more than 0.

```
> x110030[which(x110030$expRet>0),]
               close    value stock convetPrice   Premium    expRet
ret vol
2014-12-29 100.0082 105.0242 21.95        20.9 -4.776010 5.015554
0.006599760  NA
2014-12-30 100.0099 102.1533 21.35        20.9 -2.098262 2.143233
0.008299718  NA
2014-12-31 100.0115 105.2634 22.00        20.9 -4.989294 5.251296
0.009899680  NA
2015-01-05 100.0197 105.5983 22.07        20.9 -5.282876 5.577530
0.018099213  NA
           loss sharpe
2014-12-29   NA     NA
2014-12-30   NA     NA
2014-12-31   NA     NA
2015-01-05   NA     NA
```

There are four records where the return is above 0. Its conversion time is from June 30, 2015 to December 24, 2019, so these opportunities appear before the conversions. According to the overall result, Gree Real Estate Convertible Bond is not fit for the negative premium rate arbitrage.

5.3.3.4 Visualized Data Interpretation

In the following, we will draw a plot to interpret the data. It could be more intuitive and clear.

```
> library(ggplot2)
> library(scales)

# Draw a plot of premium rate
> g<-ggplot(aes(x=Index,y=Value, colour=Series),data=fortify(x110030[,c('
close','value','Premium')],melt=TRUE))
```

```
> g<-g+geom_line(size=1)
> g
```

In Figure 5.14, the Gree Real Estate Convertible Bond price (close), the conversion value (value) and the premium rate (Premium) are put together, so it is obvious that Gree Real Estate Convertible Bond was pushed high quickly after its issuance and its premium rate remains high, which can be resulted from some human factors to prevent conversion. I have checked that up to 99.80% of the Gree Real Estate Convertible Bonds is still not converted. With such a high premium rate, if the market allows reverse operation, we can short sell the convertible bonds and go long on the stock, which will help us make a bundle and quickly level the premium rate.

```
> g<-ggplot(aes(x=Index,y=Value, colour=Series),data=fortify(x110030[,c('
value','stock','convetPrice')],melt=TRUE))
> g<-g+geom_line(size=1)+scale_y_log10()
> g
```

In Figure 5.15, the Gree Real Estate price (stock), the conversion price (convetPrice) and the conversion value (value) are put together. In May of 2016, Gree Real Estate distributed its profits and increased its share capital from the accumulation fund; it lowered its stock price and adjusted the conversion price, but it did not change the conversion value.

```
> g<-ggplot(aes(x=Index,y=Value, colour=Series),data=fortify(x110030[,c('
expRet','ret','vol','loss','sharpe')],melt=TRUE))
> g<-g+geom_line(size=1)
> g
```

In Figure 5.16, all the derivative indicators are put together. We find that Gree Real Estate Convertible Bond has a very large volatility in July 2015 and the volatility became small and stable in 2016. The return of the convertible bond price has continued falling since July 2015. The

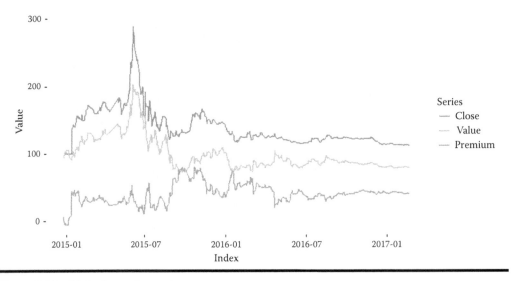

Figure 5.14 Plot of premium rate.

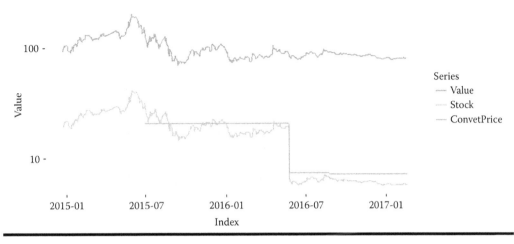

Figure 5.15 Conversion price adjustment.

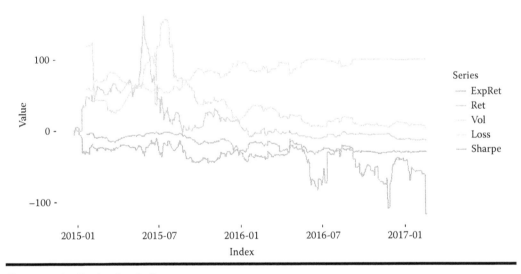

Figure 5.16 Derivative indicators.

expected return of the convertible bond arbitrage strategy was negative, the Sharpe ratio was negative and the loss probability varied from 50% to 100%.

To sum up the result of the data analysis, this convertible bond is beyond our understanding! So do not touch it.

Having gone through a convertible bond analysis, the following will be easy. We need to obtain the data of all the convertible bonds and calculate them with the method above. If the data you've got is realtime, you can realtime monitor the whole convertible bond market. In the end, update the result on the website and inform the users by text, email, etc. Congratulations! Your self-defined convertible bond trading monitoring system is completed. As an IT guy, haven't you started making money yet?

5.4 Reverse Repurchase with Risk-Free Financial Instruments

Question
How do you make reverse repos?

Introduction
In June 2013, Yu'e bao, launched by Alipay, attracted much attention from people, as well as a great amount of petty funds. Yu'e bao raised the annualized return of petty funds' interest to around 4.0%, which was much higher than the bank demand deposit interest rate of 0.3%. This shook the status of banks which used to earn money by just doing nothing.

In the financial market, it is not difficult to earn an annualized return of 4%–5%. Reverse repos can make that happen as well. If there is a currency tension, i.e. the banks are lack of money, an overnight repo rate of 50% can be possible. At that time, we can happily count money at home.

5.4.1 An Brief Introduction to Reverse Repos
Repo transaction is a form of collateralized loans, usually used in the government bonds. Bond brokers temporarily sell a certain amount of bonds to investors and meanwhile sign contracts, promising to buy the bonds back at a slightly higher price in a certain period of time. The bond brokers would get funds for investment, while the investors would profit from the bond price difference. A repo transaction can be as long as a few months, but for most of the time, it lasts only 24 hours, which makes it a super short-term financial instrument. Repo transactions can be classified into treasury bond repo transaction, bond repo transaction, security repo transaction, pledge-style repo transaction, etc. Repo transactions mostly have the attribute of short-term financing. If referring to the operation mode, they combine the characteristics of spot trading and forward trading and they are usually used in the bond trading.

A treasury bond repo transaction, also known as a treasury bond spot trading, refers to the process that a seller sells bonds in the treasury bond market with attached provisions that the seller will repurchase the bonds at a pre-agreed price in a certain period of time. It also includes the reverse repo process that a buyer purchases the bonds and then sells them back to the seller at a pre-agreed price in a certain period of time.

A pledge-style repo is a short-term financing activity which uses bond as the rights pledge. In pledge-style repo transaction, the borrower of cash (the repurchase party) pledges the bonds to the lender of cash (the counterparty) to borrow money, meanwhile, these two parties agree that on a certain date in the future, the repurchase party will return the principal and give interest by an agreed repo rate to the counterparty and the counterparty will return the original pledged bonds to the repurchase party.

A reverse repo is a transaction in which People's Bank of China purchases securities from the primary dealers and agrees to sell the securities back to the primary dealers on a certain date in the future. A reverse repo is an operation that the central bank injects liquidity to the market. A reverse repo at maturity is an operation that the central bank withdraws liquidity from the market. In simple terms, a transaction in which a party takes initiative to lend money and gets the pledged bonds is called a reverse repo transaction. In the transaction, the investor is the lender of cash who receives the pledged bonds and lends the money.

Generally speaking, A lends money to B and B will give back the principal to A and pay A the interest by an agreed interest rate when it is due. A reverse repo is risk-free, for it is pledged by treasury bonds and the pledge value is usually higher than the borrowed fund.

A reverse repo is one of the cash management financial products and its return basically levels off those of the cash management products like monetary fund, Yu'e bao, bank financing ect.

5.4.2 Varieties of Reverse Repos

There are eighteen varieties of reverse repos, with nine in Shanghai Stock Exchange (SSE) and nine in Shenzhen Stock Exchange (SZSE). The reverse repos in SSE are new pledge-style repos, with capital requirement of not less than 100,000 *yuan* and each deal an integer multiple of 100,000 *yuan*. The nine varieties are of 1, 2, 3, 4, 7, 14, 28, 91 and 182 days. The reverse repos in SZSE are treasury bond repos, with capital requirement of not less than 1000 *yuan* and each deal an integer multiple of 1000 *yuan*. The nine varieties are of 1, 2, 3, 4, 7, 14, 28, 91 and 182 days. See Figure 5.17.

Bond Repo Type	Repo Days	Repo Code	Annual Rates of Current Tradeable Repos
1-Day Shanghai New Pledge-style Repo	1	204001	4.685
2-Day Shanghai New Pledge-style Repo	2	204002	1.95
3-Day Shanghai New Pledge-style Repo	3	204003	1.3
4-Day Shanghai New Pledge-style Repo	4	204004	1.02
7-Day Shanghai New Pledge-style Repo	7	204007	3.805
14-Day Shanghai New Pledge-style Repo	14	204014	4.07
28-Day Shanghai New Pledge-style Repo	28	204028	4.2
91-Day Shanghai New Pledge-style Repo	91	204091	4.62
182-Day Shanghai New Pledge-style Repo	182	204182	4.51
1-Day Shenzhen T-bond Repo	1	131810	3.2
2-Day Shenzhen T-bond Repo	2	131811	0.2
3-Day Shenzhen T-bond Repo	3	131800	0
4-Day Shenzhen T-bond Repo	4	131809	2
7-Day Shenzhen T-bond Repo	7	131801	3.5
14-Day Shenzhen T-bond Repo	14	131802	3.6
28-Day Shenzhen T-bond Repo	28	131803	2.85
91-Day Shenzhen T-bond Repo	91	131805	4.22
182-Day Shenzhen T-bond Repo	182	131806	4.101

Figure 5.17　Reverse repos in stock exchanges.

The most common reverse repos we operate are of 1 and 7 days.

For example, we make a reverse repo. We sell the 1-day Shanghai new pledge-style reverse repo GC001 issued in SSE with 204001 as its repo code. The borrowing time is 1 day, the fund amount is 100,000 *yuan* and the interest rate is 4.685%, so after 1 calendar day, the interest received = 100000 * 4.685%/365 = 12.83562. Principal + Interest = 100000 + 12.83562 = 100012.83562. After deducting the 0.001% commission, i.e. 1 *yuan*, the final amount received should be 100012.83562 − 1 = 100011.83562 *yuan*.

If we trade the Shanghai new pledge-style repo GC007 (204007), the after-tax income after seven calendar days = 100000 * 3.805% * 7/365 − 5 = 67.9726 *yuan*.

This transaction is risk-free, i.e. the interest is the profit.

For reference of the reverse repo commission, please see Chart 5.1.

Chart 5.1   Reverse Repo Commission

Code	Abbr.	Varieties	Commission Rate	Trading Commission for Every 100,000 yuan
204001	GC001	1-Day new pledge-style repo	0.001% of the trading amount	1
204002	GC002	2-Day new pledge-style repo	0.002% of the trading amount	2
204003	GC003	3-Day new pledge-style repo	0.003% of the trading amount	3
204004	GC004	4-Day new pledge-style repo	0.004% of the trading amount	4
204007	GC007	7-Day new pledge-style repo	0.005% of the trading amount	5
204014	GC014	14-Day new pledge-style repo	0.010% of the trading amount	10
204028	GC028	28-Day new pledge-style repo	0.020% of the trading amount	20
204091	GC091	91-Day new pledge-style repo	0.030% of the trading amount	30
204182	GC182	182-Day new pledge-style repo	0.030% of the trading amount	30
131810	R-001	1-Day treasury bond repo	0.001% of the trading amount	1
131811	R-002	2-Day treasury bond repo	0.002% of the trading amount	2
131800	R-003	3-Day treasury bond repo	0.003% of the trading amount	3

131809	R-004	4-Day treasury bond repo	0.004% of the trading amount	4
131801	R-007	7-Day treasury bond repo	0.005% of the trading amount	5
131802	R-014	14-Day treasury bond repo	0.010% of the trading amount	10
131803	R-028	28-Day treasury bond repo	0.020% of the trading amount	20
131805	R-091	91-Day treasury bond repo	0.030% of the trading amount	30
131806	R-182	182-Day treasury bond repo	0.030% of the trading amount	30

5.4.3 Reverse Repo Transaction

Usually, if you have a trading account in a securities company, you can make reverse repos. You don't have to make any extra application. There are usually two operation interfaces for reverse repo transactions. One is for viewing the market situation and the other is for transactions.

5.4.3.1 Market Situation Interface of Reverse Repos

Figure 5.18 is the time-sharing plans of GC001 for observing the market situation of the reverse repo. The market situation interfaces are similar for most of the software. I have made annotations to explain the functions of every part in the figure.

Annotations:

1. Annualized return: It is represented with the y-axis and it is measured with the percentage of annualized return. On y-axis, the red values are higher than the closing price of yesterday, the white value is the closing price of yesterday and the green values are under the closing price of yesterday.
2. Annualized return curve: It is built with the x-axis (trading timeline) and the y-axis (annualized return).
3. Trading volume: It is represented with the y-axis, showing the trading amount of funds, with round block as its unit.
4. Trading volume histogram: It is built with the x-axis (trading timeline) and the y-axis (trading volume).
5. Trading timeline: It is represented with the x-axis. The trading time in China's stock market is 9:30–11:30 and 13:00–15:00.
6. Separator: It is used to separate the upper figure from the lower figure. The upper figure is the time-sharing annualized return curve and the lower figure is the trading volume histogram.
7. Percentage change: It is the y-axis on the right side. It takes the opening price as its origin. The red values are the increased percentage and the green values are the declined percentage.
8. Trading code and name: To show the trading code (204001) and the name (GC001).

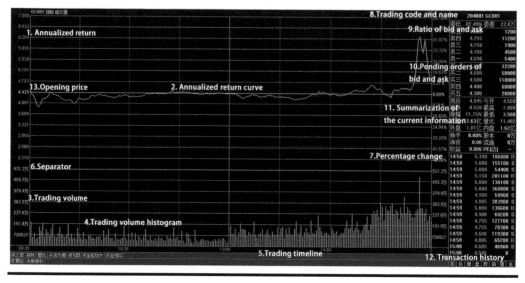

Figure 5.18   Time-sharing plans of GC001 market situation.

9. Ratio of bid and ask:
   – Ratio of bid and ask = (Bid size − Ask size)/(Bid size + Ask size) * 100%.
   – Difference of bid and ask = Bid volume − Ask volume.
   – Red means bid size > ask size. Green means the opposite.
10. Pending orders of bid and ask:
   – Bid one: Refers to the collection of the bid volume with the highest bid price at the current moment.
   – Ask one: Refers to the collection of the ask volume with the lowest ask price at the current moment.
   – Bid one, bid two, bid three, bid four and bid five are bid prices sorted from high to low.
   – Ask one, ask two, ask three, ask four and ask five are ask prices sorted from low to high.
11. Summarization of the current information:
   – Current price: Refers to the price of the last deal.
   – OpenPrice: Refers to the opening price.
   – Ups and downs: The closing price of today − The closing price of yesterday.
   – Percentage: (Current price − Closing price of the last trading day)/Closing price of the last trading day * 100%
   – Highest: The highest price of today.
   – Lowest: The lowest price of today.
   – Total amount: The sum of bid volume and ask volume. Total amount = Outer disk + Inner disk.
   – Quantity relative ratio: Refers to the ratio of today's total trading volume and the recent average trading volume. If the quantity relative ratio is larger than 1, it means the total amount of trading at the moment is enlarged; if it is smaller than 1, it means the total trading amount at the moment is shrunk.
   – Outer disk: When the trading price is the ask price, the total trading volume is called the outer disk.

– Inner disk: When the trading price is the bid price, the total trading volume is called the inner disk. If the accumulated volume of the outer disk is much larger than that of the inner disk and the stock price is rising, it means that many people are trying to buy in stocks. If the accumulated volume of the inner disk is significantly larger than that of the outer disk and the stock price is dropping, it means that many people are trying to sell stocks.

12. Transaction history: The journal account of every transaction, including trading time, trading price (accounting annualized return), trading amount of fund and trading operations (buy and sell).

13. Opening price: The price when the market opens today. Today's opening price is not equal to yesterday's closing price. It depends on the call auction.

5.4.3.2 Reverse Repo Transaction Interface

There is an interface for the transaction of reverse repos. For the sake of security, it is completely separate from the market situation interface. Users need to log in for transaction. The transaction interfaces can be different for different securities companies. The screenshot of Figure 5.19 is the operation interface of the reverse repo transaction in China Merchants Securities' transaction software.

Annotations:

1. Service option: The software of China Merchants Securities makes reverse repo a separate option. We can select this option if we want to make a reverse repo.

2. Transaction operations:
   – Shareholder code: Choose SSE or SZSE.
   – Security code: 204001.
   – Annualized repo rate: It can be interpreted as the loan interest.
   – Available fund: The balance of the current account.
   – Maximum financing volume: The block (1000 *yuan* for 1 block) amount that can be lent from the balance. It is equal to the available fund/1000.
   – Ask volume: The block amount for lending.
   – Returned money predict: To calculate the total money received when it is due. The returned money = Ask volume * 1000 * annual repo rate/365.
   – Place order: To confirm the payment.

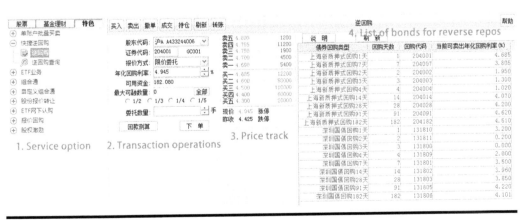

Figure 5.19 Reverse repo transaction interface.

3. Price track: Shows bid one, bid five, ask one and ask five in real time.
4. List of bonds for reverse repos: It lists all the bonds for reverse repos. It is not automatically updated and it needs to be manually refreshed.

Now we can view the market situation and meanwhile trade the reverse repos. Most of the reverse repos need manual operations.

5.4.4 Repo Operations

A repo is the reverse operation of a reverse repo. A reverse repo is to lend money to others and to profit from the interest. A repo is to borrow money from those who make reverse repos and to pay interest. It needs to make applications and pass some exams to be qualified of making repos, i.e. only professional investors can trade repos (Figure 5.20).

The varieties traded in repos are the same as those of reverse repos, so are their trading codes. The characteristics of repo operations are as follows:

1. The bond holder (the financing party) pledges the bonds and converts them to the standardized bonds.
2. They can borrow money from the investor (the funding party) within the quota of the standardized bonds and redeem them on a specified date.
3. The usage of standardized bond on the repo financing party should not surpass 90% and the hedging multiples should not exceed five.
4. Individual investors can take part in the pledge-style repo of bonds as funding parties.

Repos should be pledged with bonds. Usually, treasury bonds, convertible bonds, financial bonds, convertible bonds, etc. can be used as pledges. According to the risk rating, different pledges have

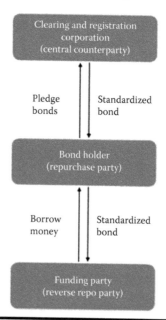

Figure 5.20 Repos and reverse repos.

different conversion rates. The conversion rate of treasury bond is 99%, usually the highest; it can be down to 40% for the financial bonds.

```
Code, Abbr., Standardized bond conversion rate

010107.SH,21 T-bond (7),1.0200
010213.SH,02 T-bond (13),0.9900
010303.SH,03 T-bond (3),0.9800
010504.SH,05 T-bond (4),1.0300
010512.SH,05 T-bond (12),1.0000
010609.SH,06 T-bond (9),0.9800
010619.SH,06 T-bond (19),0.9600
010703.SH,07 T-bond 03,0.9700
010706.SH,07 T-bond 06,0.9800
010710.SH,07 T-bond 10,0.9900
010713.SH,07 T-bond 13,0.9800
019002.SH,10 T-bond 02,0.9400
019003.SH,10 T-bond 03,0.9800
019005.SH,10 T-bond 05,1.0000
019007.SH,10 T-bond 07,0.9500
019009.SH,10 T-bond 09,0.9800
019010.SH,10 T-bond 10,1.0000
019012.SH,10 T-bond 12,0.9900
```

We can pledge treasury bonds and raise fund by repo, and purchase treasury bonds with the fund raised, then raise fund again by the newly purchased treasury bonds. The process of adding leverage to the repo can be seen in Figure 5.21.

In Figure 5.21, we purchase bonds with 1 million *yuan* cash and then add leverages by repo pledges. The fund doesn't have any loss, for the conversion rate of treasury bond is 100%. After three leverage addings, we actually buy bonds of $1 + 0.9 + 0.81 + 0.73 = 3.44$ million *yuan*. The leverage ratio is 3.4. The daily interest we need to pay is $109.589 + 98.63014 + 87.67123 + 76.71233 = 372.6027$ *yuan*.

We can continuously add leverage like this to magnify the fund, so that we can make the most of a small fund to lever a big market. However, the risk level is elevated as well. If the bond price drops or the repo rate rises, the loss will be amplified. To prevent infinite magnification, the securities company systems usually set a limit of not more than 5 multiples. We need to be especially careful when adding leverages!

5.4.5 Operations of Central Bank's Open Market

We usually learn the operations of central bank's open market from news. It is said that the transaction operation of reverse repos and repos is one of the monetary policies.

See the news on the operations of central bank's open market on February 15, 2017.

In this week, reverse repos of 900 billion *yuan* will be at maturity in the open market of China's central bank, of which 151.5 billion *yuan* of MLF will be due on Wednesday and about 630 billion *yuan* of TLF will mature on Friday. The central bank has re-launched reverse repos in this week. From Monday till now, reverse repos have been made for three consecutive days. However, the daily liquidity is still at the state of net return. The net return on Monday was 90 billion *yuan*, on Tuesday 100 billion *yuan* and today 140 billion *yuan*. The tight liquidity is also reflected in the market. The shibor interest rates all rise today. The overnight shibor reports an interest rate of 2.2658% by gaining 0.48 basis point; the 7-day shibor reports an interest rate of 2.6270% with 0.62 rising basis point;

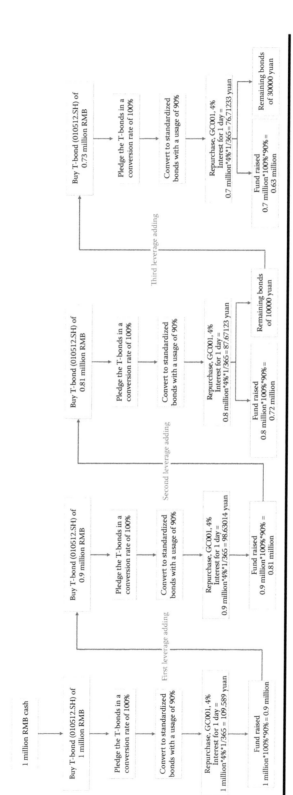

Figure 5.21 Process of adding leverage to repos.

and the 3-month shibor reports an interest rate of 4.2394% with 2.46 newly added basis points. Up to 11:25, the 7-day pledge-style repo rate of China has risen by 26 basis points to 2.7819%, which is the highest since July 2015; the 14-day interest rate has also gone up by 25 basis points to 3.3556%.

The central bank shrinks the market funds by purchasing the reverse repos; while it releases the funds and injects liquidity to the market by purchasing the repos. In this way, the country can macro control the market.

In the following, let's analyze the data and discover some rules of the interest rate in the repo market.

The system environment used here:

- Win10 64bit
- R: 3.2.3 x86_64-w64-mingw32/x64 b4bit

Organize the data of GC001 and save it to GC001.csv. The data format is as follows and the file includes seven columns.

```
Date, Opening price, Highest price, Lowest price, Closing price, Trading
amount, Trading volume
```

```
1 2006-05-08   1.5   1.5 1.5    1.5 1e+05   1e+05
2 2006-05-09   1.5   1.5 1.5    1.5 0e+00   0e+00
3 2006-05-10   1.5   1.5 1.5    1.5 0e+00   0e+00
4 2006-05-11   1.5   1.5 1.5    1.5 0e+00   0e+00
5 2006-05-12   1.5   1.5 1.5    1.5 0e+00   0e+00
6 2006-05-15   1.5   1.5 1.5    1.5 0e+00   0e+00
```

First of all, let's visualize the data in a simple way and then observe the data as a whole. Draw a curve chart of the highest closing prices, take the interest rate of 10% as a marking and see when the repo rate is above 10%.

```
> library(xts)
> library(ggplot2)
> library(scales)
# Read data
> gc<-read.csv(file="GC001.csv",header=TRUE)
> names(gc)<-c('date','open','high','low','close','value','volumn')

# Convert to the time series type
> gcx<-xts(gc[,-1],order.by=as.Date(gc$date,formate='%Y-%m-%d'))

# Draw a plot of the interest rate trend
> g<-ggplot(aes(x=Index,y=Value, colour=Series),data=fortify(gcx[,c('high
','close')],melt=TRUE))
> g<-g+geom_line(size=1)
> g<-g+geom_hline(yintercept=10,col='blue',size=1)
> g<-g+scale_y_continuous(breaks=seq(0,100,10))
> g
```

In Figure 5.22, the red line stands for the highest prices of repo rate, the blue line represents the closing prices of repo rate and the straight line parallel to x-axis is the interest rate of 10%. It is obvious that the repo rate was extraordinarily high in 2007 and 2008 and it means that the market

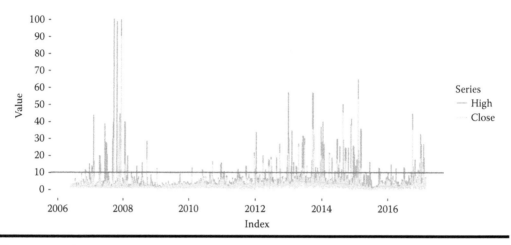

Figure 5.22   Repo rate.

was suddenly in need of money, which has never happened before. From 2012 to 2015, the interest rate of the repo market frequently exceeded 10%, which means it became normal that the market lacked money. This figure should deliver more information, but I cannot tell because of the limitation of my knowledge for now. Hope I can make clear the underlying message 1 day.

Now let's analyze the rules of the repo market and figure out when the interest rate would usually rise. Find the data with the highest prices above 6% and then calculate what day and what date the prices are most likely to appear.

```
> library(lubridate)

# The return rate is over 6
> gcx6<-gcx[which(gcx$high>6),]

# The day it is most likely to happen
> table(wday(index(gcx6))-1)

  1   2   3   4   5
 59  63  72 190  84

# the date it is most likely to happen
> table(days_in_month(index(gcx6)))

 28  29  30  31
 32  11 175 250
```

According to the data structure, the GC001 highest prices over 6% appear on Thursday for the most of times, because on Thursday the fund borrowed by the repurchase party can be used for 3 days, including weekends. From the monthly analysis, we find that in the end of the months, the highest prices over 6% appear more than other dates, because many short-term notes are at maturity and the banks are temporarily lack of money. Here is an opportunity we can actually seize. We can trade the reverse repos when the repo rate rises and earn more risk-free income.

In the end, let's get back to our real lives. I suggest that everyone open an account for reverse repo transaction. We may not make a big money out of reverse repos, but it is still a good choice to transfer the bank savings deposit.

Chapter 6

Quantitative Investment Strategy Cases

Comprehensive cases will be introduced in this chapter. The cases will study the financial market, the mathematical equations, the modeling with R language, the historical data backtesting and finally find the investment opportunities. It is a whole set of learning methods from theory to practice. I hope everyone can have fun in the financial market.

6.1 Mean Reversion, the Investment Opportunity against Market

Question
How do you implement mean reversion models with R language?

Mean Reversion
The Investment Opportunity Against Market
http://blog.fens.me/finance-mean-reversion

Introduction

There are two typical investment strategies in the stock market: trend following and mean reversion. The strategy of trend following is to find the effective trading signals in the fluctuation ranges of the market. In Section 2.3 of *R for Programmers: Advanced Techniques*, the trend following strategy is mentioned. However, the mean reversion strategy is a counter-trend strategy. A soar is usually followed by a drop, and a plummet is usually followed by a rising, which is especially effective in a volatile market. In this section, the strategy to seize this type of tiny opportunity will be introduced.

6.1.1 Mean Reversion Theory

In finance, mean reversion is a law that the price will tend to move to the average price after a certain degree of deviation from it. Essentially, mean reversion is the philosophical idea that things will go to the opposite direction when they reach the extreme. In simple terms, it is a rule that rising too much will definitely bring a fall and vice versa.

Mean reversion means that the stock price is likely to revert to the mean value when it is higher or lower than the mean, or the average price. According to this theory, the stock price should always fluctuate around its mean. The trend of rising or falling, no matter how long it lasts, cannot last forever. In the end, the mean reversion law will rule: the price will fall to the mean if it rises markedly and will increase to the mean if it plummets. When we think that everything will always tend to return to normalcy and we make all decisions based on it, we are actually applying the mean reversion theory.

In the following, the candlestick chart of PAB (000001.SZ) will be taken as an example to intuitively illustrate the phenomenon of mean reversion. The stock data from January 2005 to July 2015 is intercepted and the stock price is the price after forward restoration of rights.

There are three curves in Figure 6.1. The black one is the daily stock price of PAB after forward restoration of rights, the red one is the 20-day moving average and the blue one is the 60-day moving average. And the dotted red horizontal line in the figure, i.e. 7.14 *yuan*, is the stock average price in the past 10 years. In these 10 years, the stock price of PAB has gone through several ups and downs and traversed the mean value 7.14 many times. And this is the phenomenon of mean reversion that we are going to discuss.

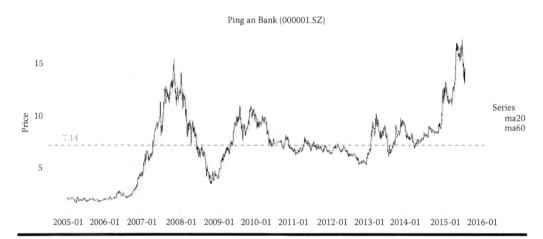

Figure 6.1   PAB stock price.

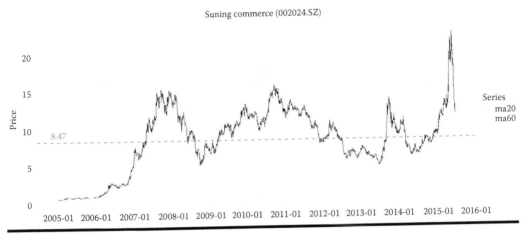

Figure 6.2 Stock price of Suning Commerse.

6.1.1.1 Three Characteristics of Mean Reversion

Mean reversion is one of the core theories to support value investment theory. It has three characteristics: necessity, asymmetry and government regulations.

For necessity, the stock price cannot always go up or down. A trend may last long, but not forever. A tendency where the stock price continuously rises or falls is called mean aversion, the opposite of which is mean reversion. However, the reversion period is random and unpredictable. The reversion periods differ in different stock markets; even in the same market, the periods can be different for different stocks.

Let's take the candlestick chart of another stock, Suning Commerse (002024.SZ) for example. Intercept the stock price data after forward restoration of rights from 2005 to July 2015. See the Figure 6.2. We see that the stock price of Suning Commerse rose dramatically from 2006 to 2007 and fell after that. Another price rising happened from 2009 to 2010; a third rising happened in the latter half of 2013; and a fourth rising happened from the latter half of 2014 to 2015. We can intuitively see in the figure that the rising of 2015 was the most dramatical and the most volatile; we can conclude from the figure that a phenomenon cannot last forever, though it may last long.

For asymmetry, the amplitudes and speeds of the stock price volatility can be different and the reversion amplitudes and speeds can be random. A symmetrical mean reversion is actually abnormal and incidental, which can be verified by the stocks.

We can put together the candlestick charts of PAB (000001.SZ) and Suning Commerse (002024.SZ), as shown in Figure 6.3. These two stocks rose significantly in the middle of 2007; their curves matched mostly. They both saw dramatical falls in 2008, yet with different volatilities and speeds. From 2010 to 2012, they followed completely different and irregular trends, which showed the randomness and the asymmetry of mean reversion.

Government behavior: The stock return would not deviate from the mean for too long. The inner power of the market would revert it back to its intrinsic value. If there is no government policy influence, the stock price will naturally revert to the mean value under the market mechanism. This is not a denial to the endeavor that the government has made to stimulate the market efficiency. Mean reversion doesn't mean that the market will immediately revert to the intrinsic value after the deviation. Mean aversion can last for long. The government behavior can be a

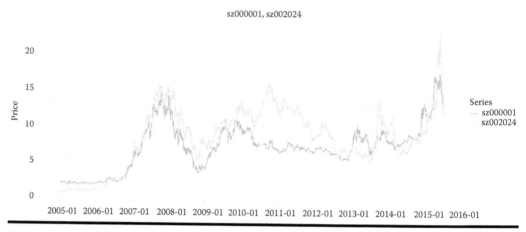

Figure 6.3   Stock prices of PAB and Suning Commerse.

restraint to the market mechanism, but it is a necessity for the market. A market failure can be the immediate result of regulations participated by the government.

The government behavior of policies like an RRR (reserve requirement ratio) raise, an RRR reduction, an interest raise and an interest reduction will significantly manifest themselves in the stock market. The real estate stocks and the banking stocks will be directly impacted by the state macro regulations. As shown in Figure 6.4, the share of VANKE A (000002.SZ) has been added and the three curves respectively represent the stocks of PAB, VANKE A and Suning Commerse. We can find that the stock price trends of real estate and banking are similar, while that of e-commerce is different.

Besides, auxiliary lines of two colors are adopted. The red ones represent the time and the interest variation of an interest raise and the yellow ones those of interest reduction. In 2007, when the stock price rose substantially, the state macro regulation raised the interest rate to encourage deposits and to suppress the over-high stock price. In 2015, due to the financial reform, the government continuously reduced the interest rate to stimulate the stock market. It is obvious in Figure 6.4 that VANKE A and PAB were sensitive to the interest raise or reduction regulations, while Suning Commerse was not.

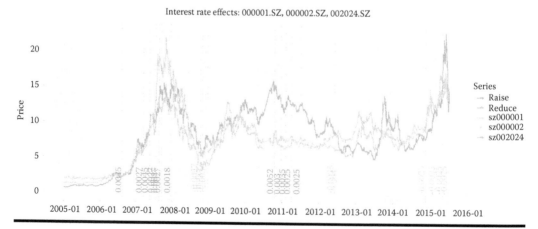

Figure 6.4   Stock prices influenced by the interest rate.

By reviewing the market, we have basically testified that the mean reversion theory is consistent with the market behavior. So how should we apply the theory and find the breakthrough point of investment?

6.1.1.2 Calculation Principle and Formula

From the perspective of value investing, we find that the stock price fluctuates around the mean; yet if taking the time costs into consideration, it is not cost-effective to invest money in the stock market and wait for a big take-off. Then we need to re-define the average. Replace the long-term average with 20-day average and then find a short-term investment method.

Calculation principle: Get a candlestick chart, take the N day moving average as the short-term average price (the mean) of mean reversion and then calculate the difference of the stock price and the mean and its N day average standard deviation, from which we can find how much the difference deviates from the mean. If the deviation is more than two multiples of the standard deviation, we consider the stock price overextended or oversold and the stock price will ceaselessly restore itself to the mean according to the mean reversion theory.

Calculation formula:

```
N day average = [T day stock price + (T-1) day stock price + ... + (T-(N-
1)) day stock price]/N
Difference = N day average - N day stock price
Average of N day difference = [T day difference + (T-1) day difference +
... + (T-(N-1)) day difference]/N
Standard deviation of N day difference = sqrt([(T day difference -
Average of T day difference)^2 + ... + ((T-(N-1)) day difference -
Average of (T-(N-1)) day difference)^2 ]/N)
```

If N is 20,

```
20-Day average = [T day stock price + (T-1) day stock price + ... +
(T-19) day stock price]/20
```

Calculate the deviation point.

```
T day difference > Standard deviation of T day difference * 2
```

We take the deviation point as the buy signal and the next intersection point of the moving average and the stock price as the sell signal. Now we have turned the mean reversion investment theory into a mathematical model.

6.1.2 Mean Reversion Theory and Its Implementation

In the following, we will process the stock data with R language, so as to implement a mean reversion model and to testify whether our investment theory can discover opportunities of making money.

The system environment used here:

■ Win10 64bit
■ R: 3.2.3 x86_64-w64-mingw32/x64 b4bit

6.1.2.1 Data Preparation

R language itself provides abundant financial function toolkits, such as the time series package zoo and xts, the index calculation package TTR, the data processing package plyr, the visualization package ggplot2, etc. We will use all these packages together for modeling, calculation and visualization. For detailed usage of the zoo package and the xts package, please refer to Section 2.1, "Basic Time Series Library of R: zoo," and Section 2.2, "Extensible Time Series: xts," of *R for Programmers: Mastering the Tools*.

The data used here can be the data you crawl from Sina Finance or the data you download from Yahoo Finance with the quantmod package.

The data used here includes the candlestick chart data of A-share (after forward restoration of rights) from July 2014 to July 2015 and it is saved to a local file stock.csv in CSV (Concurrent Versions System) format.

The data format is as follow:

```
000001.SZ,2014-07-02,8.14,8.18,8.10,8.17,28604171
000002.SZ,2014-07-02,8.09,8.13,8.05,8.12,40633122
000004.SZ,2014-07-02,13.9,13.99,13.82,13.95,1081139
000005.SZ,2014-07-02,2.27,2.29,2.26,2.28,4157537
000006.SZ,2014-07-02,4.57,4.57,4.50,4.55,5137384
000010.SZ,2014-07-02,6.6,6.82,6.5,6.73,9909143
```

The data contains seven columns:

- Column1: The stock code, code, 000001.SZ
- Column2: The trading date, date, 2014-07-02
- Column3: The opening price, Open, 8.14
- Column4: The highest price, High, 8.18
- Column5: The lowest price, Low, 8.10
- Column6: The closing price, Close, 8.17
- Column7: The trading volume, Volume, 28604171

Load the stock data with R language. We need to do the calculation by each stock, but the information of all stocks is mixed in the data, so I have made some conversion during the loading. The data has been grouped by the stock codes, a list object has been generated and meanwhile, the stock object type has been converted from data.frame to XTS time series for the convenience of the subsequent data processing.

```
#Load the packages
> library(plyr)
> library(xts)
> library(TTR)
> library(ggplot2)
> library(scales)

# Read the CSV data file
> read<-function(file){
+    df<-read.table(file=file,header=FALSE,sep = ",", na.strings = "NULL")
# Read the file
+    names(df)<-c("code","date","Open","High","Low","Close","Volume")
# Set the column name
```

```
+    dl<-split(df[-1],df$code)                          # Group by code
+
+    lapply(dl,function(row){                # Convert the data to xts type
+      xts(row[-1],order.by = as.Date(row$date))
+    })
+ }

# Load the data
> data<-read("stock.csv")

# View the data type
> class(data)
[1] "list"

# View the data index values
> head(names(data))
[1] "000001.SZ" "000002.SZ" "000004.SZ" "000005.SZ" "000006.SZ" "000007.SZ"

# View the numbers of the stocks included
> length(data)
[1] 2782

# View stock 000001.SZ
> head(data[['000001.SZ']])
               Open     High      Low    Close   Volume
2014-07-02 8.146949 8.180000 8.105636 8.171737 28604171
2014-07-03 8.171737 8.254364 8.122162 8.229576 44690486
2014-07-04 8.237838 8.270889 8.146949 8.188263 34231126
2014-07-07 8.188263 8.204788 8.097374 8.146949 34306164
2014-07-08 8.130424 8.204788 8.072586 8.204788 34608702
2014-07-09 8.196525 8.196525 7.915596 7.973434 58789114
```

The data is ready, now we can start modeling.

6.1.2.2 Mean Reversion Model

In order to get a closer understanding of the market, we have obtained the data since January 1, 2015 to build the mean reversion model. Take PAB (000001.SZ) as an example and draw its candlestick chart and its moving average since 2015.

```
# Obtain the time scope
> dateArea<-function(sDate=Sys.Date()-365,eDate= Sys.Date(),before=0){
#Starting Date, Ending Date, Bill of Lading Date
+      if(class(sDate)=='character') sDate=as.Date(sDate)
+      if(class(eDate)=='character') eDate=as.Date(eDate)
+      return(paste(sDate-before,eDate,sep="/"))
+ }

# Calculate the moving average
> ma<-function(cdata,mas=c(5,20,60)){
+      if(nrow(cdata)<=max(mas)) return(NULL)
+      ldata<-cdata
+      for(m in mas){
```

```
+           ldata<-merge(ldata,SMA(cdata,m))
+       }
+       names(ldata)<-c('Value',paste('ma',mas,sep=''))
+       return(ldata)
+ }
# The candlestick chart and the moving average
> title<-'000001.SZ'
> SZ000011<-data[[title]]                           # Obtain the stock
                                                      data
> sDate<-as.Date("2015-01-01")                      # The starting date
> eDate<-as.Date("2015-07-10")                      # The ending date
> cdata<-SZ000011[dateArea(sDate,eDate,360)]$Close  # Obtain the
                                                      closing price
> ldata<-ma(cdata,c(5,20,60))                       # Select a moving
average indicator
# Print the moving average indicator
> tail(ldata)

            Value    ma5     ma20      ma60
2015-07-03 13.07 13.768 15.2545 15.84355
2015-07-06 13.88 13.832 15.1335 15.82700
2015-07-07 14.65 13.854 15.0015 15.79850
2015-07-08 13.19 13.708 14.8120 15.74267
2015-07-09 14.26 13.810 14.6910 15.70867
2015-07-10 14.86 14.168 14.6100 15.67883
```

Let's set three moving averages: 5-day moving average, 20-day moving average and 60-day moving average. Of course, we can set the moving averages of other periods to meet our own needs. Draw the candlestick charts and the moving averages.

```
> drawLine<-function(ldata,titie="Stock_MA",sDate=min(index(ldata)),eDate
=max(index(ldata)),breaks="1 year",avg=FALSE,out=FALSE){
+       if(sDate<min(index(ldata))) sDate=min(index(ldata))
+       if(eDate>max(index(ldata))) eDate=max(index(ldata))
+       ldata<-na.omit(ldata)
+
+       g<-ggplot(aes(x=Index, y=Value),data=fortify(ldata[,1],melt=TRUE))
+       g<-g+geom_line()
+       g<-g+geom_line(aes(colour=Series),data=fortify(ldata[,-1],
        melt=TRUE))
+
+       if(avg){
+           meanVal<<-round(mean(ldata[dateArea(sDate,eDate)]$Value),2) #
            The mean
+           g<-g+geom_hline(aes(yintercept=meanVal),color="red",alpha=0.8,
            size=1,linetype="dashed")
+           g<-g+geom_text(aes(x=sDate, y=meanVal,label=meanVal),color="red",
            vjust=-0.4)
+       }
+       g<-g+scale_x_date(labels=date_format("%Y-%m"),breaks=date_
        breaks(breaks),limits = c(sDate,eDate))
+       g<-g+ylim(min(ldata$Value), max(ldata$Value))
+       g<-g+xlab("") + ylab("Price")+ggtitle(title)
```

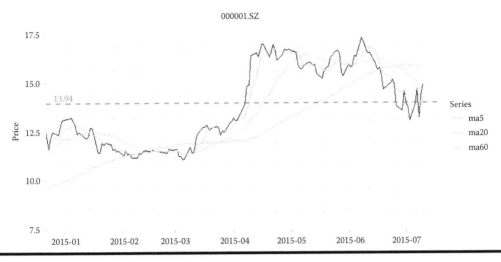

Figure 6.5   PAB stock price and moving averages.

```
+       g
+ }

> drawLine(ldata,title,sDate,eDate,'1 month',TRUE)      # Draw the plot
```

As shown in Figure 6.5, the 60-day moving average is the smoothest, while the 5-day moving average is the most volatile. The stock price has more cross points with the 5-day moving average than with the 60-day. That is, in the same period of time, the characteristic of mean reversion manifests itself more significantly in the short-term moving average than in the long-term.

Let's calculate the average standard deviations of the differences between the stock prices and the moving averages for different periods.

```
> getMaSd<-function(ldata,mas=20,sDate,eDate){
+       if(is.null(ldata) || nrow(ldata)<= max(mas)) return(NULL)
+       col<-paste('ma',mas,sep='')
+       ldata<-ldata[,c("Value",col)]
+       ldata$dif<-ldata[,col]-ldata$Value
+       ldata$sd<-runSD(ldata[,"dif"],mas)
+       ldata$rate<-round(ldata$dif/ldata$sd,2)
+       ldata[dateArea(sDate,eDate)]
+ }

# The difference between 5-day moving average and its average standard
deviation
> ldata5<-getMaSd(ldata,5,sDate,eDate)
> head(ldata5)
               Value      ma5        dif         sd   rate
2015-01-05 13.23673 12.78724 -0.4494869 0.1613198 -2.79
2015-01-06 13.03842 12.89961 -0.1388121 0.1909328 -0.73
2015-01-07 12.79055 12.99215  0.2016081 0.3169068  0.64
2015-01-08 12.36089 12.90292  0.5420283 0.4472248  1.21
2015-01-09 12.46004 12.77733  0.3172848 0.3910700  0.81
2015-01-12 12.20390 12.57076  0.3668606 0.2533165  1.45
```

```
# The difference between 20-day moving average and its average standard
deviation
> ldata20<-getMaSd(ldata,20,sDate,eDate)
> head(ldata20)
              Value     ma20          dif         sd  rate
2015-01-05 13.23673 12.18613 -1.05059293 0.6556366 -1.60
2015-01-06 13.03842 12.23778 -0.80064848 0.6021093 -1.33
2015-01-07 12.79055 12.24810 -0.54244141 0.4754686 -1.14
2015-01-08 12.36089 12.29975 -0.06114343 0.5130410 -0.12
2015-01-09 12.46004 12.33651 -0.12352626 0.5150453 -0.24
2015-01-12 12.20390 12.37163  0.16773131 0.5531618  0.30


# The difference between 60-day moving average and its average standard
deviation
> ldata60<-getMaSd(ldata,60,sDate,eDate)
> head(ldata60)
              Value     ma60         dif        sd  rate
2015-01-05 13.23673 10.06939 -3.167340 1.264792 -2.50
2015-01-06 13.03842 10.14678 -2.891644 1.271689 -2.27
2015-01-07 12.79055 10.22087 -2.569677 1.269302 -2.02
2015-01-08 12.36089 10.28752 -2.073368 1.258813 -1.65
2015-01-09 12.46004 10.35527 -2.104766 1.247967 -1.69
2015-01-12 12.20390 10.41821 -1.785691 1.233989 -1.45
```

The difference between 5-day moving average and its average standard deviation is the smallest, while the difference between 60-day moving average and its average standard deviation is the largest. If we take the 5-day moving average as the mean, we may trade frequently but profit little from every trade, so little that it may not even cover the commission. On the other hand, if we take the 60-day moving average as the mean, we may trade less frequently, but may suffer tight cash position for the stock price may go trending up or down and it may not convert to the mean for long. Summing up the two circumstances above, we can choose the 20-day moving average as the mean.

According to the calculation formula of the model, when the difference is more than twice the average standard deviation, we consider it a stock price deviation, with the deviation point as the buy signal of the model and the intersection of the moving average and the stock price as the sell signal.

In the previous step, we have calculated the deviation and saved it to the rate column. In the following, we are going to find the points with the differences larger than twice their standard deviations and draw a plot.

```
# The points with the differences larger than twice their standard
deviations
> buyPoint<-function(ldata,x=2,dir=2){
+     idx<-which(ldata$rate>x)
+     if(dir==2){
+         idx<-c(idx,which(ldata$rate<x*-1))
+     }
+     return(ldata[idx,])
+ }
```

```
# Draw the trading signal points
> drawPoint<-function(ldata,pdata,titie,sDate,eDate,breaks="1 year"){
+       ldata<-na.omit(ldata)
+       g<-ggplot(aes(x=Index, y=Value),data=fortify(ldata[,1],melt=TRUE))
+       g<-g+geom_line()
+       g<-g+geom_line(aes(colour=Series),data=fortify(ldata[,-1],
        melt=TRUE))
+
+       if(is.data.frame(pdata)){
+       g<-g+geom_point(aes(x=Index,y=Value,colour=op),data=pdata,size=4)
+       }else{
+       g<-g+geom_point(aes(x=Index,y=Value,colour=Series),data=na.omit
        (fortify(pdata,melt=TRUE)),size=4)
+       }
+       g<-g+scale_x_date(labels=date_format("%Y-%m"),breaks=date_
        breaks(breaks),limits = c(sDate,eDate))
+       g<-g+xlab("") + ylab("Price")+ggtitle(title)
+       g
+ }

> buydata<-buyPoint(ldata20,2,2)                                    #
going long or short signals
> drawPoint(ldata20[,c(1,2)],buydata$Value,title,sDate,eDate,'1 month')
# Draw a plot
```

The blue points in Figure 6.6 are the buy signals. The stock trade is one-way trade, i.e. to buy low and sell high, and it cannot be short sold, so we should filter the points with the stock price higher than the moving average, keep those with the stock prices lower than the moving average, which are our buy signals.

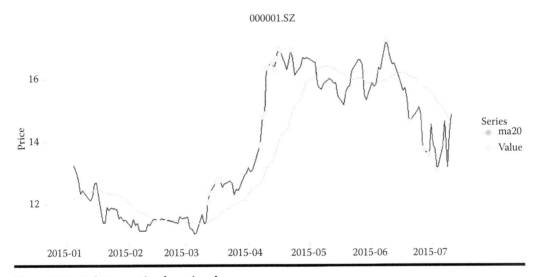

Figure 6.6 Points meeting buy signals.

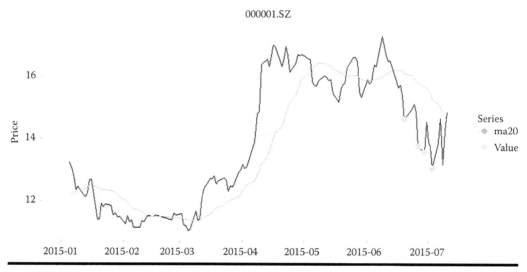

Figure 6.7   Buy signal points filtering.

Draw the buy signal points and keep only the points with the stock prices lower than the moving average (Figure 6.7).

```
> buydata<-buyPoint(ldata20,2,1)           # Going long signal points
> drawPoint(ldata20[,c(1,2)],buydata$Value,title,sDate,eDate,'1 month')
# Draw a plot
```

Calculate the sell signal points. After buying, the next intersection point of the stock price and the moving average is the sell signal point. Let's see if we can make any money from it.

```
# Calculate the sell signal points
> sellPoint<-function(ldata,buydata){
+       buy<-buydata[which(buydata$dif>0),]
+
+       aidx<-index(ldata[which(ldata$dif<=0),])
+       sellIdx<-sapply(index(buy),function(ele){
+           head(which(aidx>ele),1)
+       })
+       ldata[aidx[unique(unlist(sellIdx))]]
+ }

> selldata<-sellPoint(ldata20,buydata)

# Sell signal points
> selldata
           Value  ma20    dif        sd  rate
2015-07-10 14.86 14.61  -0.25 0.7384824 -0.34
```
Put the buy signals and sell signals together in a plot, as shown in the figure.

```
> bsdata<-merge(buydata$Value,selldata$Value)
> names(bsdata)<-c("buy","sell")
```

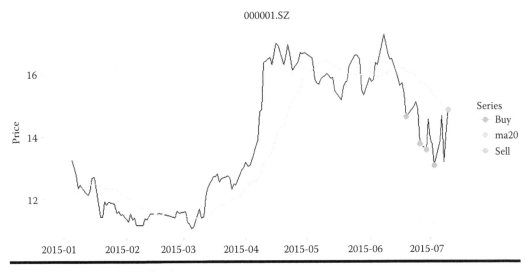

Figure 6.8   Add sell signal points.

```
> drawPoint(ldata20[,c(1,2)],bsdata,title,sDate,eDate,'1 month') #Draw a
plot
```

The figure shows that we do make money from buying on the green points and selling on the blue point. And how much do we earn? Here we need accurate calculation (Figure 6.8).

```
# Merge the trading signals
> signal<-function(buy, sell){
+      selldf<-data.frame(sell,op=as.character(rep("S",nrow(sell))))
+      buydf<-data.frame(buy,op=as.character(rep("B",nrow(buy))))
+      sdata<-rbind(buydf,selldf)
+      sdata[order(as.Date(row.names(sdata))),]
+ }

# Data of trading signals
> sdata<-signal(buydata,selldata)
> sdata
            Value    ma20      dif        sd rate op
2015-06-19 14.63 16.0965   1.4665 0.6620157  2.22  B
2015-06-26 13.77 15.7720   2.0020 0.8271793  2.42  B
2015-06-29 13.56 15.6840   2.1240 0.9271735  2.29  B
2015-07-03 13.07 15.2545   2.1845 1.0434926  2.09  B
2015-07-10 14.86 14.6100  -0.2500 0.7384824 -0.34  S
```

Make a mock trading with the data of trading signals. Let's set the trading parameters and the rules.

- ■ Take 100,000 *yuan* as the principal.
- ■ When the buy signal appears, buy the stock at the closing price. Buy the stock of 10,000 *yuan* every time. If the buy signals continue to appear, buy the stock continuously. If the capital is less than 10,000 *yuan*, ignore the buy signals.

■ When the sell signal appears, sell the stock at the closing price and close all the positions of
the stock at once.
■ This is commission-free.

```
# Make a mock trading
> trade<-function(sdata,capital=100000,fixMoney=10000){
+     amount<-0
+     cash<-capital
+
+     ticks<-data.frame()
+     for(i in 1:nrow(sdata)){
+         row<-sdata[i,]
+         if(row$op=='B'){
+             if(cash<fixMoney){
+                 print(paste(row.names(row),"No enough cash"))
+                 next
+             }
+             amount0<-floor(fixMoney/row$Value)
+             amount<-amount+amount0
+             cash<-cash-amount0*row$Value
+         }
+
+         if(row$op=='S'){
+             cash<-cash+amount*row$Value
+             amount<-0
+         }
+
+         row$cash<-round(cash,2)
+         row$amount<-amount
+         row$asset<-round(cash+amount*row$Value,2)
+         ticks<-rbind(ticks,row)
+     }
+
+     ticks$diff<-c(0,round(diff(ticks$asset),2))
+     rise<-ticks[intersect(which(ticks$diff>0),which(ticks$op=='S')),]
+     fall<-ticks[intersect(which(ticks$diff<0),which(ticks$op=='S')),]
+     return(list(ticks=ticks,rise=rise,fall=fall))
+ }
```

```
# The trading result
> result<-trade(sdata,100000,10000)
```

Let's check the details of every trade.

```
> result$ticks
            Value    ma20     dif         sd rate op     cash amount     asset     diff
2015-06-19 14.63 16.0965  1.4665 0.6620157  2.22  B 90007.71    683 100000.00     0.00
2015-06-26 13.77 15.7720  2.0020 0.8271793  2.42  B 80010.69   1409  99412.62  -587.38
2015-06-29 13.56 15.6840  2.1240 0.9271735  2.29  B 70016.97   2146  99116.73  -295.89
2015-07-03 13.07 15.2545  2.1845 1.0434926  2.09  B 60018.42   2911  98065.19 -1051.54
2015-07-10 14.86 14.6100 -0.2500 0.7384824 -0.34  S 103275.88     0 103275.88  5210.69
```

There are five transactions in all, of which four are buying and one is selling. In the end, the remaining capital is 103275.88 *yuan*, with an income of 3275.88 *yuan*, which makes 3.275% the rate of return.

There is one transaction of selling that makes money.

```
> result$rise
           Value ma20   dif        sd  rate op   cash amount   asset    diff
2015-07-10 14.86 14.61 -0.25 0.7384824 -0.34  S 103275.9      0 103275.9 5210.69
```

No loss happened during a selling.

```
> result$fall
 [1] Value  ma20   dif    sd     rate   op   cash   amount asset  diff
<Row 0> (or any row.names with 0 length)
```

Next, let's compare the net asset value (NAV) with the stock price.

```
# The net asset value curve
> drawAsset<-function(ldata,adata,sDate=FALSE,capital=100000){
+     if(!sDate) sDate<-index(ldata)[1]
+     adata<-rbind(adata,as.xts(capital,as.Date(sDate)))
+
+     g<-ggplot(aes(x=Index, y=Value),data=fortify(ldata[,1],melt=TRUE))
+     g<-g+geom_line()
+     g<-g+geom_line(aes(x=as.Date(Index), y=Value,colour=Series),data=
+       fortify(adata,melt=TRUE))
+     g<-g+facet_grid(Series ~ .,scales = "free_y")
+     g<-g+scale_y_continuous(labels=dollar_format(prefix = "￥"))
+     g<-g+scale_x_date(labels=date_format("%Y-%m"),breaks=date_breaks
+       ("2 months"),limits = c(sDate,eDate))
+     g<-g+xlab("") + ylab("Price")+ggtitle(title)
+     g
+ }

> drawAsset(ldata20,as.xts(result$ticks['asset']))  # The net asset value
  curve
```

We have just tested one stock and found it is possible to make money from it. Let's take another stock and see if it comes to the same result. Let's encapsulate the process of the data processing we've done just now to a unified quick function, with which we can quickly testify the performance of other stocks in mean reversion (Figure 6.9).

```
> quick<-function(title,sDate,eDate){
+     stock<-data[[title]]
+     cdata<-stock[dateArea(sDate,eDate,360)]$Close
+     ldata<-ma(cdata,c(20))
+     ldata<-getMaSd(ldata,20,sDate,eDate)
+     buydata<-buyPoint(ldata,2,1)
+     selldata<-sellPoint(ldata,buydata)
+     sdata<-signal(buydata,selldata)
+     return(trade(sdata))
+ }
```

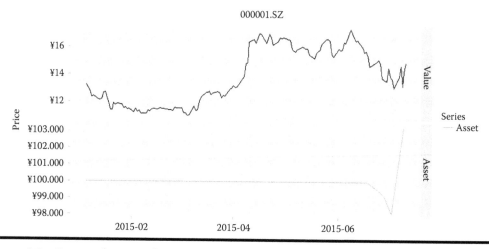

Figure 6.9 Curves of asset and net value.

Try to test it with LESHI INTERNET INF 'A'CNY1 (300104.SZ) and see if we can find any opportunities of making money.

```
> title<-"300104.SZ"
> sDate<-as.Date("2015-01-01") #The starting date
> eDate<-as.Date("2015-07-10") #The ending date

> quick(title,sDate,eDate)
$ticks
           Value   ma20     dif       sd rate op      cash amount     asset     diff
2015-06-19 55.04 69.9095 14.8695 5.347756 2.78  B  90037.76    181 100000.00     0.00
2015-06-23 54.30 68.8075 14.5075 5.477894 2.65  B  80046.56    365  99866.06  -133.94
2015-06-24 56.21 67.8735 11.6635 5.404922 2.16  B  70097.39    542 100563.21   697.15
2015-06-25 51.80 66.8775 15.0775 5.770806 2.61  B  60099.99    735  98172.99 -2390.22
2015-06-26 46.79 65.9830 19.1930 6.580622 2.92  B  50133.72    948  94490.64 -3682.35
2015-06-29 47.05 64.9445 17.8945 7.096230 2.52  B  40159.12   1160  94737.12   246.48
2015-07-07 47.86 58.8150 10.9550 5.401247 2.03  B  30204.24   1368  95676.72   939.60
2015-07-10 57.92 57.3520 -0.5680 5.625309 -0.10 S 109438.80      0 109438.80 13762.08

$rise
           Value   ma20    dif       sd rate op      cash amount    asset     diff
2015-07-10 57.92 57.352 -0.568 5.625309 -0.1  S 109438.8      0 109438.8 13762.08
$fall

[1] Value ma20    dif    sd     rate   op     cash    amount asset diff
<Row 0> (or the row.names with 0 length)
```

The result of the data tells us that we make money again. There are eight transactions in all, of which seven are buying and one is selling. In the end, the remaining capital is 109438.80 *yuan*, with an income of 9438.80 *yuan*, which makes 9.43% the rate of return.

Draw a plot of trading signals.

```
> title<-"300104.SZ"
> sDate<-as.Date("2015-01-01") #The starting date
> eDate<-as.Date("2015-07-10") #The ending date
```

```
> stock<-data[[title]]
> cdata<-stock[dateArea(sDate,eDate,360)]$Close
> ldata<-ma(cdata,c(20))
> ldata<-getMaSd(ldata,20,sDate,eDate)
> buydata<-buyPoint(ldata,2,1)
> selldata<-sellPoint(ldata,buydata)
> bsdata<-merge(buydata$Value,selldata$Value)
> drawPoint(ldata[,c(1,2)],bsdata,title,sDate,eDate,'1 month') #Draw a plot
```

In the distressed June, when people are held up by at least 30%, we can still earn a positive return of 9%. How pleasant that can be (Figure 6.10).

6.1.3 Quantitative Stock Picking

We have tested two stocks in the above and found that the mean reversion model is fit for the stock trading. If we scan all the stocks in the market with the model, more trading signals will be generated, thus more investing opportunities will appear and more profits will be made.

Now let's quantitatively pick the stocks according to the mean reversion theory.

We have learned from the previous experience that the larger the deviation between the stock price and its average standard deviation, the bigger a profit can be made. Therefore, every day we can use the quantification method to find the stocks with the largest deviations among 2,700 stocks in the market for trading and to efficiently distribute our capital for more effective investment. We need to try and see whether the market works the way we think.

Scan all the stocks in the market. Calculate the difference, the average and the average standard deviation.

```
> sDate<-as.Date("2015-01-01")          # The starting date
> eDate<-as.Date("2015-07-10")          # The ending date

# Calculate the difference, the average and the average standard
  deviation
```

Figure 6.10 Trading signals of LESHI INTERNET INF 'A'CNY1.

```
> data0<-lapply(data,function(stock){
+     cdata<-stock[dateArea(sDate,eDate,360)]$Close
+     ldata<-ma(cdata,c(5,20))
+     getMaSd(na.omit(ldata),20,sDate,eDate)
+ })

# Remove the null data
> data0<-data0[!sapply(data0, is.null)]

# All the stocks in the market
> length(data)
[1] 2782

# Effective stocks
> length(data0)
[1] 2697

# View the 1st stock
> head(data0[[1]])
               Value     ma20          dif          sd  rate
2015-01-05 13.23673 12.18613 -1.05059293 0.6556366 -1.60
2015-01-06 13.03842 12.23778 -0.80064848 0.6021093 -1.33
2015-01-07 12.79055 12.24810 -0.54244141 0.4754686 -1.14
2015-01-08 12.36089 12.29975 -0.06114343 0.5130410 -0.12
2015-01-09 12.46004 12.33651 -0.12352626 0.5150453 -0.24
2015-01-12 12.20390 12.37163  0.16773131 0.5531618  0.30
```

After the first scanning, we've found 2,697 stocks meeting the requirements and 85 stocks were removed for insufficient data sample.

Next, let's continue to screen the 2,697 stocks for the buy signals that meet the requirements.

```
# Calculate the buy signals
> buys<-lapply(data0,function(stock){
+     if(nrow(stock)==0) return(NULL)
+     buy<-buyPoint(stock,2,1)
+     if(nrow(buy)>0) {
+         return(buy)
+     }
+ })

# Remove the null data
> buys<-buys[!sapply(buys, is.null)]

# View the stocks with buy signals
> length(buys)
[1] 1819

# View the buy signals
> head(buys)
$`000001.SZ`
           Value     ma20    dif          sd rate
2015-06-19 14.63 16.0965 1.4665 0.6620157 2.22
2015-06-26 13.77 15.7720 2.0020 0.8271793 2.42
2015-06-29 13.56 15.6840 2.1240 0.9271735 2.29
2015-07-03 13.07 15.2545 2.1845 1.0434926 2.09
```

```
$`000002.SZ`
           Value     ma20    dif        sd rate
2015-03-05 11.90 12.568 0.668 0.2644101 2.53
2015-03-06 11.94 12.509 0.569 0.2674732 2.13


$`000004.SZ`
           Value     ma20     dif        sd rate
2015-01-05 15.69 17.7210  2.0310 0.7395717 2.75
2015-07-06 26.03 39.1540 13.1240 6.3898795 2.05
2015-07-07 23.43 38.2025 14.7725 6.9421723 2.13
2015-07-08 22.22 37.2635 15.0435 7.4287088 2.03


$`000005.SZ`
           Value     ma20    dif        sd rate
2015-07-06  6.02 10.9600 4.9400 2.381665 2.07
2015-07-07  5.42 10.5655 5.1455 2.333008 2.21


$`000006.SZ`
              Value      ma20       dif       sd rate
2015-01-19 5.829283 6.519462 0.6901792 0.26929 2.56


$`000007.SZ`
           Value     ma20    d if        sd rate
2015-02-06 12.47 14.4200 1.9500 0.6182860 3.15
2015-02-09 12.52 14.3270 1.8070 0.7440473 2.43
2015-02-10 12.10 14.1845 2.0845 0.8484250 2.46
```

By calculation, we find that 1,819 stocks have sent out buy signals in this half year. Different stocks give out buy signals in different timing and different frequency, so we can make diversified investment in different stocks and thus diversify the risk. If there are multiple stocks sending trading signals in 1 day and our capital is limited, to maximize the profit, we can choose and trade the stock with the largest deviation.

Then we will use programs to find the stock with the largest deviation

```
# Merge the data and convert it from list to data.frame
> buydf<-ldply(buys,function(e){
+       data.frame(date=index(e), coredata(e))
+ })


# Select the stocks with the largest rate on the same day and take them
as the buy signals
> buydatas<-ddply(buydf, .(date), function(row) {
+       row[row$rate == max(row$rate ),]
+ })


# View the buy signal
> nrow(buydatas)
[1] 81


# View the details of the buy signal
> head(buydatas)
```

```
           .id        date      Value       ma20         dif        sd rate
1   002551.SZ  2015-01-05  16.573846  19.565446   2.9916000 0.74591596 4.01
2   002450.SZ  2015-01-06  18.548809  19.766636   1.2178275 0.34008453 3.58
3   300143.SZ  2015-01-07  11.480000  12.603000   1.1230000 0.32028018 3.51
4   300335.SZ  2015-01-08  12.113677  13.139601   1.0259238 0.21760484 4.71
5   300335.SZ  2015-01-09  12.243288  13.043888   0.8005994 0.22940845 3.49
6   300335.SZ  2015-01-12  11.994036  12.941694   0.9476584 0.23168313 4.09
```

Finally, we have picked 81 buy signals. Basically, there are buy signals in every trading day. Having found the buy signals, let's continue to find the sell signals.

```
# Sell signals
> selldatas<-data.frame()
> for(i in 1:nrow(buydatas)){
+     buydata<-buydatas[i,]
+     if(is.data.frame(buydata)){
+         buydata<-xts(buydata,order.by=as.Date(buydata$date))
+     }
+
+     ldata<-data0[[buydata$.id]]
+     sell<-sellPoint(ldata,buydata)
+
+     if(nrow(sell)>0){
+         sell<-data.frame(sell,.id=buydata$.id,date=index(sell))
+         selldatas<-rbind(selldatas,sell)
+     }
+ }

# Remove the duplicate sell signals
> selldatas<-unique(selldatas)
> nrow(selldatas)
[1] 33

# View the sell signals
> head(selldatas)
              Value       ma20         dif        sd  rate       .id       date op
2015-01-12  19.232308  18.848908  -0.38340000 0.9051374 -0.42 002551.SZ 2015-01-12  S
2015-01-08  19.814257  19.729006  -0.08525126 0.3782955 -0.23 002450.SZ 2015-01-08  S
2015-01-28  11.210000  11.019500  -0.19050000 0.7781848 -0.24 300143.SZ 2015-01-28  S
2015-01-21  13.190448  12.899321  -0.29112706 0.3871871 -0.75 300335.SZ 2015-01-21  S
2015-01-213  7.140000   6.989500  -0.15050000 0.2007652 -0.75 002505.SZ 2015-01-21  S
2015-01-22   5.561561   5.490668  -0.07089242 0.2127939 -0.33 600077.SH 2015-01-22  S
```

There are 33 sell signals by calculation. Merge the buy signals and the sell signals and then calculate the profits.

```
> buydatas$op<-'B'                          # The buy marks
> selldatas$op<-'S'                         # The sell marks
> sdatas<-rbind(buydatas,selldatas)         # Merge the data
> row.names(sdatas)<-1:nrow(sdatas)         # Re-set the row numbers
> sdatas<-sdatas[order(sdatas$.id),]        # Sort by the stock codes

# View the merging signals
> head(sdatas)
```

```
         .id       date Value     ma20       dif        sd rate op
36   000002.SZ 2015-03-05 11.90 12.56800  0.668000 0.26441011  2.53  B
100  000002.SZ 2015-03-16 12.49 12.38050 -0.109500 0.23702768 -0.46  S
58   000553.SZ 2015-05-06 14.35 15.50882  1.158824 0.38429912  3.02  B
110  000553.SZ 2015-05-21 16.57 15.18903 -1.380972 0.55647152 -2.48  S
26   000725.SZ 2015-02-09  2.80  3.11400  0.314000 0.07934585  3.96  B
94   000725.SZ 2015-02-16  3.09  3.06500 -0.025000 0.08182388 -0.31  S
```

In the end, group by the stock codes and respectively calculate the profit of every stock.

```
# Calculate the profit of every stock
> slist<-split(sdatas[-1],sdatas$.id)      # Group by the stock codes
> results<-lapply(slist,trade)

# View the stocks with signals
> names(results)
 [1] "000002.SZ" "000553.SZ" "000725.SZ" "000786.SZ" "000826.SZ" "002240.
SZ" "002450.SZ"
 [8] "002496.SZ" "002505.SZ" "002544.SZ" "002551.SZ" "002646.SZ" "002652.
SZ" "300143.SZ"
[15] "300335.SZ" "300359.SZ" "300380.SZ" "300397.SZ" "300439.SZ" "300440.
SZ" "300444.SZ"
[22] "600030.SH" "600038.SH" "600077.SH" "600168.SH" "600199.SH" "600213.
SH" "600375.SH"
[29] "600490.SH" "600536.SH" "600656.SH" "600733.SH" "600890.SH" "601179.
SH" "601186.SH"
[36] "601628.SH" "601633.SH" "601939.SH" "603019.SH"
```

Check the stock of VANKE A (000002.SZ).

```
> results[['000002.SZ']]$ticks
         date Value    ma20     dif       sd rate op    cash amount
asset  diff
36  2015-03-05 11.90 12.5680  0.6680 0.2644101  2.53  B  90004.0    840
100000.0   0.0
100 2015-03-16 12.49 12.3805 -0.1095 0.2370277 -0.46  S 100495.6      0
100495.6 495.6
```

With the optimized rule design, two transactions have been made, which brought a profit of 495 *yuan*. If we did not optimize the algorithm, but kept trading VANKE A, three transactions with a profit of 955.95 *yuan* could have been made.

```
> quick('000002.SZ',sDate,eDate)$ticks
           Value    ma20     dif       sd rate op     cash amount    asset   diff
2015-03-05 11.90 12.5680  0.6680 0.2644101  2.53  B  90004.00    840 100000.0   0.00
2015-03-06 11.94 12.5090  0.5690 0.2674732  2.13  B  80010.22   1677 100033.6  33.60
2015-03-16 12.49 12.3805 -0.1095 0.2370277 -0.46  S 100955.95      0 100955.9 922.35
```

This section has to end. However, there are still many things to do, such as the optimization of the model parameters, the replacement of 20-day moving average with 10-day average, the replacement of twice the standard deviation with triple the standard deviation, the normal distribution test of the sample, the signal generation combining other trend models, etc. We cannot settle all these in just one article. You can follow my blog (http://fens.me) for more diverse strategies.

In this section, we have introduced first the theory of mean reversion, then the test of market characteristics, the mathematical equations, the modeling with R language, the historical data backtesting and finally found the investment opportunities. It is a whole set of learning method from theory to practice. Though it is beset with difficulties, we are the geeks with ideals and we are able to overcome the difficulties.

This section covers the disciplines of computer science, finance, mathematics, statistics, etc. and I think these disciplines can be the future development directions for inter-disciplinary talents. If the past decade is the golden decade for real estate, the coming decade will be the golden decade for finance. If we IT guys grasp enough knowledge of finance, we will be able to have a place in the financial market.

Programmers, let's seize the moment and fight for the future!

6.2 Build Quantitative Trading Model of Upswing Buying and Downswing Selling with R

Question
How do you implement the model of buying on the upswing and selling on the downswing with R language?

Build Quantitative Trading Model of
Upswing Buying and Downswing Selling
with R
http://blog.fens.me/finance-chase-sell/

Introduction
An experienced investor usually uses a common trading strategy, i.e. to buy on the upswing and sell on the downswing. This strategy is a very important skill in the stock market operations. It means to buy the stock when the stock price is rising and to sell it when it is dropping. It can be a good method to profit if used properly. In the first half year of 2015, the method of buying on the upswing and selling on the downswing was the golden rule of trading in the bull market of China.

6.2.1 What Is Upswing Buying and Downswing Selling?

Upswing buying and downswing selling is a term in the financial market. It is the opposite of bottom fishing and top picking. The operations of buying on the upswing and selling on the downswing: buy the financial products (stocks, futures, foreign exchanges, etc.) when their prices go up in the financial market, expecting the prices to rise more and then sell them at higher prices to earn profits; sell these financial products immediately when their prices go down to stop loss and avoid greater loss, no matter how much the buying prices were.

In the following, the candlestick chart of PAB stock (000001.SZ) will be taken as an example. The stock data from 2005 to July 2015 is intercepted and the stock price is the price after forward restoration of rights. See Figure 6.11.

There are three curves in the figure above. The black one is the daily stock price of PAB after forward restoration of rights, the red one is the 20-day moving average and the blue one is the 60-day moving average. For introduction to the moving average, please refer to the article "Surfing the Stock Market with MA Model." And the dotted red horizontal line in the figure, i.e. 7.14 *yuan*, is the stock average price in the past 10 years. In these 10 years, the stock price of PAB went ups and downs and every rising lasted for a period of time, so we could buy it in the middle of a rising and sell it before a steep drop, which is called buying on the upswing and selling on the downswing.

6.2.1.1 Buying on the Upswing

There are usually two circumstances for buying on the upswing: short-upswing buying and medium-upswing buying. Take the stock trading for example.

■ Short-upswing buying: When the stock price has risen by five points or more, you buy the stock and expect it to reach its daily limit at closing and sell it at a high price on the next trading day. This is the short-upswing buying. For the particularity of trading rules in China's stock market, the A-share stock price change should not exceed 10% and the stock trading is a T+1 trading, i.e. you buy the stock today, but cannot sell it until the next trading day.

■ Medium-upswing buying: When the price of one stock has risen for a while and its upswing is strong with a rising by 30% or more, you buy it and expect it to continuously increase. This is the medium-upswing buying.

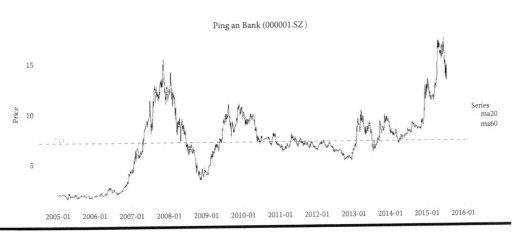

Figure 6.11   PAB stock price.

We make short-upswing buyings and medium-upswing buyings for the same reason and the same criteria, i.e. the continuously rising stock implies a strong upswing and we buy it with the expectation that the stock will continue to surge. However, the upswing buying bears great risks if we view it from another perspective, mostly because we don't know how long the rising will last, at which point the price will fall and when the market makers will sell their holdings. Therefore, the upswing buying needs great skills. It is almost an art to tell when is a good time to buy and when is not.

6.2.1.2 Selling on the Downswing

Selling on the downswing is a deep operation technique. If operated properly, it can avoid risks. However, if accurate judgments on downswing selling cannot be made, a hard and fast stop of loss may cause serious loss.

If we are going to use the strategy of selling on the downswing, we need to master the following operations:

- An important premise of downswing selling is to know the change, i.e. to accurately and speedily know the influences of information, power of strategic action and all kinds of factors around.
- To sell on the downswing, we need to be observant of the subtle changes in the situations, the forms, the information, the public opinions, etc. and we need to, based on the precise estimation on the quotations, quickly make decisions on which stock to sell and how many will be sold.
- To sell on the downswing, we need to reflect on the positive and negative correlation between the big board and the target stock, as well as the trading volume changes of the target stock, so as to find rules of changes.
- When good news of the market is released one after another, we should prepare for selling the stock. When good news ends, bad news will follow. Usually, when the bad news is confirmed, the stock price has already fallen greatly, leaving the investors unprepared.

The subsequent operation of downswing selling is comparatively simple. We should just keep a good mood and patiently wait for the signal for bottom fishing. A downswing selling success usually ends with a successful bottom fishing or a successful stock swap at the bottom; otherwise, the downswing selling will mean nothing. Be patient and don't hastily get in when the stock price rebounds, or the stock price would drop again and again after your buying, thus loss would be enlarged. We should wait till the stock price starts to rollback.

6.2.1.3 Skills of Upswing Buying and Downswing Selling

We need to know the general situation and who the winner of the duel between going long and going short in the market is.

The assumptions of upswing buying and downswing selling:

- It is a bull market. The bull market lays a solid operation foundation for individual stocks.
- Distinct hot spots or block effects with ability for sustainable rising have been formed in the market.

The objects of upswing buying:

- When a distinct and sustainable hot spot is formed in the market, buy it on the upswing. Theoretically speaking, as long as we grasp the hot spot or the block, we can profit. We should pay special attention to the leading enterprises when making an upswing buying. E.g. the constituent stocks of CSI 300 are good options.
- Those individual stocks with significant uprising trend are worth a thought of upswing buying. When an individual stock breaks through its important pressure line like the neckline of head and shoulders bottom pattern, the roof of the box pattern, etc. and the effectiveness of the breakthrough has been confirmed by the price return, we can consider an upswing buying.

Notes for upswing buying and downswing selling:

- Set the stop-profit point and the stop-loss point when buying on the upswing and set your goal of the approximate earning.
- The frequency of upswing buying and downswing selling should be reduced, for frequent operations bring about more errors, which will have a bad influence on our mood.
- During upswing buying and downswing selling, we need a peaceful mind more than a fluke mind, and we need reason more than sensibility.

6.2.1.4 Market Operations of Upswing Buying and Downswing Selling

In a weak market, short-upswing buyings are usually held up, for the quotations of most of the stocks in a weak market last for only 1 day. If you buy the stock at a high price on the day, you may not have a chance to sell it higher the other day. On the contrary, in a weak market, medium-upswing buyings are advisable. Not that many stocks can remain strong in a weak market and this strong upswing may continue. However, we may not dare to buy this kind of stocks in a weak market most of the time.

Some particularly buy stocks which reach the rising limit, some buy new stocks and some buy sub-new stocks. If someone bought the new IPO shares in the first half year of 2015, such as BAOFENG GROUP CO L 'A'CNY1, ZHONG CHAO HOLDING, etc., he/she would make a decent profit. For other times, this strategy of buying on the upswing may not work (Figure 6.12).

I will not blindly buy on the upswing in my operations. Safety is my priority. Though I have not failed any upswing buying, I do not have the ability of fundamental analysis at present and I do not have the information sources; therefore, if I make the buying just according to the technical indicators, it will be too speculative and it is easy to get caught by the market makers. It is easy to understand the reason. If I buy on the upswing, there can be large room for falling, especially in a medium upswing buying, and the makers can sell their stocks at any time. At the time, I need to make a strategy for selling on the downswing. It is common to fall by 50% if the makers sell their stocks and it will be hard to rise again if there is a fall like this, so it will be a suicide if I do not sell on the downswing at the time. We need to think before an upswing buying. We cannot just have longings for profits. A stop loss point should be set in advance.

Downswing selling is an art, too. We do not sell immediately when there is a fall, for turbulences and whipsaws are commonly seen in the stock market. The problem is how to tell a shake from a real sell. It is difficult to tell, so many stop their loss in a hard and fast way, i.e. pre-set the

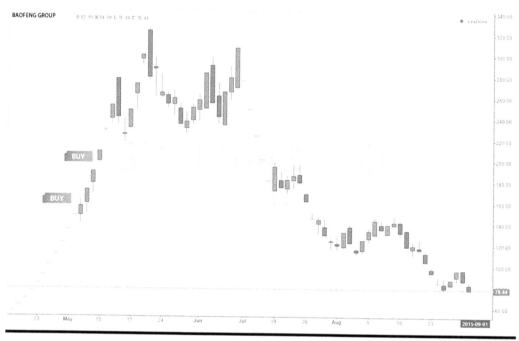

Figure 6.12 Candlestick chart of BAOFENG GROUP CO L 'A'CNY1.

price and the percentage and once the stock reaches the pre-set price, the stock will absolutely be sold, no matter it is a shake or a real sell of the market makers. If we do not have the fundamental judgment or the first-hand sources, it will be easy for us to get shaken off. Of course, I would rather be shaken off the market than bear the uncertain risks.

6.2.2 *Modeling and Implementation of Upswing Buying and Downswing Selling*

Since we are going to make upswing buyings and downswing sellings, we need to define what an upswing buying is and what a downswing selling is and then quantify the definitions for modeling and implementation. For the short-upswing buying strategy, we can simply buy the stocks with a rising of above 5% on the day, and then sell them at opening on the next day or at the time when the prices reach the stop-loss point. There are not many technical details here. After that, calculate the odds of profit and we will know if we are going to make any short-upswing buying.

In the following, we will take a medium-upswing buying for example to build the model and implement it with programs.

For a medium-upswing buying, our core strategic thinking pattern can be interpreted as: buy when the stock price (the trading volume) exceeds its highest of the recent 20 days and sell when the stock price is lower than its lowest of the recent 10 days; take the constituent stocks of CSI 300 as the target of the stock trading. We take the highest price of the recent 20 days and the lowest price of the recent 10 days because of experience; these prices can be trained and optimized as the model parameters. We take the constituent stocks of CSI 300 because these are the strong and leading stocks in all blocks, fitting the assumption of buying on the upswing and selling on the downswing.

Calculation formula:

```
Highest price in recent 20 days = max(T day stock price, T-1 day stock
price... , T-19 day stock price)
Highest trading volume in recent 20 days= max(T day trading volume, T-1
day trading volume ... , T-19 day trading volume)
Lowest price in recent 10 days = min(T day stock price,T-1 day stock
price..., T-9 day stock price)
```

It is a buy signal when the stock price is more than the highest price in recent 20 days and a sell signal when the price is below the lowest price in recent 10 days. Now we have turned the theory of upswing buying and downswing selling into a mathematical model.

In the following, we will process the stock data with R language, so as to implement a case model of upswing buying and downswing selling, to testify our investment theory and to discover opportunities of making money.

6.2.2.1 *Data Preparation*

R language itself provides abundant financial function toolkits, such as the time series package zoo and xts, the index calculation package TTR, the data processing package plyr, the visualization package ggplot2, etc. We will use all these packages together for modeling, calculation and visualization. For detailed usage of the zoo package and the xts package, please refer to Section 2.1, "Basic Time Series Library of R: zoo," and Section 2.2, "Extensible Time Series: xts," of *R for Programmers: Mastering the Tools.*

The system environment used here:

- Win10 64bit
- R: 3.2.3 x86_64-w64-mingw32/x64 b4bit

The data used here can be the data you crawl from Sina Finance or the data you download from Yahoo Finance with the quantmod package.

The data used here includes the candlestick chart data of A-share (after forward restoration of rights) from July 2014 to August 2015 and it is saved to a local file stock.csv in CSV format.

The data format is as follow:

```
000001.SZ,2014-07-02,8.14,8.18,8.10,8.17,28604171
000002.SZ,2014-07-02,8.09,8.13,8.05,8.12,40633122
000004.SZ,2014-07-02,13.9,13.99,13.82,13.95,1081139
000005.SZ,2014-07-02,2.27,2.29,2.26,2.28,4157537
000006.SZ,2014-07-02,4.57,4.57,4.50,4.55,5137384
000010.SZ,2014-07-02,6.6,6.82,6.5,6.73,9909143
```

The data contains seven columns:

- Column1: The stock code, code, 000001.SZ
- Column2: The trading date, date, 2014-07-02
- Column3: The opening price, Open, 8.14
- Column4: The highest price, High, 8.18
- Column5: The lowest price, Low, 8.10

- ◾ Column6: The closing price, Close, 8.17
- ◾ Column7: The trading volume, Volume, 28604171

Load the stock data with R language. We need to do the calculation by every stock, but the information of all stocks is mixed in the data, so I have made some conversion during the loading. The data has been grouped by stock codes, a list object has been generated and meanwhile, the stock object type has been converted from data.frame to xts time series for the convenience of the subsequent data processing.

```
#Load the packages
> library(plyr)
> library(xts)
> library(TTR)
> library(ggplot2)
> library(scales)

# Read the CSV data file
> read<-function(file){
+    df<-read.table(file=file,header=FALSE,sep = ",", na.strings = "NULL")
# Read the file
+    names(df)<-c("code","date","Open","High","Low","Close","Volume")
# Set the column names
+    dl<-split(df[-1],df$code)            # Group by ccode
+    lapply(dl,function(row){             # Convert to xts type data
+      xts(row[-1],order.by = as.Date(row$date))
+    })
+ }

# Load the data
> data<-read("stock.csv")

# View the data type
> class(data)
[1] "list"

# View the data index values
> head(names(data))
[1] "000001.SZ" "000002.SZ" "000004.SZ" "000005.SZ" "000006.SZ" "000007.SZ"

# View the number of the stocks included
> length(data)
[1] 2782

# Obtain the time scope
dateArea<-function(sDate=Sys.Date()-365,eDate= Sys.Date(),before=0){
#Starting Date, Ending Date, Bill of Lading Date
  if(class(sDate)=='character') sDate=as.Date(sDate)
  if(class(eDate)=='character') eDate=as.Date(eDate)
  return(paste(sDate-before,eDate,sep="/"))
}

# View stock 000001.SZ
> head(data[['000001.SZ']])
```

```
            Open      High       Low     Close    Volume
2014-07-02  8.146949  8.180000  8.105636  8.171737  28604171
2014-07-03  8.171737  8.254364  8.122162  8.229576  44690486
2014-07-04  8.237838  8.270889  8.146949  8.188263  34231126
2014-07-07  8.188263  8.204788  8.097374  8.146949  34306164
2014-07-08  8.130424  8.204788  8.072586  8.204788  34608702
2014-07-09  8.196525  8.196525  7.915596  7.973434  58789114
```

The data is ready, now we can start modeling.

6.2.2.2 *Model of Upswing Buying and Downswing Selling*

In order to get a closer understanding of the market, we have obtained the data since January 1, 2015 to build the model of upswing buying and downswing selling. Take LESHI INTERNET INF 'A'CNY1 (300104.SZ) for example. Draw the everyday closing price since 2015, the highest price in recent 20 days and the lowest price in recent 10 days. LESHI INTERNET INF 'A'CNY1, as the benchmarking listed company in the growth enterprises market, has been brought into the CSI 300 in July 2015 for the first time.

```
# Data of candlestick chart
> title<-'300104.SZ'
> stock<-data[[title]]              # Obtain the stock
                                      data
> sDate<-as.Date("2015-01-01")     # Starting date
> eDate<-as.Date("2015-08-24")     # Ending date
> cdata<-stock[dateArea(sDate,eDate,360)]$Close   # Obtain the
                                                    closing price
> vdata<-stock[dateArea(sDate,eDate,360)]$Volume  # Obtain the
                                                    trading volume


# Closing price
> names(cdata)<-"Value"     # Reset the column names
> tail(cdata)
            Value
2015-08-14  49.81
2015-08-17  48.30
2015-08-18  45.57
2015-08-19  46.98
2015-08-20  45.79
2015-08-21  42.14

# Trading volume
> tail(vdata)
            Volume
2015-08-14  42108324
2015-08-17  35939096
2015-08-18  52745702
2015-08-19  43447844
2015-08-20  32916746
2015-08-21  34802494
```

Define the plot functio n drawLine(), which supports multiple curves, including the closing price, the highest price and the lowest price.

```
# Plot function
> drawLine<-function(cdata,titie="Stock",sDate=min(index(cdata)),eDate=ma
x(index(cdata)),breaks="1 year"){
+     if(sDate<min(index(cdata))) sDate=min(index(cdata))
+     if(eDate>max(index(cdata))) eDate=max(index(cdata))
+     cdata<-na.omit(cdata)
+
+     g<-ggplot(aes(x=Index, y=Value),data=fortify(cdata[,1],melt=TRUE))
+     g<-g+geom_line()
+
+     if(ncol(cdata)>1){ # Multiple curves
+     g<-g+geom_line(aes(colour=Series),data=fortify(cdata[,-1],
      melt=TRUE))
+     }
+
+     g<-g+scale_x_date(labels=date_format("%Y-%m"),breaks=date_
      breaks(breaks),limits = c(sDate,eDate))
+     g<-g+ylim(min(cdata$Value), max(cdata$Value))
+     g<-g+xlab("") + ylab("Price")+ggtitle(title)
+     g
+ }


# Draw the closing price
# drawLine(cdata,title,sDate,eDate,'1 month')     # Draw a plot

# Calculate the highest price in recent 20 days and the lowest price in
recent 10 days
> minmax<-function(data,max=20,min=10){
+    d1<-na.locf(data,fromLast=TRUE)
+    d2<-merge(d1,min=runMin(d1,min),max=runMax(d1,max))
+    return(d2[,-1])
+ }


# Draw the stock price, the highest price and the lowest price
> ldata<-cbind(cdata,minmax(cdata))
> drawLine(ldata,title,sDate,eDate,'1 month')     # Draw a plot
```

There are three curves in Figure 6.13. The black curve is the daily closing price of LESHI INTERNET INF 'A'CNY1, the blue one is the highest price in recent 20 days and the red one is the lowest price in recent 10 days.

Let's calculate the buy signals by the formula of the model. We will buy when the stock price is higher than the highest price in recent 20 days.

```
# The function of buy signal
> buyPoint<-function(ldata){
+   idx<-which(ldata$Value == ldata$max)
+   return(ldata[idx,])
+ }


# Calculate the buy points
> buydata<-buyPoint(ldata)
> buydata
```

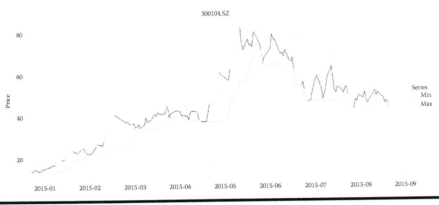

Figure 6.13 **Closing price, highest price and lowest price of LESHI INTERNET INF 'A'CNY1.**

```
              Value      min       max
2015-01-08  17.43721  13.70164  17.43721
2015-01-09  17.98709  13.74254  17.98709
2015-01-12  19.53222  13.74254  19.53222
2015-01-15  20.21389  14.74232  20.21389
2015-01-16  22.23619  16.08749  22.23619
2015-01-19  23.04056  16.36016  23.04056
2015-01-20  23.89947  16.36016  23.89947
2015-01-26  24.77656  19.22774  24.77656
2015-01-27  25.16284  19.40043  25.16284
2015-02-05  26.91247  21.99533  26.91247
2015-02-10  28.68482  21.99533  28.68482
2015-02-11  31.55239  21.99533  31.55239
2015-02-12  31.87960  21.99533  31.87960
2015-02-13  35.06983  22.72245  35.06983
2015-02-16  38.57817  24.22213  38.57817
2015-02-17  40.99130  24.46753  40.99130
2015-03-16  41.07764  34.32453  41.07764
2015-03-18  41.94564  34.32453  41.94564
2015-03-24  45.34946  37.17393  45.34946
2015-04-23  46.27199  37.06031  46.27199
2015-04-24  50.89829  37.06031  50.89829
2015-04-27  50.90283  37.06031  50.90283
2015-04-28  55.44277  37.06031  55.44277
2015-04-29  60.98705  37.06031  60.98705
2015-05-06  62.25497  45.19495  62.25497
2015-05-07  66.20413  46.27199  66.20413
2015-05-08  67.23573  50.89829  67.23573
2015-05-11  73.96157  50.90283  73.96157
2015-05-12  81.36000  55.44277  81.36000
2015-05-13  82.49000  57.16514  82.49000
```

Draw a plot of buy signals to get an intuitive result.

```
# Plot function
> drawPoint<-function(ldata,pdata,titie,sDate,eDate,breaks="1 year"){
+     ldata<-na.omit(ldata)
+     g<-ggplot(aes(x=Index, y=Value),data=fortify(ldata[,1],melt=TRUE))
```

```
+       g<-g+geom_line()
+       g<-g+geom_line(aes(colour=Series),data=fortify(ldata[,-1],
        melt=TRUE))
+
+       if(is.data.frame(pdata)){
+       g<-g+geom_point(aes(x=Index,y=Value,colour=op),data=pdata,size=4)
+       }else{
+       g<-g+geom_point(aes(x=Index,y=Value,colour=Series),data=na.omit
        (fortify(pdata,melt=TRUE)),size=4)
+       }
+       g<-g+scale_x_date(labels=date_format("%Y-%m"),breaks=date_
        breaks(breaks),limits = c(sDate,eDate))
+       g<-g+xlab("") + ylab("Price")+ggtitle(title)
+       g
+ }

> drawPoint(ldata,buydata$Value,title,sDate,eDate,'1 month')  # Draw a plot
```

As shown in Figure 6.14, the blue points are the points with stock prices equal to or higher than the highest price in recent 20 days and they are the buy signals. All the buy signals appear on the bullish upswing, so in the first half year of 2015, the signals of buying on the upswing would be triggered massively.

Then let's continue to calculate the sell signals. It is a sell signal when the price is below the lowest price in recent 10 days.

```
> # Calculate the sell signals
> stopPoint<-function(ldata,buydata){
+       idx<-which(ldata$Value == ldata$min)
+       idx<-idx[which(c(0,diff(idx))!=1)]    # Represent the first point
with 0
+
+       selldata<-ldata[idx,]                 # All points below the lowest
value
```

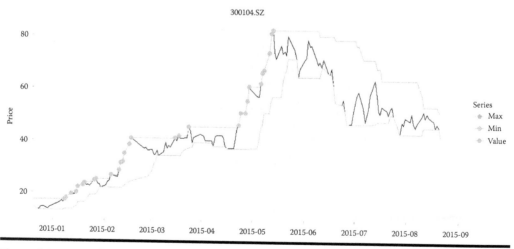

Figure 6.14 Buy signals.

```
+       idx2<-sapply(index(buydata),function(e){  # Sell signal points
after buying
+           head(which(index(selldata)>e),1)
+       })
+
+       return(selldata[unique(idx2),])
+ }

# Sell signals
> selldata<-stopPoint(ldata,buydata)
> selldata
              Value      min      max
2015-01-30 21.99533 21.99533 25.16284
2015-03-06 34.32453 34.32453 40.99130
2015-04-08 38.01011 38.01011 45.34946
2015-05-28 64.68000 64.68000 82.49000
```

There are four sell signals. To make the data intuitive, we merge the buy signals with the sell signals and visualize them with a plot.

```
# Draw a plot of the buy and sell signals
> bsdata<-merge(buydata$Value,selldata$Value)
> names(bsdata)<-c("buy","sell")
> drawPoint(ldata,bsdata,title,sDate,eDate,'1 month') #Draw a plot
```

In Figure 6.15, the purple points are the sell signals and the red ones are the buy signals. It is obvious that if, according the trading signals, we buy on the red points and sell on the purple points, we will make money. So how much exactly can we earn? I need to calculate that.

```
> # Merge the trading signals
> signal<-function(buy, sell){
+       selldf<-data.frame(sell,op=as.character(rep("S",nrow(sell))))
+       buydf<-data.frame(buy,op=as.character(rep("B",nrow(buy))))
+       sdata<-rbind(buydf,selldf)
# Data of trading signals
```

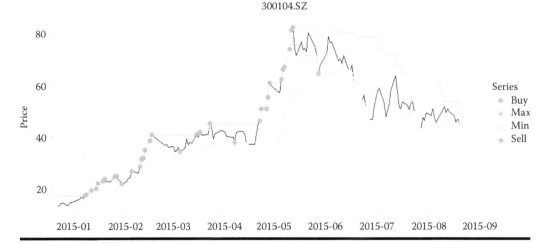

Figure 6.15 Add sell signal points.

```
+      sdata[order(as.Date(row.names(sdata))),]
+ }

# Merge the trading signals
> sdata<-signal(buydata,selldata)
> sdata
               Value      min      max op
2015-01-08 17.43721 13.70164 17.43721  B
2015-01-09 17.98709 13.74254 17.98709  B
2015-01-12 19.53222 13.74254 19.53222  B
2015-01-15 20.21389 14.74232 20.21389  B
2015-01-16 22.23619 16.08749 22.23619  B
2015-01-19 23.04056 16.36016 23.04056  B
2015-01-20 23.89947 16.36016 23.89947  B
2015-01-26 24.77656 19.22774 24.77656  B
2015-01-27 25.16284 19.40043 25.16284  B
2015-01-30 21.99533 21.99533 25.16284  S
2015-02-05 26.91247 21.99533 26.91247  B
2015-02-10 28.68482 21.99533 28.68482  B
2015-02-11 31.55239 21.99533 31.55239  B
2015-02-12 31.87960 21.99533 31.87960  B
2015-02-13 35.06983 22.72245 35.06983  B
2015-02-16 38.57817 24.22213 38.57817  B
2015-02-17 40.99130 24.46753 40.99130  B
2015-03-06 34.32453 34.32453 40.99130  S
2015-03-16 41.07764 34.32453 41.07764  B
2015-03-18 41.94564 34.32453 41.94564  B
2015-03-24 45.34946 37.17393 45.34946  B
2015-04-08 38.01011 38.01011 45.34946  S
2015-04-23 46.27199 37.06031 46.27199  B
2015-04-24 50.89829 37.06031 50.89829  B
2015-04-27 50.90283 37.06031 50.90283  B
2015-04-28 55.44277 37.06031 55.44277  B
2015-04-29 60.98705 37.06031 60.98705  B
2015-05-06 62.25497 45.19495 62.25497  B
2015-05-07 66.20413 46.27199 66.20413  B
2015-05-08 67.23573 50.89829 67.23573  B
2015-05-11 73.96157 50.90283 73.96157  B
2015-05-12 81.36000 55.44277 81.36000  B
2015-05-13 82.49000 57.16514 82.49000  B
2015-05-28 64.68000 64.68000 82.49000  S
```

Next, let's make a mock trading with the data of trading signals. Let's set the trading parameters and the rules.

- Take 100,000 *yuan* as the principal.
- When the buy signal appears, buy the stock at the closing price. Buy the stock of 10,000 *yuan* every time. If the buy signals continue to appear, buy the stock continuously. If the capital is less than 10,000 *yuan*, ignore the buy signals.
- When the sell signal appears, sell the stock at the closing price and close all the positions of the stock at once.
- This is commission free.

In the following, let's make a mock trading.

```
> # Make a mock trading
> trade<-function(sdata,capital=100000,fixMoney=10000){ # Trading
signals, total fund, fixed investment fund for every time
+       amount<-0
+       cash<-capital
+
+       ticks<-data.frame()
+       for(i in 1:nrow(sdata)){
+           row<-sdata[i,]
+           if(row$op=='B'){
+               if(cash<fixMoney){
+                   print(paste(row.names(row),"No enough cash"))
+                   next
+               }
+               amount0<-floor(fixMoney/row$Value) # The volume of this
trading
+               amount<-amount+amount0
+               cash<-cash-amount0*row$Value
+           }
+
+           if(row$op=='S'){
+               cash<-cash+amount*row$Value
+               amount<-0
+           }
+
+           row$cash<-round(cash,2)
+           row$amount<-amount
+           row$asset<-round(cash+amount*row$Value,2)
+           ticks<-rbind(ticks,row)
+       }
+
+
+       ticks$diff<-c(0,round(diff(ticks$asset),2))
+
+       rise<-ticks[intersect(which(ticks$diff>0),which(ticks$op=='S')),]
# Tradings of profit
+       fall<-ticks[intersect(which(ticks$diff<0),which(ticks$op=='S')),]
# Tradings of loss
+
+       return(list(
+           ticks=ticks,
+           rise=rise,
+           fall=fall
+       ))
+ }

# Trading result
> result<-trade(sdata,100000,10000)
```

Let's view the details of every trading.

```
> result$ticks
```

	Value	min	max	op	cash	amount	asset	diff
2015-01-08	17.43721	13.70164	17.43721	B	90008.48	573	100000.0	0.00
2015-01-09	17.98709	13.74254	17.98709	B	80025.65	1128	100315.1	315.08
2015-01-12	19.53222	13.74254	19.53222	B	70044.68	1639	102058.0	1742.91
2015-01-15	20.21389	14.74232	20.21389	B	60059.02	2133	103175.2	1117.26
2015-01-16	22.23619	16.08749	22.23619	B	50074.97	2582	107488.8	4313.56
2015-01-19	23.04056	16.36016	23.04056	B	40075.37	3016	109565.7	2076.90
2015-01-20	23.89947	16.36016	23.89947	B	30085.39	3434	112156.2	2590.46
2015-01-26	24.77656	19.22774	24.77656	B	20100.44	3837	115168.1	3011.92
2015-01-27	25.16284	19.40043	25.16284	B	10110.79	4234	116650.2	1482.16
2015-01-30	21.99533	21.99533	25.16284	S	103239.02	0	103239.0	-13411.23
2015-02-05	26.91247	21.99533	26.91247	B	93254.49	371	103239.0	0.00
2015-02-10	28.68482	21.99533	28.68482	B	83272.17	719	103896.6	657.54
2015-02-11	31.55239	21.99533	31.55239	B	73301.62	1035	105958.3	2061.78
2015-02-12	31.87960	21.99533	31.87960	B	63323.30	1348	106297.0	338.66
2015-02-13	35.06983	22.72245	35.06983	B	53328.40	1633	110597.4	4300.43
2015-02-16	38.57817	24.22213	38.57817	B	43336.66	1892	116326.6	5729.13
2015-02-17	40.99130	24.46753	40.99130	B	33375.77	2135	120892.2	4565.63
2015-03-06	34.32453	34.32453	40.99130	S	106658.65	0	106658.6	-14233.54
2015-03-16	41.07764	34.32453	41.07764	B	96676.78	243	106658.6	0.00
2015-03-18	41.94564	34.32453	41.94564	B	86693.72	481	106869.6	210.92
2015-03-24	45.34946	37.17393	45.34946	B	76716.83	701	108506.8	1637.24
2015-04-08	38.01011	38.01011	45.34946	S	103361.92	0	103361.9	-5144.89
2015-04-23	46.27199	37.06031	46.27199	B	93367.17	216	103361.9	0.00
2015-04-24	50.89829	37.06031	50.89829	B	83391.11	412	104361.2	999.28
2015-04-27	50.90283	37.06031	50.90283	B	73414.15	608	104363.1	1.87
2015-04-28	55.44277	37.06031	55.44277	B	63434.45	788	107123.4	2760.29
2015-04-29	60.98705	37.06031	60.98705	B	53493.56	951	111492.2	4368.89
2015-05-06	62.25497	45.19495	62.25497	B	43532.77	1111	112698.0	1205.79
2015-05-07	66.20413	46.27199	66.20413	B	33535.95	1262	117085.6	4387.51
2015-05-08	67.23573	50.89829	67.23573	B	23585.06	1410	118387.4	1301.88
2015-05-11	73.96157	50.90283	73.96157	B	13600.25	1545	127870.9	9483.44
2015-05-12	81.36000	55.44277	81.36000	B	3674.33	1667	139301.5	11430.58
2015-05-28	64.68000	64.68000	82.49000	S	111495.89	0	111495.9	-27805.56

There are 34 transactions in all, of which 30 are buying and 4 are selling. In the end, the remaining capital is 111,495.9 *yuan*, with an income of 11,495 *yuan*, which makes 11.5% the rate of return.

6.2.3 Model Optimization

We see that we can profit well by the upswing buying in a big bull market. In fact, we can further optimize the model. When building the sell signals, we can take better sell points than those resulting from the lowest price in recent 10 days. Let's optimize the model, for example, with the conditions: sell the stock when the stock price is lower than the previous buy point and set the lowest price in recent 10 days as the stop-loss point. Can we profit more by this optimization strategy?

This optimizing pattern makes our strategy more sensitive to the volatility, so it is easier for us to get shaken off; yet the advantage is that it will, more speedily, trigger the stop-profit or the stop-loss conditions, which will earn us more profit in a bull market.

```
# The optimization conditions: sell the stock when the stock price is
lower than the previous buy point and set the lowest price in recent 10
days as the stop-loss point.
> # Calculate the sell signals
```

```
> sellPoint<-function(ldata,buydata){
+
+     arr<-c()
+     for(i in 1:nrow(buydata)){
+
+         if(i>1){ # Jump to the 1st point
+             date<-index(buydata[i,])#;print(date)
+
+             # Sell when the price is lower than the last buy price
+             last<-as.vector(buydata[i-1,]$Value) # The last buy price
+             lst<-ldata[paste(date,"/",sep="")]$Value
+             idx<-head(which(lst < last),1)
+
+             if(length(idx)>0){
+                 arr<-rbind(arr,index(lst[idx]))
+             }
+         }
+     }
+     selldata<-ldata[as.Date(unique(arr)),]
+
+     # Filter the redundant sell points
+     bsdata<-merge(buydata$Value,selldata$Value)
+     names(bsdata)<-c("buy","Value")
+     idx1<-which(!is.na(bsdata$Value))
+     idx2<-idx1[which(c(0,diff(idx1))==1)]
+     bsdata$Value[idx2]<-NA
+     return(bsdata$Value[which(!is.na(bsdata$Value))])
+
+ }

# Sell signals
> selldata<-sellPoint(ldata,buydata)
> selldata
                Value
2015-01-21 22.81788
2015-01-28 23.60408
2015-02-25 36.89217
2015-03-17 39.97333
2015-03-19 40.96858
2015-03-26 39.25985
2015-05-14 74.24000
```

The result has been regenerated. There are seven sell signals, three more than last time. Draw a plot of trading signals (Figure 6.16).

```
> sdata<-signal(buydata$Value,selldata$Value)
# Merge the trading signals
> sdata
                Value op
2015-01-08 17.43721  B
2015-01-09 17.98709  B
2015-01-12 19.53222  B
```

Figure 6.16   Trading signals after optimization.

```
2015-01-15 20.21389   B
2015-01-16 22.23619   B
2015-01-19 23.04056   B
2015-01-20 23.89947   B
2015-01-21 22.81788   S
2015-01-26 24.77656   B
2015-01-27 25.16284   B
2015-01-28 23.60408   S
2015-02-05 26.91247   B
2015-02-10 28.68482   B
2015-02-11 31.55239   B
2015-02-12 31.87960   B
2015-02-13 35.06983   B
2015-02-16 38.57817   B
2015-02-17 40.99130   B
2015-02-25 36.89217   S
2015-03-16 41.07764   B
2015-03-17 39.97333   S
2015-03-18 41.94564   B
2015-03-19 40.96858   S
2015-03-24 45.34946   B
2015-03-26 39.25985   S
2015-04-23 46.27199   B
2015-04-24 50.89829   B
2015-04-27 50.90283   B
2015-04-28 55.44277   B
2015-04-29 60.98705   B
2015-05-06 62.25497   B
2015-05-07 66.20413   B
2015-05-08 67.23573   B
2015-05-11 73.96157   B
2015-05-12 81.36000   B
```

Figure 6.17 Merged trading signals.

```
2015-05-13  82.49000  B
2015-05-14  74.24000  S
```

There are 37 transactions in all, of which 30 are buying and 7 are selling. In the end, the remaining capital is 137,483.8 *yuan*, with an income of 37,483 *yuan*, which makes 37.5% the rate of return.

Let's merge the trading signals and the stop-loss signals and draw a plot of them.

```
# Stop-loss signals
> stopdata<-stopPoint(ldata,buydata)

# Merge the trading signals and the stop-loss signals
> bsdata<-merge(buydata$Value,selldata$Value,stopdata$Value)
> names(bsdata)<-c("buy","sell","stop")
> drawPoint(ldata,bsdata,title,sDate,eDate,'1 month') #Draw a plot
```

In Figure 6.17, the red points represent the buy points, the blue the sell points and the purple the stop-loss points. It is clear in the figure that the blue sell points are higher than the purple stop-loss points. So we have attained our goal of model optimization. It is a small optimization, but will bring a better income.

The strategy of upswing buying and downswing selling will profit us well in a bull market. However, the bull market has gone in the latter half year of 2015 and the volatility would continue. Then we can adopt the strategy of mean reversion to find investment opportunities in a reverse tendency.

In conclusion, in this article, we have implemented an investment idea in our head by the process of introducing the idea of upswing buying and downswing selling, testing its market characteristics, building the mathematical formula, modeling with R and backtesting with historical data. Everyone bears an investment idea similar to this, but IT guys are those who can use their technical advantages to actually implement the ideas with real practice.

This makes it easier to realize the conversion from IT technology to real value. Fighting, IT guys!

6.3 Build Quantitative Model of Pairs Trading with R

Question

How do you implement the model of pairs trading with R language?

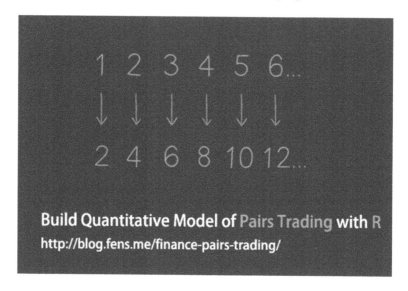

Build Quantitative Model of Pairs Trading with R
http://blog.fens.me/finance-pairs-trading/

Introduction

Individual investors go through the ups and downs of China's stock market every day. It is lucky for them to make money, and it is also normal for them to lose money. Well, is there any way to turn money making to normalcy for them?

We can find the ineffectiveness of the market by statistical arbitrage. Pairs trading is one of the statistical arbitrage strategies. It hedges most of the market risks, takes advantages of the arbitrage opportunities and accumulates small profits to big ones.

6.3.1 What Is Pairs Trading?

The idea of pairs trading originates from the sister stocks trading strategy of Jesse Livermore, a legendary trader in Wall Street in 1920s. The basic principle of pairs trading is to find a pair of highly correlated stocks or other financial products, then go short on the over-performer while simultaneously go long on the under-performer, and finally close the positions when both of the prices return to its norm to derive the profit from the price difference.

Assume that these two financial products stay in a good correlation. Once there is a weakness in the correlation, i.e. the price tendencies of the two products deviate and this deviation will be repaired in the future, there can be an opportunity for arbitrage. This kind of opportunities for statistical arbitrage is what pairs trading is looking for.

6.3.1.1 Characteristics of Pairs Trading

The biggest difference between a pairs trading and a traditional stock trading lies in that the investment target of pairs trading is the price difference between the two stocks, which is a relative value instead of an absolute value. A pairs trading opens a long position and a short position at the same time, which hedges most of the market risks, so it is a neutral strategy. No matter the market is rising or falling, the return of a pairs trading is comparative stable. It is lowly correlated to the market tendency.

When there is no trending opportunity in the market, pairs trading can be used to avoid the systematic risks in the stock market and to obtain the absolute value of Alpha. For trend trading strategy, please refer to Section 2.3, "Surfing the Stock Market with MA Model," of *R for Programmers: Advanced Techniques*.

3.3.1.2 Operations of Pairs Trading

1. Pairs screening: Find the matching financial products or portfolios in the market, check their trends of historical prices and judge whether they are fit for pairs trading. We screen for pairs mainly with these indicators: the correlation coefficient, the mean reversion rate by model calculation, the co-integration test, the fundamental factors, etc. With these factors, we will find the stably correlated pairs.
2. Risk measurement and dynamic pairs building: Calculate the expected return, the expected risks and the trading costs for each one of the pair; find the way in which the price difference is distributed; figure out whether correlation is short-term or long-term; find out the frequency of the price difference leap, etc.
3. Trading rules confirmation: Determine the trading frequency (high frequency trading or low frequency trading), the trading trigger conditions and the rules of closing positions, etc. by the characteristics of price difference.
4. Trading execution and risk control: In addition to the trading rules, we still need to execute the trading according to the dynamic tendency of price difference. If an abrupt change is detected, we should adjust the arbitrage mode and the trading frequency in time.

3.3.1.3 Disadvantages of Pairs Trading

- The rules of statistical arbitrage are all calculated based on the historical data. However, the history does not represent the future. If there is any change in the market, the model will become invalid.
- We can hardly estimate accurately the time of the price correction in the market. We can only make a rough estimation according to the history. If the reversion time is too long, it will be a test for the arbitrager's fund costs and it may lead to an arbitrage failure as well.

6.3.2 Model of Pairs Trading

We can design the model of pairs trading according to its principle. First, we need to quantify all the indicators concerned, such as how to screen for the highly correlated financial instrument pairs, when to go long, when to go short, and when to close positions, etc.

The system environment used here:

- Win10 64bit
- R: 3.2.3 x86_64-w64-mingw32/x64 b4bit

According to the concepts, we can generate two virtual instruments X and Y, including the fields of time and price. Make the prices of X and Y normally distributed and generate the data of 100 dates. It is a test program, in which the field of date means the calendar days, so here the dates refer to successive dates.

The codes of R implementation are as follow:

```
> set.seed(1)                          #Set the random seed
> dates<-as.Date('2010-01-01')+1:100   #100 dates
> x<-round(rnorm(100,50,40),2)         #Randomly generate 100 normally
distributed closing prices of X
> y<-round(rnorm(100,50,40),2)         #Randomly generate 100 normally
distributed closing prices of Y
> df<-data.frame(dates,x,y)
> head(df,20)
           dates        x       y
1      2010-01-02    24.94   25.19
2      2010-01-03    57.35   51.68
3      2010-01-04    16.57   13.56
4      2010-01-05   113.81   56.32
5      2010-01-06    63.18   23.82
6      2010-01-07    17.18  120.69
7      2010-01-08    69.50   78.67
8      2010-01-09    79.53   86.41
9      2010-01-10    73.03   65.37
10     2010-01-11    37.78  117.29
11     2010-01-12   110.47   24.57
12     2010-01-13    65.59   31.53
13     2010-01-14    25.15  107.29
14     2010-01-15   -38.59   23.97
15     2010-01-16    95.00   41.70
16     2010-01-17    48.20   34.29
17     2010-01-18    49.35   37.20
18     2010-01-19    87.75   38.84
19     2010-01-20    82.85   69.77
20     2010-01-21    73.76   42.91
```

Visualize the data, so that we can understand the data more intuitively.

```
# Load the class library of R
> library(ggplot2)
> library(scales)
> library(reshape2)

# Data type conversion
> df2<-melt(df,c('dates'))

# Draw a plot
> g<-ggplot(data=df2,aes(x=dates,y=value,colour=variable))
> g<-g+geom_line()
> g<-g+scale_x_date(date_breaks = "1 week",date_labels='%m-%d')
> g<-g+labs(x='date',y='Price')
> g
```

In Figure 6.18, *X*-axis is the time and *Y*-axis is the price; the red line is the price of *X* and the blue one is the price of *Y*. It is obvious that there is no correlation between *X* and *Y*.

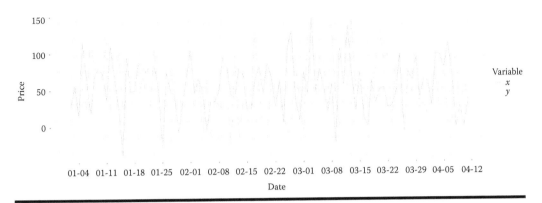

Figure 6.18   Price line chart of *X* and *Y*.

According to the assumptions of pairs trading, if the price difference, or the spread, of the two instruments is convergent, we will test the spread convergence next. Let's subtract the price of *Y* from that of *X*. If the difference is a positive value, we consider the price of *X* too high and the price of *Y* too low, then go short on *X* and go long on *Y*; if the difference is a minus, we consider the price of *X* too low and that of *Y* too high, then go long on *X* and go short on *Y*; if the difference is 0, which means the prices have been corrected by the market, we will close the positions of both.

To make the difference more obvious, we define the formula as follow:

```
Spread Z = Price of X - Price of Y
When Z > 10, go short on X and go long on Y; when Z < 0, close the
positions
When Z < -10, go long on X and go short on Y; when Z > 0, close the
positions
```

Calculate the spread and then calculate the trading statistics.

```
# Calculate the spreads
> df$diff<-df$x-df$y

# Find the points with spreads higher than 10
> idx<-which(df$diff>10)
> idx<-idx[-which(diff(idx)==1)-1]

# Print the spread index values
> idx
 [1]  4 11 15 23 25 30 34 36 38 43 48 53 55 59 61 68 76 81 83 86 88 92 95
98
```

Next, let's make a mock trading. Take the point of the first index value, and then go short on *X* and go long on *Y* on January 4, 2010. When the spread is less than 0, i.e. on January 6, 2010, close the positions.

```
# Print the first 20 data
> head(df,20)
```

```
        dates      x      y    diff
1   2010-01-02  24.94  25.19   -0.25
2   2010-01-03  57.35  51.68    5.67
3   2010-01-04  16.57  13.56    3.01
4   2010-01-05 113.81  56.32   57.49
5   2010-01-06  63.18  23.82   39.36
6   2010-01-07  17.18 120.69 -103.51
7   2010-01-08  69.50  78.67   -9.17
8   2010-01-09  79.53  86.41   -6.88
9   2010-01-10  73.03  65.37    7.66
10  2010-01-11  37.78 117.29  -79.51
11  2010-01-12 110.47  24.57   85.90
12  2010-01-13  65.59  31.53   34.06
13  2010-01-14  25.15 107.29  -82.14
14  2010-01-15 -38.59  23.97  -62.56
15  2010-01-16  95.00  41.70   53.30
16  2010-01-17  48.20  34.29   13.91
17  2010-01-18  49.35  37.20   12.15
18  2010-01-19  87.75  38.84   48.91
19  2010-01-20  82.85  69.77   13.08
20  2010-01-21  73.76  42.91   30.85
```

```
# When the spread is larger than 10, go short on X; when the spread is
less than 0, close the positions.
# Go short in the 4th row and close positions in the 6th row
> xprofit<- df$x[4]-df$x[6];xprofit
[1] 96.63
```

```
# When the spread is larger than 10, go long on Y; when the spread is
less than 0, close positions.
# Go short in the 4th row and close positions in the 6th row
> yprofit<- df$y[6]-df$y[4];yprofit
[1] 64.37
```

According to the trading result, our first pairs trade makes us money.

And why?

According to the assumptions of pairs trading, if the spread of the two instruments is convergent, we can test the spread convergence with co-integration tests. If the data is convergent, it will embody the characteristics of mean reversion. For the usage of mean reversion model, please refer to Section 6.1, "Mean Reversion, the Investment Opportunity Against Market," of this book.

Draw the spread plot of *X* and *Y*. We can see from the plot that the spread fluctuates around 0, which is obviously convergent and meanwhile accords with the characteristics of mean reversion (Figure 6.19).

```
> plot(df$diff,type='l')
```

This is the rule of the market. We can find the ineffectiveness of the market with pairs trading and then make profit by the arbitrage.

6.3.3 R Implementation of Pairs Trading

We may be excited to know the above way of making money. However, most of the data in the market will not serve our purpose of profiting as easily as our assumptions. We can scan the whole

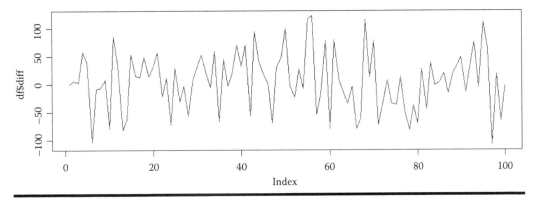

Figure 6.19   Spread line chart.

market by computer programs to find out the transaction opportunities. Of course, you can also use your own eyes to observe.

There are instruments with inborn high correlations, which can be our targets of arbitrage.

- Stocks: The individual stocks of the same industry with similar market values and similar fundamentals, such as BANK OF CHINA (601988) and AGRICULTURAL BANK OF CHINA (601288).
- Funds: The funds taking the same index as their target, such as Security B (150172) and Securities Company B (150201).
- Futures: Different contracts of the same futures, such as Copper (cu1605) and Copper (cu1606).
- Mixed: The instruments targeted at crossing markets, such as CSI 300 and the future contracts of IF.

In the following, we will take different contracts of the same futures variety as an example. We will pairs trade the contracts of cu1605 and cu1606. See what will happen. Futures support T+0 trading and the arbitrage operations usually do not hold a position overnight, so we try to operate in a short period, and close the positions in a day. We will take 1 minute as our trading period.

6.3.3.1 Data Preparation

R language itself provides abundant financial function toolkits, such as the time series package zoo and xts, the index calculation package TTR, the visualization package ggplot2, etc. We will use all these packages together for modeling, calculation and visualization.

The data used in this section is the day trading data of Copper in 1 minute line from February 1, 2016 to February 29, 2016. It is saved to local files named cu1605.csv and cu1606.csv in CSV format. There are 3-day trading time periods for commodity futures: 09:00:00–10:14:59, 10:30:00–11:29:59 and 13:30:00–14:59:59. It is a test, so we do not take night trading into consideration.

The data format is as follow:

```
2016-02-01 09:00:00,35870,35900,35860,35880
2016-02-01 09:01:00,35890,35890,35860,35870
2016-02-01 09:02:00,35870,35870,35860,35870
2016-02-01 09:03:00,35870,35900,35870,35900
```

```
2016-02-01 09:04:00,35900,35900,35870,35870
2016-02-01 09:05:00,35870,35880,35860,35870
2016-02-01 09:06:00,35880,35880,35860,35870
```

The data contains five columns:

- Column1: The trading date, date, 2016-02-01 09:00:00
- Column2: The opening price, Open, 35870
- Column3: The highest price, High, 35900
- Column4: The lowest price, Low, 35860
- Column5: The closing price, Close, 35880

Load the Copper data of 1-minute line with R language. We need to do an intraday trading, so I have made some conversion during the loading. The data has been grouped by dates, a list object of R has been generated and meanwhile, the object type has been converted from data.frame to XTS time series for the convenience of the subsequent data processing.

```
#Load the packages
> library(xts)
> library(TTR)

# Read the CVS data files
> read<-function(file){
+     df<-read.table(file=file,header=FALSE,sep = ",", na.strings =
"NULL")   # Read the files
+     names(df)<-c("date","Open","High","Low","Close")      # Set the
column names
+     dl<-split(df,format(as.POSIXct(df$date),'%Y-%m-%d'))  # Group by
dates
+
+     lapply(dl,function(item){                    # Convert the data
to xts type
+         xts(item[-1],order.by = as.POSIXct(item$date))
+     })
+ }

# Load the data
> cu1605<-read(file='cu1605.csv')
> cu1606<-read(file='cu1606.csv')

# View the data type
> class(cu1605)
[1] "list"

# View the index value of dates
> names(cu1605)
 [1] "2016-02-01" "2016-02-02" "2016-02-03" "2016-02-04" "2016-02-05"
 [6] "2016-02-15" "2016-02-16" "2016-02-17" "2016-02-18" "2016-02-19"
[11] "2016-02-22" "2016-02-23" "2016-02-24" "2016-02-25" "2016-02-26"
[16] "2016-02-29"

# View the everyday data volume
> nrow(cu1605[[1]])
[1] 223
```

```
# View the contract data of cu1605
> head(cu1605[['2016-02-01']])
                       Open  High   Low Close
2016-02-01 09:00:00 35870 35900 35860 35880
2016-02-01 09:01:00 35890 35890 35860 35870
2016-02-01 09:02:00 35870 35870 35860 35870
2016-02-01 09:03:00 35870 35900 35870 35900
2016-02-01 09:04:00 35900 35900 35870 35870
2016-02-01 09:05:00 35870 35880 35860 35870
```

The data is ready, now we can start modeling.

6.3.3.2 *Model of Pairs Trading*

Take the trading of February 1, 2016 for example. Calculate the difference between the closing prices of cu1605 and cu1606 in 1 minute line. We will process the data in the following: merge the data of these two contracts on February 1, 2016, process the null values and finally calculate the spread between the contracts.

```
# Merge data
> xdf<-merge(cu1605[['2016-02-01']]$Close,cu1606[['2016-02-01']]$Close)
> names(xdf)<-c('x1','x2')

# Replace the null value with its previous value
> xdf<-na.locf(xdf)

# Calculate the spreads
> xdf$diff<-xdf$x1-xdf$x2

# Print the data of the first 20 rows
> head(xdf,20)
                       x1    x2  diff
2016-02-01 09:00:00 35880 35900  -20
2016-02-01 09:01:00 35870 35920  -50
2016-02-01 09:02:00 35870 35910  -40
2016-02-01 09:03:00 35900 35940  -40
2016-02-01 09:04:00 35870 35910  -40
2016-02-01 09:05:00 35870 35920  -50
2016-02-01 09:06:00 35870 35910  -40
2016-02-01 09:07:00 35860 35910  -50
2016-02-01 09:08:00 35840 35880  -40
2016-02-01 09:09:00 35790 35840  -50
2016-02-01 09:10:00 35800 35840  -40
2016-02-01 09:11:00 35790 35830  -40
2016-02-01 09:12:00 35820 35860  -40
2016-02-01 09:13:00 35810 35850  -40
2016-02-01 09:14:00 35790 35830  -40
2016-02-01 09:15:00 35780 35830  -50
2016-02-01 09:16:00 35770 35810  -40
2016-02-01 09:17:00 35760 35820  -60
2016-02-01 09:18:00 35750 35800  -50
2016-02-01 09:19:00 35760 35810  -50
```

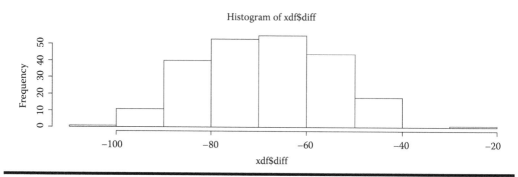

Figure 6.20 Histogram of spread distribution.

Data explanations:

- Column x1: The first leg, corresponding to the cu1605 contract.
- Column x2: The second leg, corresponding to the cu1606 contract.
- Column diff: Refers to cu1605-cu1606.

According to the spread result, the price of cu1605 contract is lower than that of cu1606 contract in every minute. The difference differs from –110 to –20 and fluctuates around the mean of –63.

```
# Calculate the spread scope
> range(xdf$diff)
[1] -110  -20

# Calculate the spread mean
> mean(xdf$diff)
[1] -63.90135

# Draw a histogram of spread distribution
> hist(xdf$diff,10)
```

Draw a histogram of spread distribution (Figure 6.20).

Let's assume that the mean value of mean reversion is –63. If the difference is larger than –45, we consider the price of X too high and the price of Y too low, then go short on X and go long on Y; if the difference is smaller than –75, we consider the price of X too low and that of Y too high, then go long on X and go short on Y; if the difference is –63, which means the prices have been corrected by the market, we will close both the positions. Hold the positions of cu1605 and cu1606 contracts at the ratio of 1:1, and go long on one round block of a contract while go short on one round block of the other.

First of all, load DanQuant, the R package developed by myself and used for pairs trading. It is still a test version, so it is only released on github.

```
> library(devtools)
> install_github('bsspirit/DanQuant')
> library(DanQuant)
```

Define the model indicators. The diff column is for the differences, the mid column for the mean value of mean reversion, the top column for the maximum thresholds and the bottom column for the minimum thresholds.

```
# Define the model indicators
> target.pair<-function(xdf){
+     xdf$diff<-xdf$x1-xdf$x2      #The difference
+     xdf$mid<- -63                #The mean value of mean reversion
+     xdf$top<- -45                #The maximum threshold
+     xdf$bottom<- -75             #The minimum threshold
+     return(xdf)
+ }
```

After the indicator definitions, we can start building the model of pairs trading.

```
# Define the strategy model
> mod.pair<-function(tXTS,params){
+
+     #The basic model
+     ops<-model.boll(tXTS)
+
+     #Signal processing
+     ops<-model.filter.pingLatest(ops)                      #Reserve the
latest point of closing a position
+     ops<-model.filter.pingStrat(ops)                       #Remove the
points smaller than the position opening point after closing a position
+     ops<-model.filter.closeOut(ops,tXTS,1)                 #Intraday
position closing
+     ops<-model.filter.stopProfit.pair(ops,tXTS,500,params) #Stop profit
and stop loss
+     ops<-model.filter.posPeriod(ops,40)                    #Time of
holding positions
+     ops<-model.filter.contineKai(ops)
#Continuously open positions
+     ops<-model.filter.continePing(ops)
#Continuously close positions
+     return(ops)
+ }
```

Next, backtest the pairs dataset. Make a mock trading with the trading signals generated, output a list and visualize the trading result.

```
# Initialize the trading varieties
> params<-newParams('cu1605','cu1606','2016-02-03')

# Execute the trading model
> tXTS<-target.pair(xdf)

# Generate the trading signals
> ops<-mod.pair(tXTS,params)
```

The trading signals generated are as follows:

```
> ops
                   date    x1    x2 diff mid top bottom op
21 2016-02-01 09:00:00 35880 35900  -20 -63 -45    -75 ks
1  2016-02-01 09:25:00 35740 35810  -70 -63 -45    -75 pb
22 2016-02-01 09:40:00 35690 35730  -40 -63 -45    -75 ks
2  2016-02-01 09:47:00 35700 35770  -70 -63 -45    -75 pb
```

```
13 2016-02-01 10:00:00 35690 35770  -80 -63 -45    -75 kb
5  2016-02-01 10:01:00 35710 35760  -50 -63 -45    -75 ps
23 2016-02-01 10:02:00 35710 35750  -40 -63 -45    -75 ks
3  2016-02-01 10:07:00 35680 35750  -70 -63 -45    -75 pb
14 2016-02-01 10:37:00 35720 35800  -80 -63 -45    -75 kb
6  2016-02-01 10:42:00 35740 35790  -50 -63 -45    -75 ps
15 2016-02-01 11:20:00 35700 35780  -80 -63 -45    -75 kb
7  2016-02-01 11:21:00 35710 35750  -40 -63 -45    -75 ps
24 2016-02-01 11:21:00 35710 35750  -40 -63 -45    -75 ks
4  2016-02-01 11:23:00 35690 35760  -70 -63 -45    -75 pb
16 2016-02-01 11:29:00 35690 35770  -80 -63 -45    -75 kb
8  2016-02-01 13:36:00 35660 35720  -60 -63 -45    -75 ps
17 2016-02-01 13:45:00 35660 35740  -80 -63 -45    -75 kb
9  2016-02-01 13:46:00 35670 35730  -60 -63 -45    -75 ps
18 2016-02-01 13:52:00 35650 35730  -80 -63 -45    -75 kb
10 2016-02-01 13:53:00 35650 35710  -60 -63 -45    -75 ps
19 2016-02-01 13:56:00 35640 35720  -80 -63 -45    -75 kb
11 2016-02-01 14:49:00 35600 35660  -60 -63 -45    -75 ps
20 2016-02-01 14:52:00 35610 35700  -90 -63 -45    -75 kb
12 2016-02-01 14:58:00 35610 35690  -80 -63 -45    -75 ps
```

Data explanations:

- Column date: The trading time.
- Column x1: The first leg, corresponding to the cu1605 contract.
- Column x2: The second leg, corresponding to the cu1606 contract.
- Column diff: Refers to cu1605-cu1606.
- Column mid: The mean value of mean reversion.
- Column top: The maximum threshold.
- Column bottom: The minimum threshold.
- Column op: The trading signals.

There are four types of trading signals.

- ks: To open a position and go short (sell). The reverse operation is pb.
- kb: To open a position and go long (buy). The reverse operation is ps.
- ps: To close a position and go short (sell). The reverse operation is kb.
- pb: To close a position and go long (buy). The reverse operation is ks.

In all, 24 trading signals appeared. We are trying to make pairs tradings here, so when there is a ks (to close a position and go long) signal, there are actually two trades going on: opening a position for the first leg and going short on it; opening a position for the second leg and going long on it.

Next, take a backtesting and calculate the trading list.

```
# Calculate the trading signals
> sigs<-backtest.signal(tXTS,ops)

# Take a backtesting and generate the trading list
> report<-backtest.pair(sigs,params)
```

```
# Print the backtesting report
> report
$day
[1] "2016-02-03"
$capital
[1] 2e+05
$cash
[1] 201417.8
$x1
                      code op price pos    fee  value  margin balance    cash
2016-02-01 09:00:00 cu1605 ks 35880   1 8.9700 179400 26910.0      NA 173081.0
2016-02-01 09:25:00 cu1605 pb 35740   0 8.9350      0     0.0     700 173748.1
2016-02-01 09:40:00 cu1605 ks 35690   1 8.9225 178450 26767.5      NA 173437.7
2016-02-01 09:47:00 cu1605 pb 35700   0 8.9250      0     0.0     -50 173339.9
2016-02-01 10:00:00 cu1605 kb 35690   1 8.9225 178450 26767.5      NA 173552.0
2016-02-01 10:01:00 cu1605 ps 35710   0 8.9275      0     0.0     100 173574.2
2016-02-01 10:02:00 cu1605 ks 35710   1 8.9275 178550 26782.5      NA 173651.3
2016-02-01 10:07:00 cu1605 pb 35680   0 8.9200      0     0.0     150 173753.4
2016-02-01 10:37:00 cu1605 kb 35720   1 8.9300 178600 26790.0      NA 173758.1
2016-02-01 10:42:00 cu1605 ps 35740   0 8.9350      0     0.0     100 173780.2
2016-02-01 11:20:00 cu1605 kb 35700   1 8.9250 178500 26775.0      NA 173887.3
2016-02-01 11:21:00 cu1605 ps 35710   0 8.9275      0     0.0      50 173859.4
2016-02-01 11:21:001 cu1605 ks 35710  1 8.9275 178550 26782.5      NA 174044.1
2016-02-01 11:23:00 cu1605 pb 35690   0 8.9225      0     0.0     100 174096.2
2016-02-01 11:29:00 cu1605 kb 35690   1 8.9225 178450 26767.5      NA 174173.3
2016-02-01 13:36:00 cu1605 ps 35660   0 8.9150      0     0.0    -150 173945.5
2016-02-01 13:45:00 cu1605 kb 35660   1 8.9150 178300 26745.0      NA 174260.1
2016-02-01 13:46:00 cu1605 ps 35670   0 8.9175      0     0.0      50 174232.3
2016-02-01 13:52:00 cu1605 kb 35650   1 8.9125 178250 26737.5      NA 174331.9
2016-02-01 13:53:00 cu1605 ps 35650   0 8.9125      0     0.0       0 174254.1
2016-02-01 13:56:00 cu1605 kb 35640   1 8.9100 178200 26730.0      NA 174403.8
2016-02-01 14:49:00 cu1605 ps 35600   0 8.9000      0     0.0    -200 174125.9
2016-02-01 14:52:00 cu1605 kb 35610   1 8.9025 178050 26707.5      NA 174490.6
2016-02-01 14:58:00 cu1605 ps 35610   0 8.9025      0     0.0       0 174405.3


$x2
                      code op price pos    fee  value  margin balance    cash
2016-02-01 09:00:00 cu1606 kb 35900   1 8.9750 179500 26925.0      NA 146147.1
2016-02-01 09:25:00 cu1606 ps 35810   0 8.9525      0     0.0    -450 200214.2
2016-02-01 09:40:00 cu1606 kb 35730   1 8.9325 178650 26797.5      NA 146631.3
2016-02-01 09:47:00 cu1606 ps 35770   0 8.9425      0     0.0     200 200328.4
2016-02-01 10:00:00 cu1606 ks 35770   1 8.9425 178850 26827.5      NA 146715.6
2016-02-01 10:01:00 cu1606 pb 35760   0 8.9400      0     0.0      50 200442.7
2016-02-01 10:02:00 cu1606 kb 35750   1 8.9375 178750 26812.5      NA 146829.8
2016-02-01 10:07:00 cu1606 ps 35750   0 8.9375      0     0.0       0 200557.0
2016-02-01 10:37:00 cu1606 ks 35800   1 8.9500 179000 26850.0      NA 146899.1
2016-02-01 10:42:00 cu1606 pb 35790   0 8.9475      0     0.0      50 200671.2
2016-02-01 11:20:00 cu1606 ks 35780   1 8.9450 178900 26835.0      NA 147043.4
2016-02-01 11:21:00 cu1606 pb 35750   0 8.9375      0     0.0     150 200835.5
2016-02-01 11:21:001 cu1606 kb 35750  1 8.9375 178750 26812.5      NA 147222.6
2016-02-01 11:23:00 cu1606 ps 35760   0 8.9400      0     0.0      50 200949.8
2016-02-01 11:29:00 cu1606 ks 35770   1 8.9425 178850 26827.5      NA 147336.9
2016-02-01 13:36:00 cu1606 pb 35720   0 8.9300      0     0.0     250 201014.1
2016-02-01 13:45:00 cu1606 ks 35740   1 8.9350 178700 26805.0      NA 147446.2
2016-02-01 13:46:00 cu1606 pb 35730   0 8.9325      0     0.0      50 201078.4
2016-02-01 13:52:00 cu1606 ks 35730   1 8.9325 178650 26797.5      NA 147525.5
2016-02-01 13:53:00 cu1606 pb 35710   0 8.9275      0     0.0     100 201142.7
2016-02-01 13:56:00 cu1606 ks 35720   1 8.9300 178600 26790.0      NA 147604.8
```

```
2016-02-01 14:49:00     cu1606 pb 35660   0  8.9150      0     0.0    300 201207.0
2016-02-01 14:52:00     cu1606 ks 35700   1  8.9250 178500 26775.0     NA 147706.7
2016-02-01 14:58:00     cu1606 pb 35690   0  8.9225      0     0.0     50 201221.4
```

Data explanations:

- Part $x1: The trading list of the first leg.
- Part $x2: The trading list of the second leg.
- code: The contract code.
- op: The trading signals.
- price: The price of the transaction.
- pos: The trading volume.
- fee: The commission.
- value: The underlying value.
- margin: The cash deposit.
- balance: The balance when closing a position.
- cash: The capital of the account.

Sum up the trading result in the trading list.

```
# Backtesting report statistics
> page<-backtest.reading.pair(report)
> page
$day       # Trading date
[1] "2016-02-01"

$capital    # Initial fund
[1] 2e+05

$cash       # Balance of the account
[1] 201221.4

$num        # Number of trading signals
[1] 24

$record    # Balance when closing the positions of pairs trading
                    x1    x2 balance
2016-02-01 09:25:00  700 -450     250
2016-02-01 09:47:00  -50  200     150
2016-02-01 10:01:00  100   50     150
2016-02-01 10:07:00  150    0     150
2016-02-01 10:42:00  100   50     150
2016-02-01 11:21:00   50  150     200
2016-02-01 11:23:00  100   50     150
2016-02-01 13:36:00 -150  250     100
2016-02-01 13:46:00   50   50     100
2016-02-01 13:53:00    0  100     100
2016-02-01 14:49:00 -200  300     100
2016-02-01 14:58:00    0   50      50
```

```
$balance    # The total balance when closing positions, the balance of the
first leg, the balance of the second leg
[1] 1650   850   800

$fee        # The total commission, the commission of the first leg, the
commission of the second leg
[1] 429 214 215

$profit     # The net income of the account, the rate of return (in the
margin)
[1] 1221.000    0.023

$wins       # The win rate, the number of profit trades, the number of
loss trades
[1]   1 12   0
```

Finally, visualize and output the trading signals (Figure 6.21).

Illustrations:

- Brown line: The price difference, diff
- Purple line: The maximum threshold, top
- Red line: The minimum threshold, bottom
- Blue line: The line of the mean value (mid), parallel to the top and bottom lines
- Light blue line: The trades of ks (opening positions and going short)
- Green line: The trades of kb (opening positions and going long)

It is obvious in the figure that we made 12 trades, 4 deals for each trade, with a win rate of 100%.

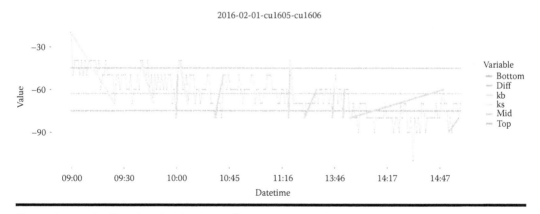

Figure 6.21   Trading signals of pairs trading.

Then let's run a backtesting for the data of the whole February. The backtesting result is as follow:

```
> cu_cu
      date      profit ret  balance fee winRate win fail maxProfit maxLoss avgProfit avgLoss
1  2016-02-01  1221 0.023  1650 429    1.00  12    0       250      50       138      NaN
2  2016-02-02  1077 0.020  1650 573    1.00  15    0       150       0       110      NaN
3  2016-02-03    64 0.001   100  36    1.00   1    0       100     100       100      NaN
4  2016-02-04   113 0.002   150  37    1.00   1    0       150     150       150      NaN
5  2016-02-05   926 0.017  1400 474    1.00  13    0       150     100       108      NaN
6  2016-02-15  1191 0.022  1550 359    1.00  10    0       250     100       155      NaN
7  2016-02-16    78 0.001   150  72    1.00   1    0       150       0       150      NaN
8  2016-02-17   179 0.003   250  71    1.00   2    0       200      50       125      NaN
9  2016-02-18    14 0.000    50  36    1.00   1    0        50      50        50      NaN
10 2016-02-19   -36 -0.001    0  36     NaN   0    0         0       0       NaN      NaN
11 2016-02-22    64 0.001   100  36    1.00   1    0       100     100       100      NaN
12 2016-02-23   632 0.012   850 218    1.00   6    0       200     100       142      NaN
13 2016-02-24   470 0.009   650 180    1.00   4    0       200       0       162      NaN
14 2016-02-25   114 0.002   150  36    1.00   1    0       150     150       150      NaN
15 2016-02-26   178 0.003   250  72    1.00   2    0       150     100       125      NaN
16 2016-02-29   511 0.009   800 289    0.88   7    1       150     -50       121      -50
```

Data explanations:

- date: The trading date
- profit, the net income
- ret: The daily return
- balance: The balance when closing a position
- fee: The commission
- winRate: The win rate
- win: The number of profit trades
- fail: The number of loss trades
- maxProfit: The maximum profit for one trade
- maxLoss: The maximum loss for one trade
- avgProfit: The average profit
- avgLoss: The average loss

This shows that cu1605 and cu1606 are two instruments highly correlated. The intertemporal arbitrage we usually refer to is implemented based on this idea. The pairs trading introduced in this section is a basic model of statistical arbitrage. After we all have grasped its principle, which is simple, the trading speed will be crucial for our winning.

To profit from the market ineffectiveness is the goal of every arbitrage strategy. We can find the market ineffectiveness by statistical methods, use hedge operation to avoid most of the market risks and then make profit from the self-repairing of the market. It seems easy. However, the market ineffectiveness can be repaired in an extremely short period of time.

"Only the fastest kung fu is impenetrable." We can use quantification and computer to discover opportunities, then trade and make profits. Everything would be in harmony.

6.4 Fund Accounting System Design and Implementation

Question

How do you implement the fund accounting?

**Fund Accounting System
Design and Implementation**
http://blog.fens.me/finance-fund-accounting/

Introduction

People make investments every day, but most of us mind only our own money. When you get steady and sustainable returns from investments, your friends may want you to help manage their money. So how to help others manage their money? You need a management system of fund accounting here.

You may receive 10 million or 100 million investment funds one day. To manage it in a scientific and standardized way will make the fund grow larger.

6.4.1 An Introduction to Fund Accounting System

It is necessary for a financial practitioner to have an understanding of the fund accounting knowledge. In the financial market, any error can lead to a loss and any grasp of details may bring along a big profit.

You may not be sensitive to the commission when making individual transactions. However, for the futures transactions of high frequency, the commission in one month may exceed your principal. Sometimes, one extra amount of income may appear in your record sheet. It is not a large amount. It can be the interest of your stocks or bonds. A few days later, you may find a deduction of a smaller amount, which can be the interest tax of the stocks or bonds. When the funds flow in and out of the account more frequently, you may find it difficult to tell which fund is for which trade. However, to be a little extreme, if you cannot tell whether you have made a profit or a loss after a daily trading, you can quit finance.

In this section, I will introduce how to implement a fund accounting system mainly used for the meeting bookkeeping service of a private equity in financial trading. The system includes the core computing functions, such as asset accounting, NAV accounting, fund shares accounting, etc. The fund accounting system mainly deals with the situation in which there are large amounts of money from more than one investor, there are various instruments from all over the market for trading and there can be frequent transactions. The system is needed in this situation to help us with accounts management.

Note: The system design introduced here is merely my understanding of the market. If there is any definition that does not conform to the textbooks, the textbooks should prevail.

I will introduce the fund accounting system from the following aspects, so as to reveal its full view and its business logic.

1. Financial transaction
2. Trading record
3. Position holding record
4. Investors
5. Fund inflow and outflow
6. Investor shares
7. Traders
8. NAV
9. NAV per share

6.4.1.1 Financial Transaction

Financial transactions refer to the transactions of financial products or financial derivatives. Take China's market for example. There are many kinds of instruments we can trade. They include but are not limited to the instruments listed in the following.

■ Stock: It is a security issued by a joint-stock company to the shareholders to raise funds. It is a certificate of title to shares, by which the shareholders can get stock dividend and bonus.

■ Futures: Different from the spot commodities which are real tradeable goods (commodities), futures are not goods, but standardized tradeable contracts targeting at certain bulk commodities like cotton, soybean, petroleum, etc. and financial assets like stocks, bonds, etc.

■ Futures margin: In the futures market, the traders only need to pay a certain ratio of the future contract price, i.e. a small amount of money, as a financial guarantee for the contract implementation, and then they can participate in the trading of the future contract. Here the money that the traders pay is the futures margin.

■ Convertible bond: A convertible bond is a type of bond which can be converted to the stock shares of the issuing company and it bears a comparatively lower coupon rate. Essentially, a convertible bond is a corporate bond with attached options and its holder is allowed to convert the bond into the underlying stock within a period of time.

■ Reverse repo: It refers to the transaction where a party takes initiative to lend money and gets income from the interest.

■ Repo: A repo is to borrow money from those who make reverse repos and to pay the interest.

■ Bitcoin: It is a virtual currency, constituted with a series of complicated codes generated by computers. It is not issued by the government, but can be traded on the bitcoin trading platforms and can be exchanged with RMB, US dollars and Euros.

■ Financing: It is an operation that the investor pledges with funds or securities to borrow money from securities companies, in order to buy securities. The investor will return the principal and pay interest in an agreed period.

■ Securities loan: It is an operation that the investor pledges with funds or securities to borrow securities from securities companies and then sells the borrowed securities. The investor will, in an agreed period, return the securities of the same amount and the same type and pay the securities loan fee to the securities companies.

We divide the financial transactions into two types: asset type and debt type.

- Asset type: Stocks, futures, convertible bonds, reverse repos, bitcoin, etc.
- Debt type: Repos, financing, securities loan, futures margins, etc.

6.4.1.2 Trading Record

Trading record means the written record of the related materials of the successful transactions in the exchanges. Different exchanges provide different detail fields and different recording methods in their trading list. To run a unified accounting management, we need a standardized recording format for these trading records. First, we need to have some knowledge of the trading records of the securities market, the futures market and the bitcoin market provided in different exchanges.

The futures trading record is as shown in Chart 6.1.

A futures trading is a T+0 trading. We can go long or go short on it. The corresponding operation of buying in and going long is to buy | open a position. The corresponding operation of buying in and going short is to sell | open a position. If we buy the contract and sell it on the same day, the operation should be close today's position; if we sell it the next day, it is close a position. It is commission free for some instruments if we close today's position, so many investors do not hold the position overnight. They sell all their products before the closing of the day and keep only cash. This is what we called the intraday trading.

The securities trading record is as shown in Chart 6.2.

Most of the securities tradings are T+1 tradings and we can only go long on them, never go short. Securities trading can trade more instruments than futures, so the record keeping becomes complicated. Its operations can be stocks trading, convertible bonds trading, conversion from the convertible bonds to stocks, conversion from the treasury bonds to the standardized bonds, repo pledged by standardized bonds, margin trading with credit accounts, etc. And there are more transaction fees than futures. The fees include commission, stamp duty, transfer fee and clearing fee.

- Commission: It is charged by the exchanges and the securities companies. When you opened the account, your securities company may have said that the transaction fee can be lowered. By the transaction fee, they referred to the commission.
- Stamp duty: It is charged by the country and will directly go to the fiscal levy. Only stock selling will be charged for stamp duty and other tradings of other instruments will be stamp duty free.
- Transfer fee: It is charged only by Shanghai Stock Exchange. Those traded in Shenzhen Stock Exchange will not be charged for transfer fee separately.
- Clearing fee: It is charged in securities tradings when there is a warrant exercise or a warrant trade.

The bitcoin trading record is as shown in Chart 6.3.

The bitcoin trading is comparatively simple. It can be traded in 24 hours. We can only go long on it, never go short. There is only one product for bitcoin and its operations include only recharge, withdraw, buy and sell. The commission is usually charged by the platform and for most of the time, it is commission free. Only when the bitcoin value is exceedingly high, the commission will be charged to suppress the price.

We have seen that there are different trading rules for different financial assets in different financial markets, so there are different fields to record tradings for futures, securities and bitcoin. We need to combine these trading records of different assets and output a standardized one.

Chart 6.1   Futures Trading Record

Date	Time	Contract	Transaction	Open or Close a Position	Transaction Price	Volume	Commission	Investors' Security Fund	Transaction Code
20140217	10:49:34 PM	ag1406	Sell	Close a position	4,325.000	2	9.73	Speculation	4404
20140217	10:50:31 PM	ag1406	Buy	Open a position	4,328.000	2	9.74	Speculation	4541
20140217	11:51:57 PM	ag1406	Sell	Close today's position	4,316.000	2	0.00	Speculation	6464
20140217	1:44:08 AM	ag1406	Buy	Open a position	4,327.000	1	4.87	Speculation	7616
20140217	1:49:50 AM	ag1406	Sell	Close today's position	4,326.000	1	0.00	Speculation	7632
20140218	9:00:37 PM	au1406	Buy	Open a position	261.000	1	15.00	Speculation	214
20140218	9:39:07 PM	au1406	Sell	Close today's position	260.900	1	0.00	Speculation	2441
20140218	9:40:51 PM	au1406	Buy	Open a position	261.000	1	15.00	Speculation	2469
20140218	9:41:38 PM	au1406	Sell	Close today's position	261.000	1	0.00	Speculation	2556
20140218	9:41:47 PM	au1406	Sell	Open a position	261.000	1	15.00	Speculation	2561
20140218	9:45:33 PM	au1406	Buy	Close today's position	260.950	1	0.00	Speculation	2632
20140218	10:10:32 PM	au1406	Buy	Open a position	261.000	1	15.00	Speculation	3033
20140218	11:01:03 PM	au1406	Sell	Close today's position	261.000	1	0.00	Speculation	3463
20140220	10:21:02 AM	TF1403	Buy	Open a position	91.904	1	4.50	Speculation	41511
20140220	10:22:03 AM	TF1403	Sell	Close today's position	91.902	1	0.00	Speculation	41512

Chart 6.2 Securities Trading Record

Currency	Security Name	Trading Date	Trading Price	Trading Volume	Incurred Amount	Fund Balance	Contract No.	Service Name	Commission	Stamp Duty	Transfer Fee	Clearing Fee	Security Code	Shareholder Code
RMB	CMBC	20140122000930	7.100	1200.00	−8527.39	−7528.24	911	Buy securities (CMBC)	6.39	0.00	1.00	0.00	600016	A433244006
RMB	R-001	20140122000931	7.320	−300.00	30006.02	22477.78	872	Lend and repo (R-001)	0.00	0.00	0.00	0.00	131810	0155546324
RMB	ANHUI ZHONGDING convertible bond	20140122000932	117.500	−60.00	7042.95	29520.73	918	Sell securities (ANHUI ZHONGDING convertible bond)	7.05	0.00	0.00	0.00	125887	0155546324
RMB	CMBC convertible bond	20140123000954	93.000	5.00	−4651.00	24869.73	940	Buy securities (CMBC convertible bond)	1.00	0.00	0.00	0.00	110023	A433244006
RMB	CMBC convertible bond	20140124001080	94.320	8.00	−7547.11	17322.62	987	Buy securities (CMBC convertible bond)	1.51	0.00	0.00	0.00	110023	A433244006
RMB	CMBC	20140124001081	7.050	1000.00	−7056.29	10266.33	990	Buy securities (CMBC)	5.29	0.00	1.00	0.00	600016	A433244006
RMB	CMBC convertible bond	20140124001082	93.950	5.00	−4698.50	5567.83	996	Buy securities (CMBC convertible bond)	1.00	0.00	0.00	0.00	110023	A433244006
RMB	CMBC convertible bond	20140124001083	94.520	5.00	−4727.00	840.83	1018	Buy securities (CMBC convertible bond)	1.00	0.00	0.00	0.00	110023	A433244006
RMB	CMBC convertible bond	20140124001084	94.660	10.00	−9467.89	−8627.06	1025	Buy securities (CMBC convertible bond)	1.89	0.00	0.00	0.00	110023	A433244006

Chart 6.3 Bitcoin Trading Record

Transaction ID	Transaction Type	Transaction Time	Bitcoin	RMB	For One Bitcoin
6760639	Commission	December 17, 2013 4:17:26 PM	฿−0.00003000	¥0.00	
6760638	Buy bitcoin	December 17, 2013 4:17:26 PM	฿0.01000000	¥−41.00	¥4,100.00
6749278	Commission	December 17, 2013 3:10:01 PM	฿0.00000000	¥−0.13	
6749277	Sell bitcoin	December 17, 2013 3:10:01 PM	฿−0.01000000	¥42.00	¥4,200.00
6690344	Commission	December 17, 2013 9:51:43 AM	฿−0.00000900	¥0.00	
6690343	Buy bitcoin	December 17, 2013 9:51:43 AM	฿0.00300000	¥−10.41	¥3,470.00
6684013	Commission	December 17, 2013 9:33:23 AM	฿−0.00003000	¥0.00	
6684012	Buy bitcoin	December 17, 2013 9:33:23 AM	฿0.01000000	¥−37.30	¥3,730.00
5278051	Sell bitcoin	December 7, 2013 9:37:29 PM	฿−0.01000000	¥50.09	¥5,009.00
5276581	Recharge bitcoin	December 7, 2013 9:30:04 PM	฿0.01000000	¥0.00	

6.4.1.3 Position Holding Record

Position holding record refers to the record of financial products when we are buying or short selling (short futures). If a position is held till closing after buying on the same day, the value of the product will be accounted and recorded by the closing price. Usually, the positions of an intraday arbitrage will be reflected on the holding cash, if no position is held.

The position holding record is different from the trading record. The position holding record will make realtime accounting for the accounts and will reflect the realtime influence of the financial product price changes in the market on the accounts.

The position holding record of securities is shown as below:

```
RMB: Balance:757.63 | Available:36763.70 | Withdrawable:757.63 | Market
value reference:51183.93 | Asset:51183.93 | Profit and loss:3119.30
----------------------------------------------------------------------
--------------------------------

Securities name | Securities volume | Sellable volume | Cost price |
Floating profit and loss | Ratio of profit and loss (%) | Latest market
```

```
value |  Current price | Today's buy volume | Today's sell volume |
Security code | Shareholder code | Cost amount |  Note |
CMBC | 3300 | 3300 | 7.078 | 898.50 | 3.84 |  24255.00 |  7.35 | 0 |  0 |
600016 | A433244006 | 23356.50 |
New standardized bond | 100 | 0 |  1000.000 | 0.00 | 0.00 |  100000.00 |
1000.000 | 0 |  0 |  888880 | A433244006 | 100000.00 |
```

For position holding record, we can assess the financial asset's profit by the floating profit and loss and the ratio of profit and loss and we can have knowledge of the funds in the account by balance, available and withdrawal.

- Balance: Represents the cash balance, which is withdrawable.
- Available: Represents the tradeable fund, but not necessarily withdrawable. For example, if you sell a stock today, the fund paid to your account becomes available and it will become withdrawable tomorrow.
- Withdrawable: Represents the withdrawable fund in the account.

6.4.1.4 Investors

Investors are those who provide funds. They are the capital sources of investments. There are investors for publicly offered funds and privately offered funds.

- Publicly offered funds are the funds targeted at the general public, i.e. the unspecific investors in society.
- While privately offered funds are the funds targeted at a minority of specific high-net-worth investors, including institutions and high-net-worth individuals. Privately offered funds require every capital investment higher than one million.

6.4.1.5 Fund Inflow and Outflow

Fund inflow and outflow is the process of subscription and redemption by investors. The redemption provisions can be found in the fund prospectus.

- Subscription: Purchase the fund with capital and hold shares.
- Redemption: Withdraw the capital and withdraw the shares.

Below are the rules of subscribing for and redeeming a fund in Alipay. See Figure 6.22.

6.4.1.6 Investor Shares

After the investors have subscribed for a fund, they hold shares of the fund. The holding shares of the investors will be calculated by the subscription amounts and the NAVs.

The shares change only if the investors make subscriptions or redemptions. In the following, the formula to calculate the investor shares will be introduced.

6.4.1.7 Traders

A trader is a person usually entrusted by institutions or individuals to implement the transactions in the financial market. Banks, securities companies (investment banks), listed companies,

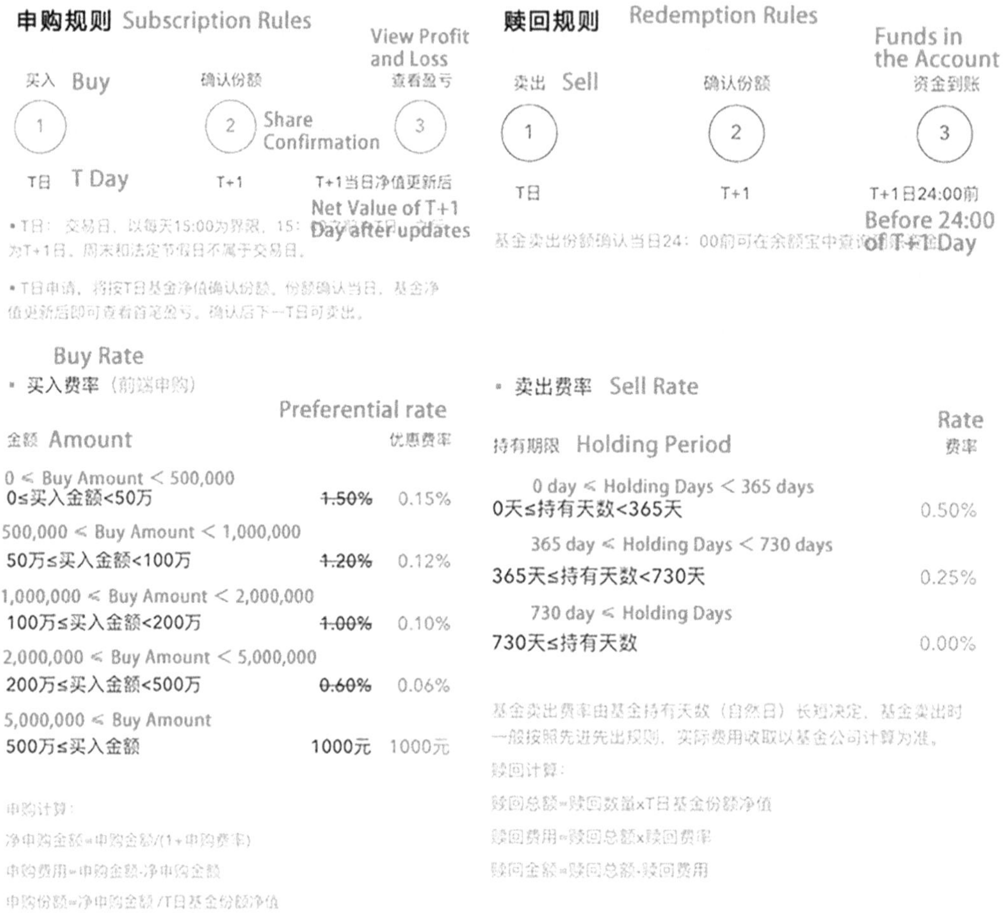

申购规则 Subscription Rules | **赎回规则** Redemption Rules

View Profit and Loss 查看盈亏

Funds in the Account 资金到账

Buy Rate
■ 买入费率（前端申购）

金额 Amount	Preferential rate 优惠费率
0 < Buy Amount < 500,000 0≤买入金额<50万	~~1.50%~~ 0.15%
500,000 < Buy Amount < 1,000,000 50万≤买入金额<100万	~~1.20%~~ 0.12%
1,000,000 < Buy Amount < 2,000,000 100万≤买入金额<200万	~~1.00%~~ 0.10%
2,000,000 < Buy Amount < 5,000,000 200万≤买入金额<500万	~~0.60%~~ 0.06%
5,000,000 < Buy Amount 500万≤买入金额	1000元 1000元

■ 卖出费率 Sell Rate

持有期限 Holding Period	Rate 费率
0 day < Holding Days < 365 days 0天≤持有天数<365天	0.50%
365 day < Holding Days < 730 days 365天≤持有天数<730天	0.25%
730 day < Holding Days 730天≤持有天数	0.00%

Figure 6.22 Rules of subscribing for and redeeming a fund in Alipay.

funds and professional trading companies are willing to pay great money to recruit excellent traders, because the traders' trading proficiency have an immense influence on the company's performance.

Traders can be classified into day traders and interday traders. The day trades aim at the T+0 market, where financial instruments can be bought and sold on the same trading day and can be gone long or short on. The interday traders, similar to the stock traders in China, can only go long. They buy on the day and sell on the next day in the T+1 market.

The accounting system records every operation of traders and gives the performances of traders according to their tradings.

6.4.1.8 Net Asset Value (NAV)

NAV, also called net value of portfolio, refers to the net asset value of a fund. NAV is the total asset value calculated by the closing price after closing in a trading day minus various costs and fees of the day. In short, NAV is the aggregate market value of the fund asset minus the liabilities at a certain time. It represents the rights and interests of the fund holders.

6.4.1.9 NAV per Share

NAV per share is the NAV for one share of fund. NAV per share = (Total asset of the fund − Total liabilities of the fund) / Total shares of the fund issued. The formula of NAV accounting will be discussed later.

- For subscriptions, the NAV will be calculated before 15:00 of the day (T day).
- For redemptions, the NAV will be calculated before 24:00 of the day (T day).

The accumulated NAV refers to the sum of the latest NAV and the total dividends after the fund establishing, that is, the accumulated NAV per share = NAV per share + the accumulated dividend per share after the fund establishing. For instance, Fund A has distributed the dividend, 3 *yuan* for each 10 shares. If the NAV per share published of the day is 1.02 *yuan*, the accumulated NAV per share will be 1.02 + (3 / 10) = 1.32 *yuan*.

6.4.2 Asset Accounting

According to the financial accounting requirements, we need to divide the financial products into assets and liabilities. An asset is an economic resource built by an enterprise's past transactions or other past events, owned or controlled by the enterprise and expected to produce economic benefits. A liability is a present obligation of the enterprise arising from its past transactions or events, the settlement of which is expected to result in an outflow of its resources embodying economic benefits.

Note: All the items below are settled in RMB.

The formulas to calculate the total assets and the total liabilities are as follows:

```
Assets = Cash + Stocks + Treasury bonds + Convertible bonds + Reverse
repos + Bitcoin + Other receivables
Liabilities = Repos + Financing + Securities loan + Futures liabilities +
Other payables
```

The owner's equity is the fund holder's equity. By the accounting equation, Owner's equity = Assets−Liabilities, where the owner's equity can be defined as the NAV of the fund.

6.4.3 NAV per Share Accounting

For NAV accounting, many items need to be calculated. Let's walk through it by an example.

For instance, we prepare to establish a fund and we need to raise funds at the beginning. We get investments from two initial investors and the NAV per share is 1.0. The fund has performed well in the first 2 months, so new investors come. At the time, the NAV per share of the fund has raised from 1.0 to 1.15. After 6 months, the fund has enlarged from 10 million to 120 million, but the NAV per share has not increased much. Therefore, some investors cannot wait and withdraw their capital. The fund size shrinks back to 80 million. In the year after, the overweight stock holdings and the operation mistakes lead to a severe loss, thus the NAV per share falls to 0.7, reaching the winding-up line. The investors withdraw their capital in succession and the fund is wound up and dissolved.

It is an example conceived by myself, but it basically fits the reality and it includes the whole process from the establishing of the fund to the liquidation. The core of the accounting system is to calculate the holder's equity of the fund for every day.

6.4.3.1 Text Description

Let's simulate the fund establishment.

- There are three investors: A, B and C.
- On the establishment day, A subscribes for 50,000 *yuan* and B for 50,000 *yuan*.
- On the second day, the fund has a return of 500 *yuan*.
- On the third day, the fund has a return of 1,000 *yuan* and an investor C subscribes for 50,000 *yuan*.
- Neglect the commission, the trustee fee, etc.

On the establishment day, A subscribes for 50,000 *yuan* and B for 50,000 *yuan*.

```
NAV = Subscription fund of A + Subscription fund of B = 50,000 + 50,000 =
100,000 yuan
Total shares of the fund = Shares of A + Shares of B = 50,000 + 50,000 =
100,000 Shares
NAV per share = NAV/Total shares of the fund =100,000/100,000=1.0

Shares of investors:
Shares of A = 50,000
Asset value of A = Shares of A * NAV per share = 50,000 * 1.0 = 50,000
yuan

Shares of B = 50,000
Asset value of B = Shares of B * NAV per share = 50,000 * 1.0 = 50,000
yuan
```

On the second day, the fund has a return of 500 *yuan*.

```
Fund return = 500 yuan.

NAV = NAV of last day + Fund return = 100,000 + 500 = 100,500 yuan
Rate of return = Fund return / NAV of last day = 500/100,000 * 100% = 0.5%
Accumulated rate of return = (1+Rate of return) * NAV per share of last
day=(1+0.5%)*1 = 1.005
Total shares of the fund = Shares of A + Shares of B = 50,000 + 50,000 =
100,000 shares
NAV per share = NAV / Total shares of the fund = 100,500/100,000= 1.005

Shares of investors:
Shares of A = 50,000
```

Asset value of A = Shares of A * NAV per share = 50,000 * 1.005 = 50,250 *yuan*

Shares of B = 50,000
Asset value of B = Shares of B * NAV per share = 50,000 * 1.005 = 50,250 *yuan*

On the third day, the fund has a return of 1,000 *yuan* and an investor C subscribes for 50,000 *yuan*.

Fund return = 1000 *yuan*.
Subscription fund of C = 50,000 *yuan*.

NAV = NAV of 2nd day + Subscription fund of C + Fund return= 100,500 + 500,00 + 1000 = 151,500 *yuan*
Rate of return = Fund return / NAV of 2nd day = 1000 / 100,500 = 1%
Accumulated rate of return = (1 + Rate of return) * NAV per share of 2nd day = (1 + 1%) * 1.005 = 1.01505

Shares of C = Subscription fund of C / Accumulated rate of return = 50,000 / 1.01505 = 49,258.66
Total shares of the fund = Shares of A + Shares of B + Shares of C = 50,000 + 50,000 + 49258.66 = 149,258.7
NAV per share = NAV / Total shares of the fund = 151,500/149,258.7= 1.015016

Shares of investors:
Shares of A=50,000
Asset value of A = Shares of A * NAV per share = 50,000 * 1.015016 = 50,750.8 *yuan*

Shares of B=50,000
Asset value of B = Shares of B * NAV per share = 50,000 * 1.015016 = 50,750.8 *yuan*

Shares of C = 49,258.66
Asset value of C = Shares of C * NAV per share = 49258.66 * 1.015016 = 49,998.33 *yuan*

On the fund establishment day, the NAV per share is 1.0 and all the shares of investors are equal to the subscription fund. After an investment was made in the fund, the daily NAV per share changes. At this time, if you invest in it, the shares need to be calculated by the daily NAV. The fund will be settled in every trading day, of which the daily NAV, the NAV per share and the owner's equity will be calculated. Some instruments can be traded in 24 hours, without any closing time. For example, bitcoin and exchange rates can be traded every day; we can take the price of a fixed time as the closing price of the day to calculate the indicators.

Some notes for the calculation of investor's shares:

- Only when there is a subscription or a redemption do the shares of investors need to be re-calculated.
- Assume that there is no limit of the fund issue volume, the investors can subscribe for whatever volume they want.

■ Do not standardize the investor shares. Just keep records of them.
■ If the subscription fund arrives at the account before the closing of the day, the shares will be calculated by the NAV of the day. If the subscription fund arrives after the closing of the day, the shares will be calculated by the NAV of the next day.

6.4.3.2 R Implementation

We can implement the above calculation process with R language. It is not very complicated. First, we define functions for all the calculation methods. Then in the following calculations, we can just call the functions. If there is any change in the calculation method, we can directly modify it in the codes of the functions.

```
# NAV= Subscription funds of investors + Fund return
> calc_net<-function(money,revenue){
+     sum(money,revenue)
+ }

# NAV per share= NAV / Total shares of the fund
> calc_netUnit<-function(net,share){
+     net/share
+ }

# Shares of investor = Subscription fund of investor / Accumulated rate
of return
> calc_share_per<-function(money,retSum){
+     money/ifelse(retSum==0,1,retSum)
+ }
# Fund shares = Shares of investors
> calc_share<-function(share_list){
+     sum(unlist(share_list))
+ }

# Accumulated rate of return = (1+Rate of return) * NAV per share of T-1
day
> calc_retSum <- function(netUnit_last1,ret){
+     (1+ret)*netUnit_last1
+ }

# Rate of return = Fund return / NAV of T-1 day
> calc_ret <- function(net_last1,revenue){
+     revenue/net_last1
+ }

# Asset value of investor = Shares of investor * NAV per share
> calc_asset_per<-function(netUnit,share_per){
+     share_per*netUnit
+ }
```

Next, according to the assumptions of the fund, calculate all its indicators from the establishment day to the fourth day.

On the establishment day, A subscribes for 50,000 *yuan* and B for 50,000 *yuan*.

```
> Shares<-0                          # Total shares of the fund
> share_list<-list()                 # List of holers' shares
> netUnit<-1                         # Initialize it. The NAV per share
                                       is 1
> revenue<-ret<-retSum<-0            # Initialize it. The fund return,
                                       the rate of return and the
                                       accumulated rate of return are all 0
> money_A<-money_B<-50000            # A subscribes for 50000 yuan and
                                       gets 50000 shares. B subscribes for
                                       50000 yuan and gets 50000 shares.

> share_list[['A']]<-calc_share_per  # Add the subscription shares of A
  (money_A,retSum)
> share_list[['B']]<-calc_share_per  # Add the subscription shares of B
  (money_B,retSum)
> Shares<-calc_share(share_list)     # Calculate the total shares of the
                                       fund
> net<-calc_net(c(money_A,money_B),  # Calculate the NAV
  revenue)

# View rate of return, accumulated rate of return, NAV per share and NAV
> ret;retSum;netUnit;net
[1] 0
[1] 0
[1] 1
[1] 100000
```

On the second day, the fund has a return of 500 *yuan*.

```
> revenue<-500                       # Return of investment
> ret<-calc_ret(net,revenue)         # Calculate the rate of return
> retSum<-calc_retSum(netUnit,ret)   # Calculate the accumulated rate of
                                       return
> net<-calc_net(net,revenue)         # Calculate the NAV
> netUnit<-calc_netUnit(net,Shares)  # Calculate the NAV per share

# View rate of return, accumulated rate of return, NAV per share and NAV
> ret;retSum;netUnit;net
[1] 0.005
[1] 1.005
[1] 1.005
[1] 100500
```

On the third day, the fund has a return of 1,000 *yuan* and an investor C subscribes for 50,000 *yuan*.

```
> revenue<-1000                      # Return of investment
> money_C<-50000                     # C subscribes for 50000 yuan
> ret<-calc_ret(net,revenue)         # Calculate the rate of return
> retSum<-calc_retSum(netUnit,ret)   # Calculate the accumulated
                                       rate of return
> share_list[['C']]<-calc_share_per  # Add the subscription shares
  (money_C,retSum)                     of C
```

```
> Shares<-calc_share(share_list)              # Calculate the total shares
                                              of the fund
> net<-calc_net(c(net,money_C),revenue)       # Calculate the NAV
> netUnit<-calc_netUnit(net,Shares)           # Calculate the NAV per share

# View rate of return, accumulated rate of return, NAV per share and NAV
> ret;retSum;netUnit;net
[1] 0.009950249
[1] 1.015
[1] 1.015
[1] 151500

# Calculate the asset values of investors
> lapply(share_list,function(x) x* netUnit)
$A
[1] 50750
$B
[1] 50750
$C
[1] 50000
```

On the fourth day, the fund has a loss of 5,000 *yuan*.

```
> revenue<- -5000                             # Return of investment
> ret<-calc_ret(net,revenue)                  # Calculate the rate of
                                              return
> retSum<-calc_retSum(netUnit,ret)            # Calculate the
                                              accumulated rate of return
> net<-calc_net(net,revenue)                  # Calculate the NAV
> netUnit<-calc_netUnit(net,Shares)           # Calculate the NAV per
                                              share

# View rate of return, accumulated rate of return, NAV per share and NAV
> ret;retSum;netUnit;net
[1] -0.0330033
[1] 0.9815017
[1] 0.9815017
[1] 146500

# Calculate the asset values of investors
> lapply(share_list,function(x) x* netUnit)
$A
[1] 49075.08
$B
[1] 49075.08
$C
[1] 48349.83
```

Finally, we sum up a report on the everyday situations of the fund, as shown in Chart 6.4.

Every day, we need this report to have a knowledge of our fund's operation situation, so as to make corresponding management decisions.

Chart 6.4   NAV Records

Date	Daily Rate of Return	Accumulated Rate of Return	NAV per Share	Net Asset Value (NAV)
Establishment day	0	0	1.0	100,000
Second day	0.005	1.005	1.005	100,500
Third day	0.009950249	1.015	1.015	151,500
Fourth day	−0.0330033	0.9815017	0.9815017	146,500

6.4.4 A Calculation Case

In the following, we will walk through the whole process from the establishment to the liquidation of a privately offered fund with a real case. The fund in this case was spontaneously founded by several of my friends and me when I started learning quantitative investment. Though we did not invest much in it and did not earn much from it as well, the system did lay a foundation of trust among us. We needed to count our money well when doing financial quantification. Even brothers keep careful accounts, so I developed this system before starting any transaction. Counting with the system would avoid wrangles, thus everyone could focus on making transactions.

This was my first fund. It was run for about half a year. The process included the whole operations: fund establishment, subscriptions of investors, redemptions of investors, liquidation, etc. The investors were the four of us. We defined some management rules. Some of us made decisions, some made transactions and some kept accounts.

Define the investors and the events of subscription and redemption.

- 2/10/2014 A subscribed for 42,520.30 *yuan*
- 2/11/2014 B subscribed for 50,000 *yuan*
- 2/12/2014 NA
- 2/13/2014 NA
- 2/14/2014 C subscribed for 40,000 *yuan*, D subscribed for 30,000 *yuan*
- 6/30/2014 NA
- 7/1/2014 D subscribed for 50,000 *yuan*
- 7/2/2014 B planned to redeem all
- 7/3/2014 Cash met the redemption conditions
- 7/4/2014 B redeemed 50,174.39 *yuan*
- 7/7/2014 B redeemed 100,000.00 *yuan*
- 7/8/2014 A subscribed for 50,000 *yuan*, D redeemed 30,000 *yuan*
- 7/9/2014 Planned to redeem all
- 7/11/2014 Cleared
- 7/12/2014 A redeemed 93,817.31 *yuan,* B redeemed 100,088.50 *yuan,* C redeemed 41,076.38 *yuan,* D redeemed 51,469.81 *yuan*

Calculate the indicators of NAV, rate of return, NAV per share, etc., as shown in Chart 6.5.

Chart 6.5   NAV Report

Date	NAV of T Day	NAV of T-1 Day	NAV Change	Return of T Day	Rate of Return of T Day (%)	Accumulated Rate of Return	NAV per Share
2/10/2014	425,20.30&	–	42,520.30	–	–	–	1
2/11/2014	93,020.15	42,520.30	50,499.85	499.85	1.18	1.0118	1.0118
2/12/2014	93,092.15	93,020.15	72.00	72.00	0.08	1.0125	1.0125
2/13/2014	93,055.78	93,092.15	−36.37	−36.37	−0.04	1.0121	1.0121
2/14/2014	162,088.65	93,055.78	69032.87	−967.13	−1.04	1.0016	1.0016
6/30/2014	163,853.98	162,088.65	1765.33	1765.33	1.09	1.0125	1.0125
7/1/2014	214,187.65	163,853.98	50333.67	333.67	0.20	1.0146	1.0146
7/2/2014	214,250.36	214,187.65	62.71	62.71	0.03	1.0149	1.0149
7/3/2014	214,333.25	214,250.36	82.89	82.89	0.04	1.0153	1.0153
7/4/2014	165,160.86	214,333.25	−49,172.39	1002.00	0.47	1.0200	1.0215
7/7/2014	266,161.10	165,160.86	101,000.24	1000.24	0.61	1.0277	1.0277
7/8/2014	285,960.99	266,161.10	19,799.89	−200.11	−0.08	1.0269	1.0268
7/9/2014	286,500.01	285,960.99	539.02	539.02	0.19	1.0287	1.0287
7/11/2014	286,452.00	286,500.01	−48.01	−48.01	−0.02	1.0286	1.0286
7/12/2014	–	286,452	−286,452	–	–	–	–

Calculate the indicators of total shares and shares of every investor, as shown in Chart 6.6.

6.4.5 Accounting System Architecture

In the end, let's land the accounting system. I will introduce what we should do to develop the system from the perspective of software architecture. I cannot take your hand and show you exactly what to do, but those who already have experience in software development and financial transaction should not find it difficult to develop a similar system with reference to this article.

With the description above, we should have known what the accounting system is for. Next, let's make a service module plan. In my opinion, the service module planning is a process of looking for "entities," or "nouns." What does it mean? When we described the services in the above, we have used quantities of nouns like "investor," "trader," "trading record," "NAV," etc. Extract these nouns, match them with verbs and constitute the subject-predicate structure, which is the process of defining the abstract service modules.

Let's define these abstract service modules. See Figure 6.23.

We can design the table structure of the underlying database based on the definitions of the entities' relation. The fields of the table structure are actually the attribute extension of the entity nouns, such as "the name of investor," "the time of trading record," etc.

Chart 6.6   Fund Shares Report

Date	Investor A	Investor B	Investor C	Investor D	Shares Sum
2/10/2014	42,520.30	–	–	–	42,520.30
2/11/2014	42,520.30	49,419.05	–	–	91,939.35
2/12/2014	42,520.30	49,419.05	–	–	91,939.35
2/13/2014	42,520.30	49,419.05	–	–	91,939.35
2/14/2014	42,520.30	49,419.05	39,935.15	29,951.36	161,825.86
6/30/2014	42,520.30	49,419.05	39,935.15	29,951.36	161,825.86
7/1/2014	42,520.30	49,419.05	39,935.15	79,232.13	211,106.63
7/2/2014	42,520.30	49,419.05	39,935.15	79,232.13	211,106.63
7/3/2014	42,520.30	49,419.05	39,935.15	79,232.13	211,106.63
7/4/2014	42,520.30	–	39,935.15	79,232.13	161,687.58
7/7/2014	42,520.30	97,307.72	39,935.15	79,232.13	258,995.30
7/8/2014	91,210.77	97,307.72	39,935.15	50,039.81	278,493.45
7/9/2014	91,210.77	97,307.72	39,935.15	50,039.81	278,493.45
7/11/2014	91,210.77	97,307.72	39,935.15	50,039.81	278,493.45
7/12/2014	–	–	–	–	–

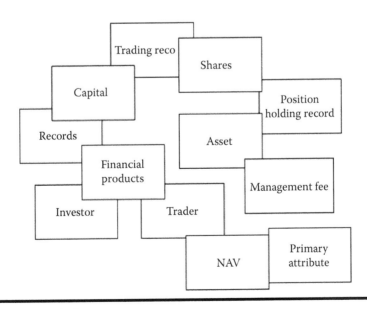

Figure 6.23   Entities' relation.

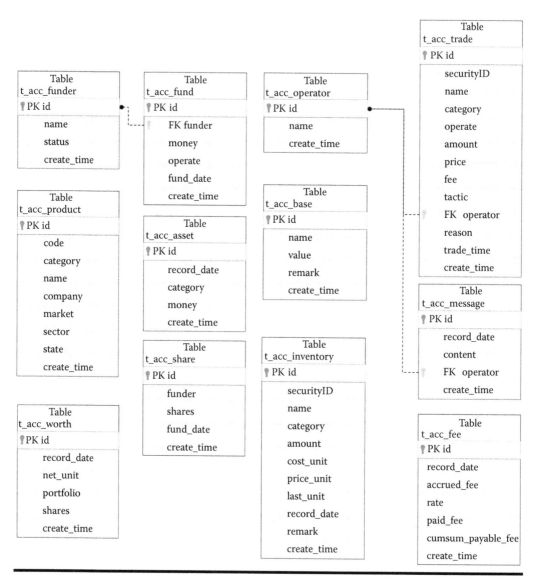

Figure 6.24   Table structure of database.

See Figure 6.24 for the table structure design of the accounting system database.

The definitions of the data dictionary corresponding to the database table structure are as show in Chart 6.7.

When we finish the database design, we have finished a half of the software architecture. Next step will be technology selection and technology implementation. This system is just a small internal system which I develop by myself and it does not require much of IO, CPU, memory or hard disk, so I will keep everything simple: simple codes, simple architecture and simple middleware. The front end and back end of Web part are all based on Javascript. The back end is Nodejs and the front end is Angular.js. The database uses open source MySQL, calculates the algorithm with R and sets daily timed tasks with crontab of Linux to initiate the R execution of algorithm. In fact, it is simple. The system structuring is as shown in Figure 6.25.

Chart 6.7 Data Dictionary

Entity	Table Name	Field Interpretation
Investor table	t_acc_funder	id: the self-incrementing key
		name: the name of the investor
		status: the status of the investor (active: it's been activated, freeze: it's been frozen and will not partake in shares calculation)
		create_time: the time of inserting the database
Operator table	t_acc_operator	id: the self-incrementing key
		name: the name of the operator
		create_time: the time of inserting the database
Basic attribute table	t_acc_base	id: self-incrementing key
		name: the name of the attribute
		value: the value of the attribute
		remark: the remark of the attribute
		create_time: the time of inserting the database
Product code table	t_acc_product	id: the self-incrementing key
		code: the code of the product
		category: the category of the product
		name: the name of the product
		company: the name of the issuer
		market: the trading market
		sector: the trading sector
		state: the trading state
		create_time: the time of inserting the database
Trading record table	t_acc_trade	id: the self-incrementing key
		securityID: the ID of the product
		name: the name of the product

(Continued)

Chart 6.7 (*Continued*) Data Dictionary

Entity	Table Name	Field Interpretation
		category: the category of the product
		operate: the operation of buy and sell (B represents buy, S sell and T convert to stock)
		amount: the amount of transactions (securities are counted by its number and bitcoin is counted from the fourth digit after the decimal point)
		price: the price of the transaction
		fee: the commission fee (including stamp duty, transfer fee, clearing fee, etc.)
		tactic: the strategy
		operator: the ID of the operator, related to t_acc_operator.id
		reason: the reason of transaction
		trade_time: the trading time
		create_time: the time of inserting the database
Position holding record table	t_acc_inventory	id: the self-incrementing key
		securityID: the ID of the product
		name: the name of the product
		category: the category of the product
		amount: the amount of positions holding (securities are counted by its number and bitcoin is counted from the fourth digit after the decimal point)
		cost_unit: the unit cost
		price_unit: the unit average price (the average value of ChinaBond)
		last_unit: the unit closing price
		record_date: the date of the trading record
		create_time: the time of inserting the database
		remark: the remark of the attribute

(*Continued*)

Chart 6.7 (*Continued*)   Data Dictionary

Entity	Table Name	Field Interpretation
Asset table	t_acc_asset	id: the self-incrementing key
		record_date: the date of record
		category: the category of the asset
		money: the amount of money
Capital flow table	t_acc_fund	id: the self-incrementing key
		funder: the ID of investor, t_acc_funder.id
		money: the amount of money for operation
		operate: the operation of recharge and withdrawal (D, W)
		fund_date: the date of operation
		create_time: the time of inserting the database
Share distribution table	t_acc_share	id: the self-incrementing key
		funder: the ID of investor, t_acc_funder.id
		share: the share of capital
		record_date: the date of record
		create_time: the time of inserting the database
NAV table	t_acc_worth	id: the self-incrementing key
		record_date: the date of record
		net_unit: the NAV per share
		portfolio: the NAV
		share: the share of capital
		create_time: the time of inserting the database
Management fee table	t_acc_fee	id: the self-incrementing key
		record_date: the date of record
		accrued_fee: the accured expenses

(*Continued*)

Chart 6.7 (*Continued*) Data Dictionary

Entity	Table Name	Field Interpretation
		rate: the rate
		paid_fee: the paid management fee
		cumsum_payable_fee: the cumulative accrued management fee
		create_time: the time of inserting the database
Information record table	t_acc_message	id: the self-incrementing key
		record_date: the date of record
		content: the content of the information
		operator: the recorder
		create_time: the time of inserting the database

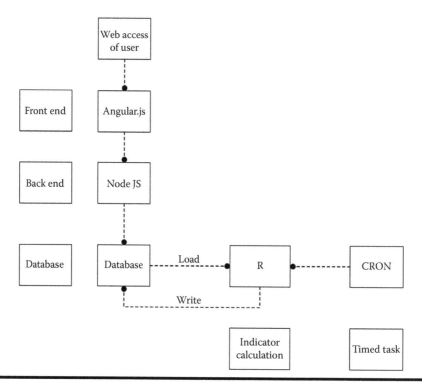

Figure 6.25 System architecture.

Finish the coding of R, SQL, Javascript and Shell and then deploy and launch the codes. In the end, let's have a look at several screenshots of the accounting system: the fund NAV page, the daily profit and loss page and the position holding record management page. Overall, the system is simple in style, but practical in use. For it is an internal system, I will not give its link (Figures 6.26–6.28).

In this section, the core services and the calculation methods of fund accounting system have been introduced in detail. With the system, we do not need to use Excel for manual calculation

Accounting System

NAV list	Asset list	Holding Record	Trading Record	Capital List	Share List	Management Fee	Investors	Trader	Product Data	Basic Data

净值表

Starting Time	Ending Time	Query

Transaction time	NAV per share	Portfolio NAV	Total Shares	Recording Time
2014-06-11	1.01764	679376	667602	2014-06-11 20:36:15
2014-06-10	1.02233	682510	667602	2014-06-10 22:12:20
2014-06-09	1.0228	682823	667602	2014-06-09 23:28:10
2014-06-08	1.01965	680720	667602	2014-06-09 22:50:26
2014-06-07	1.0197	680757	667602	2014-06-09 22:50:24
2014-06-06	1.01976	680794	667602	2014-06-06 21:52:58
2014-06-05	1.01737	679198	667602	2014-06-05 20:10:29
2014-06-04	1.01731	679157	667603	2014-06-04 21:20:01
2014-06-03	1.01409	577326	569304	2014-06-03 22:23:20
2014-06-02	1.01232	526405	519999	2014-06-03 21:45:17

Figure 6.26   Fund NAV page.

Daily Profit and Loss

2014-06-06   Query

Product Name	T Day Holding Positions	T Day Closing Price	T-1 Day Holding Positions	T-1 Day Closing Price	Change of Holding Positions	Hold_P&L	Trade_P&L	Total_P&L
期货现金(CASH_FUTURE)	211835	1	211835	1	0	0.00	0.00	0.00
证券账户现金(CASH_SEC)	4	1	2	1	2	0.00	0.00	0.00
R-001(131810)	0	0	50000	1	-50000	-50000.00	50000.00	0.00
应收利息——逆回购(interest_rec)	0	0	7	1	-7	-7.00	0.00	-7.00
双福A(150012)	11000	0.972	11000	0.973	0	-11.00	0.00	-11.00
12南糖债(112109)	120	96.019	120	95.45	0	68.28	-90.70	-22.42
12东锆债(112101)	130	90.06	130	88.93	0	146.90	-858.90	-712.00
11中孚债(122093)	130	88.49	130	87.62	0	113.10	-834.20	-721.10
12三维债(112163)	130	85.9	130	85.099	0	104.13	-852.57	-748.44
12赣昌01(112160)	130	87.85	130	87.3	0	71.50	-869.31	-797.81
13南洋债(112179)	130	86.108	130	85.596	0	66.56	-866.52	-799.96
12正邦债(112155)	130	88.779	130	88.262	0	64.61	-879.80	-815.19
12景兴债(112121)	130	90.65	130	90.39	0	33.80	-866.55	-832.75
12德豪债(112165)	120	94.58	120	93.867	0	85.56	-936.79	-851.23
11三钢02(112073)	130	89.368	130	89.112	0	33.28	-888.37	-855.09
12墨龙01(112178)	120	92.73	120	92.58	0	24.00	-885.77	-861.77
12西钢债(122077)	120	94.17	120	93.9	0	32.40	-897.00	-864.60
11云煤债(122073)	130	90.8	130	90.45	0	45.50	-912.90	-867.40
11安钢01(122107)	120	93.23	120	93	0	27.60	-896.20	-868.60
11万基债(122829)	120	95.94	120	95.7	0	28.80	-901.80	-873.00
11众和债(122110)	110	96.09	110	95.63	0	50.60	-923.80	-873.20
12亿利02(112159)	120	96.22	120	95.95	0	32.40	-907.80	-875.40

Figure 6.27   Daily profit and loss page.

Accounting System

| NAV list | Asset list | Holding Record | Trading Record | Capital List | Share List | Management Fee | Investors | Trader | Product Data | Basic Data |

Position Table Operation

Date of record	2014-06-12		
Product Code	122829		
Product Name	11万基德		
Product Classification	企业债券		▼
Position Volume	100	Cost Price	1
Average Price	10	Closing Price	10
Cost Amount	100	Market Value Amount	1000
Note	高受益债辅略交易		

Submit

Position List

Holding Date	Product Name	Product Classification	Position Volume	Cost Price	Average Price	Closing Price	Operation
2014-06-11	深市1日正回购应付利息(int_p_r001)	应付利息	3	1	1	1	删除
2014-06-11	新质押式回购(131810)	新质押式正回购	30000	1	1	1	删除
2014-06-11	证券账户现金(CASH_SEC)	现金	15	1	1	1	删除
2014-06-11	期货现金(CASH_FUTURE)	现金	211935	1	1	1	删除
2014-06-11	天天发(tiantianfa)	开放式基金	92638	1	1	1	删除
2014-06-11	双禧A(150012)	开放式基金	11000	0.974	0.974	0.976	删除
2014-06-11	11万基德(122829)	企业债券	150	94.231	94.231	95.15	删除
2014-06-11	11紫察伦(122811)	企业债券	130	95.323	95.323	95.98	删除
2014-06-11	13永泰债(122267)	企业债券	130	97.799	97.799	100.78	删除
2014-06-11	12中孚债(122162)	企业债券	130	100.122	100.122	100.65	删除

Figure 6.28 Position holding record management page.

every day, which will tremendously improve the work efficiency. Using our IT technology to reform the business of another industry is the very point of IT entering finance.

Finally, I want to restate that all the design patterns of accounting system here are the conclusion of my own experience. If there is any financial concept that does not conform to the textbooks of finance, the textbooks should prevail.

It is time to found another privately offered fund and no wrong counting will occur again.

6.5 Data Interpretation of Machine Gene Investment

Question
How do you analyze the application of Machine Gene Investment?

Introduction

In 2016, as the world economic crisis came, the Internet start-ups in China encountered a long winter. Many companies stopped running for lack of capital. On the contrary, AI technology thrived and got funds. Intelligent investment advising, as the hot spot in AI finance, has continued heating up.

In the end of 2016, China Merchants Bank (CMB) launched an application named "Machine Gene Investment," which drew many people's attention and opened a door for the banks to enter intelligent investment advising. In this section, we will interpret how Machine Gene Investment works with data.[*]

6.5.1 An Introduction to Machine Gene Investment

Machine Gene Investment is an application for mobile terminal released by CMB on December 6, 2016. It is embedded in the APP (Application) of CMB and it brings in the concept of FinTech and combines finance and artificial intelligence (AI).

According to the release introduction of CMB, Machine Gene Investment adopts the machine learning algorithm and builds the "intelligent fund portfolio allocation service" based on the publicly offered funds and the service will allocate assets globally. After the selections of investment horizon and risk-profit by a client, Machine Gene Investment will build a fund portfolio according to the client's requirement of target return. The client can choose whether to one-button purchase the service and continue to use it.

Machine Gene Investment is not a single product, but a set of asset allocation services including target and risk confirmation, portfolio building, one-button purchase, risk warning, position change prompt, one-button optimization, after service report, etc., which involve the whole fund investment service process of pre-sales, in-sales and after-sales. For example, Machine Gene Investment will realtime scan the global market and calculate the optimal portfolio according to the latest market situation. If the portfolio the client holds deviates from the optimal portfolio, it will give a dynamic advice of fund portfolio adjustment. If the client agrees on the advice, it will automatically make a one-button optimization (Figure 6.29).

After the official introduction from CMB, we are going to see how intelligent Machine Gene Investment is by data analysis.

6.5.2 Data Collection

In order to do data analysis, we need to think in the way of data. My way is to collect data first, then organize the input and output data of the application and find the relation of the data by statistics and financial knowledge.

Note: There is not any field definition in the application, so I interpret the fields by their literal meanings.

There are only two fields for the input data:

■ Rough investment horizon: The time period from the start of investment to the end of it.
■ Risk tolerance: How much risk and consequent loss you can bear.

---

[*] The analysis in this article was made in December 2016, when Machine Gene Investment was first released. Some changes have occurred to this application when the book comes out.

Figure 6.29   Boot screen of Machine Gene Investment.

There are many fields for the output data, including:

- Simulated historical annualized return (%): The annualized return rate by historical data backtesting.
- Simulated historical annualized volatility (%): The annualized volatility by historical data backtesting.
- Simulated historical return (*yuan*): The yield amount of a 1-year investment of 10,000 *yuan*.
- Loss of 95% possibility (*yuan*): The maximum loss amount in the possibility of 95%.
- Fixed income (%): The allocation ratio of the fixed-income funds.
- Cash&currency (%): The allocation ratio of the cash&currency funds.

- Stock (%): The allocation ratio of the stock funds.
- Special&others (%): The allocation ratio of special or other funds.
- Portfolio return curve: The return rate curve generated by the portfolio of a certain proportion.
- Portfolio allocation detail: The specific fund varieties corresponding to the four types of assets and their allocation ratios.

I have already marked the extraction places of input and output data on the interface of Machine Gene Investment.

The data collected from the interface has been organized in CSV format for the convenience of later analysis. The data has been saved to three CSV files.

- a.csv: For collecting the data of the 1st and 2nd interfaces and the data input and output directly by the clients, including rough investment horizon (term), risk tolerance (risk), simulated historical annualized return (ret), simulated historical annualized volatility (vol), simulated historical return (gains), loss (loss), fixed income (fixed), cash&currency (cash), stock (stock) and special&others (alter).
- B.csv: for collecting the data of the markets corresponding to all the target funds, including fund name (name), fund code (code), fund establishment time (create), fund type (type), NAV 20140101 (first2014), NAV 20150101 (first2015), NAV 20160101 (first2016) and NAV 20161208 (last).
- c.csv: For collecting the data of the 3rd interface and the allocation proportion of the target funds for every portfolio, including rough investment horizon (term), risk tolerance (risk), fund type (type), fund code (code) and allocation proportion (weight).

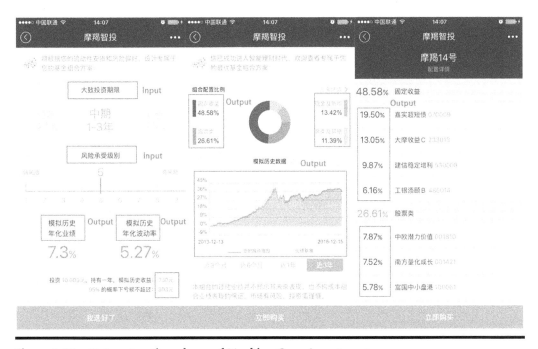

Figure 6.30   Data extraction places of Machine Gene Investment.

The first 10 data samples of a.csv are as follows:

```
> head(dfa,10)
   term risk   ret   vol fixed  cash stock alter gains loss
1     1    1  5.01  2.57 65.00 20.00  5.00 10.00   501    3
2     1    2  5.30  2.72 55.00 19.60 10.40 15.00   530    3
3     1    3  5.88  3.05 48.80 18.50 15.60 17.10   588   10
4     1    4  6.64  3.88 51.70 14.90 23.30 10.10   664   96
5     1    5  7.13  4.90 49.40 14.20 25.00 11.40   713  247
6     1    6  7.64  6.06 46.00 12.00 30.00 11.40   764  424
7     1    7  8.18  7.36 42.90 10.20 35.40 11.50   818  625
8     1    8  8.75  8.84 38.53  8.37 46.63  6.47   875  858
9     1    9  9.41 10.55 33.44  6.27 53.84  6.45   941 1127
10    1   10 10.22 12.73 31.07  0.00 62.83  6.10  1022 1473
```

The first 10 data samples of b.csv are as follows:

```
> head(dfb,10)
               name      code   create  type first2014 first2015 first2016
               last
1    ICBC Credit Suissie pure Bond Fund B OF000403 20140516 fixed
      0.0000     1.0850     1.2200 1.2520
2    E Fund Return Enhancement Fund B OF110018 20080319 fixed      1.4403
      1.8124     2.1180 2.1693
3  CCB Principal Stable Income Fund C OF530008 20080625 fixed      1.4529
    1.8030     2.0304 2.0608
4      ICBC Credit Suissie Tianyi Fund B OF485014 20110810 fixed
         1.2090     1.7300     1.9740 1.9680
5   China Southern Quantitative Growth Fund OF001421 20150629 stock
      0.0000     0.0000     1.2300 1.3400
6    ZOFund Potential Value Fund OF001810 20150930 stock      0.0000
      0.0000     1.1010 1.2170
7      Dacheng S&P 500 Index Fund OF096001 20110323 stock      1.2790
         1.4314     1.4651 1.7457
8    HuaAn AU9999 Feeder Fund C OF000217 20130822 alter      0.8800
      0.8940     0.8340 0.9965
9    China Southern Asian USD Yield Fund C OF002401 20160303 alter
      0.0000     0.0000     0.0000 1.0867
10     Morgan Stanley Huaxin Multiple Income Bond Securities Investment
        Fund C OF233013 20120828 fixed     1.0440     1.2970     1.5610 1.6350
```

The first 20 data samples of c.csv are as follows:

```
> head(dfc,20)
   term risk  type      code weight
1     1    1 fixed OF000403 0.2000
2     1    1 fixed OF110018 0.1400
3     1    1 fixed OF530008 0.1100
4     1    1 fixed OF233013 0.1000
5     1    1 fixed OF050111 0.0500
6     1    1 fixed OF485014 0.0500
7     1    1  cash OF217004 0.2000
8     1    1 alter OF002401 0.0500
9     1    1 alter OF000217 0.0500
```

```
10    1    1 stock OF001810 0.0500
11    1    2 fixed OF000403 0.2000
12    1    2 fixed OF110018 0.1289
13    1    2 fixed OF233013 0.1000
14    1    2 fixed OF485014 0.0616
15    1    2 fixed OF530008 0.0586
16    1    2  cash OF217004 0.1965
17    1    2 alter OF002401 0.0500
18    1    2 alter OF000217 0.0500
19    1    2 alter OF340001 0.0500
20    1    2 stock OF001810 0.0544
```

In particular, for the data from the application may change dynamically, the data here is collected from the Machine Gene Investment application of **December 8, 2016**.

6.5.3 Analysis of Data Modeling

Since the data has been collected, now we can start data analysis. Of course, we can analyze the data from various perspectives, such as finance, statistics, data mining, etc. or we can do the analysis by calculating some simple indicators like the maximum value, the minimum value and the average, etc. I will focus on finance and statistics. If there is any one-sidedness, please help me correct it.

In the following, I will analyze Machine Gene Investment in these six aspects.

6.5.3.1 Analysis One: Only Two Input Items

There are only two input items: rough investment horizon and risk tolerance. There are three options for rough investment horizon and ten for risk tolerance, so the actual combination numbers are 3 * 10 = 30. Only 30 options cannot totally satisfy the individualization demands. When there are 31 persons purchasing the service, at least two of them will use the same combination.

6.5.3.2 Analysis Two: Only 17 Target Funds

We have tried to allocate the 30 combinations and found that there are only 17 funds with different allocation proportions in the detail position holding plans. A small number of target funds cannot spread the risks enough, which may further lead to big withdrawals if encountering some extreme situations. The 17 funds are:

```
> paste(dfb$name,"(",dfb$code,")",sep="")
 [1] " ICBC Credit Suissie pure Bond Fund B (OF000403)"      " E Fund
     Return Enhancement Fund B (OF110018)"   " CCB Principal Stable
     Income Fund C (OF530008)"
 [4] " ICBC Credit Suissie Tianyi Fund B (OF485014)"      " CSAM
     Quantitative Growth Fund (OF001421)"  " ZOFund Potential Value Fund
     (OF001810)"
 [7] " Dacheng S&P 500 Index Fund (OF096001)"       " HuaAn AU9999 Feeder
     Fund C (OF000217)"   " China Southern Asian USD Yield Fund C
     (OF002401)"
[10] " Morgan Stanley Huaxin Multiple Income Bond Securities Investment
     Fund C (OF233013)"     " Bosera Credit Market Fixed-income Securities
     Fund C(OF050111)"   " AIFMC Convertible Bond Fund (OF340001)"
```

```
[13] " Truvalue Quantitative Multi-factor Fund (OF002210)"    "China
     Merchants Cash Accretion Money Market Fund A (OF217004)" "FullGoal
     Small and Medium Cap Hybrid Fund (OF100061)"
[16] " ICBC Credit Suissie Global Selection Fund(OF486002)"    "China
     Southern Constituent Stock Selection Hybrid Fund (OF202005)"
```

6.5.3.3 Analysis Three: Correlation Analysis

Find the correlations of the input items and the output items from the dataset of a.csv.
   R implementation

```
# Load data
> dfa<-read.csv(file="a.csv")
> names(dfa)<-c("term","risk","ret","vol","fixed","cash","stock","alter",
"gains","loss")

# Draw the pairs plot
> pairs(df)
```

Visualizing the data will help us understand it.

- The scatter plots of the term column and other columns are discretely distributed, which means there is no correlation between the term column and other columns.
- risk column is linearly correlated with all other columns except the alter column.

Let's add more elements to Figure 6.30, such as correlation coefficient, fitted curve, scatter plot, etc. Redraw the plots, as shown in Figure 6.31.

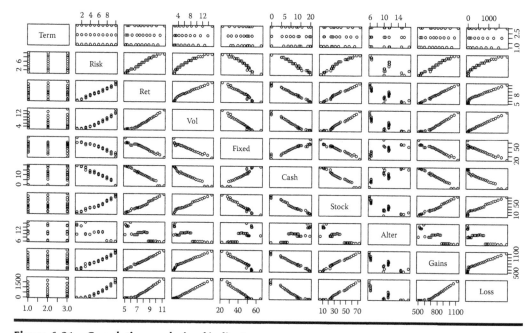

Figure 6.31   Correlation analysis of indicators.

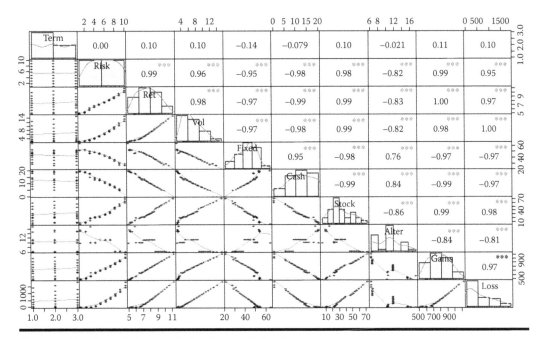

Figure 6.32 Correlation analysis after optimization.

This is much clearer.

■ risk column shows highly positive correlations with simulated historical annualized return (ret), simulated historical annualized volatility (vol) and simulated historical return (gains). Their output data is completely influenced by the value of risk.

■ risk column shows highly negative correlations with fixed income (fixed) and cash&currency (cash), a highly positive correlation with stock (stock) and a negative correlation with special&others (alter), which matches the risk return nature of the asset.

■ vol column is 100% linearly correlated with loss (loss).

■ ret column is 100% linearly correlated with simulated historical return (gains). Here is the formula: gains = 10,000 * ret.

6.5.3.4 Analysis Four: Linear Regression

By checking the correlations, we have found that risk is highly correlated with many other columns.

Then we can estimate the parameters of the columns related to risk by using the method of linear regression. If you do not have much knowledge of simple linear regression, please refer to Section 2.2, "R Interpretation of Simple Linear Regression Model," of this book.

For vol and loss are 100% linearly correlated, we take vol as x and loss as y to build a simple linear regression equation.

```
# The regression equation
> lv<-lm(loss~vol,data=dfa)
> summary(lv)
Call:
lm(formula = loss ~ vol, data = dfa)
```

```
Residuals:
Min      1Q  Median      3Q      Max
-36.119 -31.491  -6.621  27.884  67.305

Coefficients:
Estimate Std. Error t value Pr(>|t|)
(Intercept) -447.514      13.056  -34.28    <2e-16 ***
vol          149.109       1.707   87.34    <2e-16 ***
---
Signif. codes: 0 '***' 0.001 '**' 0.01 '*' 0.05 '.' 0.1 ' ' 1

Residual standard error: 34.2 on 28 degrees of freedom
Multiple R-squared: 0.9963,    Adjusted R-squared: 0.9962
F-statistic: 7629 on 1 and 28 DF,  p-value: < 2.2e-16
```

Test the statistics of linear regression: T test and F test are both extremely significant and meanwhile, the R-squared value is 0.9963, which means high correlation.

```
# Draw the scatter plot and the fitted curve
> plot(loss~vol,data=dfa)
> abline(lv)
```

In Figure 6.33, we can see that they fit well and the formula can be given as: loss = −447.514 + 149.109 * vol.

Besides, because risk determines vol, let's find the relation between risk and loss. Take risk as x and loss as y to build a simple linear regression equation.

```
# Build a simple linear regression equation
> lm(loss~risk,data=dfa)
Call:
lm(formula = loss ~ risk, data = dfa)

Coefficients:
(Intercept)       risk
    -435.8        180.0

# Detailed indicators
> summary(lr)
Call:
lm(formula = loss ~ risk, data = dfa)
```

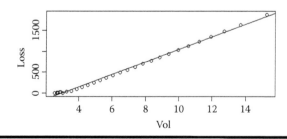

Figure 6.33　Regression fitting.

```
Residuals:
Min      1Q  Median      3Q     Max
-219.88 -136.93  -59.26  100.69  508.31

Coefficients:
Estimate Std. Error t value Pr(>|t|)
(Intercept)  -435.84      72.38  -6.021 1.73e-06 ***
risk          179.95      11.67  15.426 3.23e-15 ***
---
Signif. codes: 0 '***' 0.001 '**' 0.01 '*' 0.05 '.' 0.1 ' ' 1

Residual standard error: 183.5 on 28 degrees of freedom
Multiple R-squared: 0.8947,    Adjusted R-squared: 0.891
F-statistic: 238 on 1 and 28 DF,  p-value: 3.232e-15
```

The results of T test and F test are extremely significant; the value of R-squared is comparatively high.

Next, check the residuals. Point No. 30 deviates greatly, so it may be an outlier.

Let's remove point No. 30 and then run the significance test and the residual analysis again.

```
> dfa2<-dfa[-30,]
> lr2<-lm(loss~risk,data=dfa2)
> summary(lr2)
Call:
lm(formula = loss ~ risk, data = dfa2)

Residuals:
Min      1Q  Median      3Q     Max
-203.00 -100.98  -58.98   83.53  327.46

Coefficients:
Estimate Std. Error t value Pr(>|t|)
(Intercept)  -397.55      62.23  -6.389 7.64e-07 ***
risk          169.51      10.32  16.431 1.39e-15 ***
---
Signif. codes: 0 '***' 0.001 '**' 0.01 '*' 0.05 '.' 0.1 ' ' 1

Residual standard error: 155.3 on 27 degrees of freedom
Multiple R-squared: 0.9091,    Adjusted R-squared: 0.9057
F-statistic: 270 on 1 and 27 DF,  p-value: 1.391e-15
```

Without point No. 30, the value of R-squared is 0.9091, larger than the previous 0.8947.

We can see in Figure 6.34 that there is no obvious outlier, so it meets the optimization requirements to remove point No. 30 (Figure 6.35).

6.5.3.5 Analysis Five: Financial Thinking about Point No. 30

We find in the data that the maximum loss of point No. 30 has exceeded its return, that is, you bear great risk but profit not as much (Figure 6.36).

According to capital asset pricing model (CAPM), the return of our portfolio comes from two parts: risk-free return and risk return. The risk-free return can be obtained from the funds of cash&currency, while the risk return mainly from the funds of stock, bond and special&others.

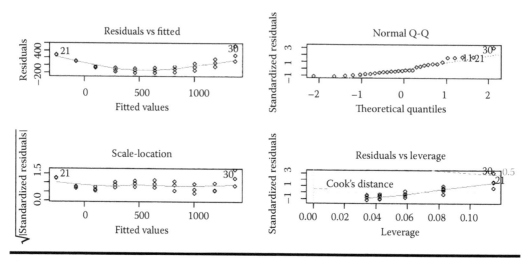

Figure 6.34   Residual analysis

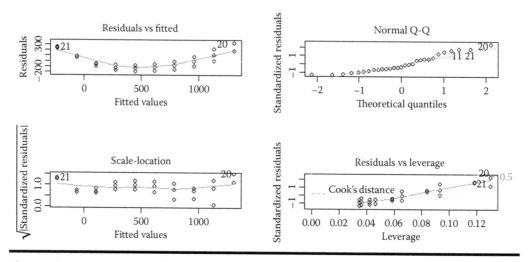

Figure 6.35   Optimized residual analysis.

For intuitive understanding, the ratio of risk and profit should be at least 1:1, i.e. if 100 *yuan* is lost, 100 *yuan* should be gained as risk compensation. For private placements, the investors may have higher requirements, for example, they may require risk:profit = 1:2. Please refer to Section 2.1 of this book for introduction to the CAPM.

The ratio of risk and profit discussed above does not involve anything about probability. I guess that 95% is a probability calculated by VaR value.

6.5.3.6 Analysis Six: Calculate the Return Rate by the Target Funds

We can get the allocation proportion of every fund in the application of Machine Gene Investment and the NAV data in an open market, so we can calculate the expected return by ourselves and see if it is in accordance with the result provided by Machine Gene Investment.

Figure 6.36   Interface screenshot of Point No. 30.

In the following, we are going to use the datasets of b.csv and c.csv introduced above.

```
# Load data
> dfb<-read.csv(file="b.csv",encoding="utf-8",fileEncoding = "utf-8")
> names(dfb)<-c("name","code","create","type","first2014","first2015","fi
  rst2016","last")

# Respectively calculate the return rates of 2014, 2015 and 2016
> dfb$ret2014<-(dfb$first2015-dfb$first2014)/dfb$first2014
> dfb$ret2015<-(dfb$first2016-dfb$first2015)/dfb$first2015
> dfb$ret2016<-(dfb$last-dfb$first2016)/dfb$first2016

# Assign 0 to invalid values
> dfb$ret2014[c(which(is.na(dfb$ret2014)),which(is.
  infinite(dfb$ret2014)))]<-0
> dfb$ret2015[c(which(is.na(dfb$ret2015)),which(is.
  infinite(dfb$ret2015)))]<-0
> dfb$ret2016[c(which(is.na(dfb$ret2016)),which(is.
  infinite(dfb$ret2016)))]<-0
```

```
# Print the first 6 items
> head(dfb)
          name       code    create   type first2014 first2015 first2016
last    ret2014      ret2015        ret2016
1       ICBC Credit Suissie pure Bond Fund B OF000403 20140516 fixed
        0.0000      1.0850    1.2200 1.2520 0.00000000   0.12442396
        0.026229508
2     E Fund Return Enhancement Fund B OF110018 20080319 fixed     1.4403
        1.8124     2.1180 2.1693 0.25834896   0.16861620   0.024220963
3   CCB Principal Stable Income Fund C OF530008 20080625 fixed     1.4529
      1.8030     2.0304 2.0608 0.24096634   0.12612313   0.014972419
4       ICBC Credit Suissie Tianyi Fund B OF485014 20110810 fixed
        1.2090     1.7300    1.9740 1.9680 0.43093466   0.14104046
        -0.003039514
5     CSAM Quantitative Growth Fund OF001421 20150629 stock    0.0000
        0.0000     1.2300 1.3400 0.00000000   0.00000000   0.089430894
6     ZOFund Potential Value Fund OF001810 20150930 stock    0.0000
        0.0000     1.1010 1.2170 0.00000000   0.00000000   0.105358765
```

China Merchants Cash Accretion Money Market Fund A (OF217004) is a cash fund, so we simply take the annualized return as its rate of return, rather than calculate it according to the method above. The average annualized return of China Merchants Cash Accretion Money Market Fund A in 2014 is 4.52%, in 2015 3.6% and in 2016 2.3%.

```
#For cash funds, find the annualized return in wind and assign it to the
 rate of return
dfb[which(dfb$code=='OF217004'),]$ret2014<-0.0452
dfb[which(dfb$code=='OF217004'),]$ret2015<-0.036
dfb[which(dfb$code=='OF217004'),]$ret2016<-0.0237
```

Then load the detailed allocation of funds from c.csv.

```
> dfc<-read.csv(file="c.csv")
> names(dfc)<-c("term","risk","type","code","weight")

# View data
> head(dfc)
  term risk  type     code weight
1    1     1 fixed OF000403   0.20
2    1     1 fixed OF110018   0.14
3    1     1 fixed OF530008   0.11
4    1     1 fixed OF233013   0.10
5    1     1 fixed OF050111   0.05
6    1     1 fixed OF485014   0.05
```

Transpose the data to a horizontal table by the type column, remove the code column and fill it with weight values. Then we have a new dataset r1.

```
> head(r1)
  term risk alter   cash  fixed  stock
1    1     1 0.1000 0.2000 0.6500 0.0500
2    1     2 0.1500 0.1965 0.5491 0.1044
3    1     3 0.1562 0.1842 0.4881 0.1715
```

```
4   1   4 0.1011 0.1490 0.5162 0.2337
5   1   5 0.1137 0.1416 0.4943 0.2504
6   1   6 0.1143 0.1208 0.4655 0.2994
```

Let's generate the allocation of plan1, with term = 1 and risk = 1.

```
# Only reserve the data with term=1 and risk=1
> plan1<-dfc[dfc$term==1 & dfc$risk==1,]

# Merge the datasets of plan1 and dfb
> plan1m<-merge(plan1[,c("term","risk","code","type","weight")],dfb[,c("c
  ode","ret2014","ret2015","ret2016")],by="code")

# Calculate the rates of returns by allocation proportions
> plan1m$ret2014w<-plan1m$weight*plan1m$ret2014
> plan1m$ret2015w<-plan1m$weight*plan1m$ret2015
> plan1m$ret2016w<-plan1m$weight*plan1m$ret2016

# The respective return rates that different funds in plan1 contribute in
  2014, 2015 and 2016
> plan1m
        code term risk  type weight     ret2014      ret2015       ret2016
ret2014w      ret2015w      ret2016w
1  OF000217    1    1 alter   0.05 0.01590909 -0.06711409  0.194844125
   0.0007954545 -0.003355705  0.0097422062
2  OF000403    1    1 fixed   0.20 0.00000000  0.12442396  0.026229508
   0.0000000000  0.024884793  0.0052459016
3  OF001810    1    1 stock   0.05 0.00000000  0.00000000  0.105358765
   0.0000000000  0.000000000  0.0052679382
4  OF002401    1    1 alter   0.05 0.00000000  0.00000000  0.000000000
   0.0000000000  0.000000000  0.0000000000
5  OF050111    1    1 fixed   0.05 0.87631433  0.12603844  0.034050727
   0.0438157167  0.006301922  0.0017025363
6  OF110018    1    1 fixed   0.14 0.25834896  0.16861620  0.024220963
0.0361688537  0.023606268   0.0033909348
7  OF217004    1    1  cash   0.20 0.04520000  0.03600000  0.023700000
   0.0090400000  0.007200000  0.0047400000
8  OF233013    1    1 fixed   0.10 0.24233716  0.20354665  0.047405509
   0.0242337165  0.020354665  0.0047405509
9  OF485014    1    1 fixed   0.05 0.43093466  0.14104046 -0.003039514
   0.0215467328  0.007052023 -0.0001519757
10 OF530008    1    1 fixed   0.11 0.24096634  0.12612313  0.014972419
   0.0265062977  0.013873544  0.0016469661
```

Merge the data and calculate the return rate of plan1 and the return contributions of different assets in plan1.

```
# The return rate of plan1
> plan1r<-ddply(plan1m,.(term,risk),summarise,ret2016=sum(ret2016w),ret20
  15=sum(ret2015w),ret2014=sum(ret2014w))
> plan1r
  term risk    ret2016     ret2015    ret2014
1    1    1 0.03632506 0.09991751 0.1621068
```

```
#Calculate the 3-year cumulative return rate curve
> plan1r$cumret<-sum(c(plan1r$ret2016,plan1r$ret2015,plan1r$ret2014))
> plan1r
  term risk    ret2016     ret2015    ret2014     cumret
1    1    1 0.03632506 0.09991751 0.1621068 0.2983493

# The return contributions of different assets in plan1
> plan1rm<-ddply(plan1m,.(term,risk,type),summarise,ret2016=sum(ret2016w)
,ret2015=sum(ret2015w),ret2014=sum(ret2014w))
> plan1rm
  term risk  type     ret2016       ret2015      ret2014
1    1    1 alter 0.009742206 -0.003355705 0.0007954545
2    1    1  cash 0.004740000  0.007200000 0.0090400000
3    1    1 fixed 0.016574914  0.096073214 0.1522713174
4    1    1 stock 0.005267938  0.000000000 0.0000000000
```

Compare my calculation result with the return rate curves of Machine Gene Investment in recent 1 year and in recent 3 years (Figure 6.37).

The 1-year curve corresponds to plan1 ret2016 = 0.03632506 = 3.63 cumret = 0.2983493 = 29.83%. I have found that there is difference between my calculation result and the curve, but the difference is not very big according to the numbers of the final result. However, for the 3-year

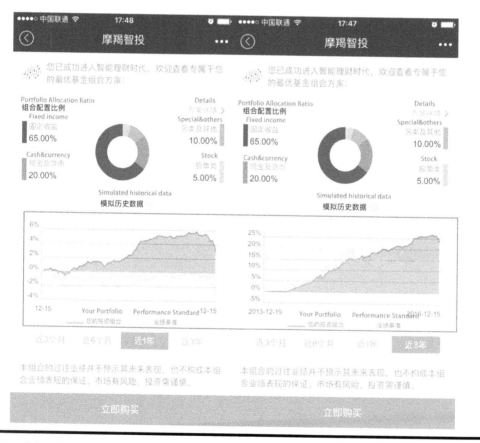

Figure 6.37   1-year and 3-year return rate curves comparison.

return rate curve, the difference is significant. The 3-year return rate curve provided by Machine Gene Investment rises evenly and stably, yet my calculation is that the rising rate of 2014 is 16%, that of 2015 is 9% and that of 2016 is 3%, which means the return rate declines year by year, so it cannot show an even and stable rising.

Let's find what causes the difference. By checking the portfolio data of plan1, we find that there is a ZOFund Potential Value Fund (OF001810), which was founded on September 30, 2015 and did not constitute the return rate of 2014 and 2015. Therefore, there are position changes for the portfolio in these 3 years. The return rate curve of Machine Gene Investment does not show the data of position changes, so it is not transparent and cannot be the basis for the user's purchase decision making.

Next, let's calculate the return rates of all 30 portfolios. Then compare them with the return rates provided by Machine Gene Investment.

In Figure 6.38, the ret column is the data collected from Machine Gene Investment interfaces; ret2016, ret2015 and ret2014 are our calculation result of percentages according to the fund data in the open market; mean is the arithmetic mean of ret2016, ret2015 and ret2014. In the data, there are some values of ret column approximate to those of mean column. Let's run the correlation analysis again (Figure 6.39).

Now it is easier to interpret the result. ret is linearly correlated with the return rate of 2016 and mean is linearly correlated with the return rate of 2014. I guess the bull markets of both stocks and bonds at the end of 2014 made the mean deviate. Therefore, there is no correlation between ret and mean. Machine Gene Investment is expected to put more weight in the returns of 1-year portfolios.

	term	risk	ret	ret2016	ret2015	ret2014	mean
1	1	1	5.01	3.632506	9.991751	16.210677	9.944978
2	1	10	10.22	6.198760	7.049315	17.743376	10.330484
3	1	2	5.30	4.449338	9.129718	14.047953	9.209003
4	1	3	5.88	4.371248	7.617743	11.439545	7.809512
5	1	4	6.64	4.296000	8.023777	10.688780	7.669519
6	1	5	7.13	4.478504	7.492688	10.262947	7.411380
7	1	6	7.64	4.538284	6.980744	10.347668	7.288898
8	1	7	8.18	4.585401	6.471830	10.817985	7.291739
9	1	8	8.75	5.916795	6.421429	6.852006	6.396743
10	1	9	9.41	6.374440	5.295182	13.085237	8.251620
11	2	1	5.12	4.457547	9.110942	14.094492	9.220993
12	2	10	10.60	6.261426	6.718109	17.889275	10.289603
13	2	2	5.39	4.609648	9.065737	13.761499	9.145628
14	2	3	6.03	4.453050	7.853433	10.372646	7.559710
15	2	4	6.80	4.316154	7.841456	10.478096	7.545235
16	2	5	7.30	5.177302	7.322135	10.233540	7.577659
17	2	6	7.81	5.495333	6.853176	10.707537	7.685349
18	2	7	8.36	5.495333	6.853176	10.707537	7.685349
19	2	8	8.96	6.104869	5.784689	12.969496	8.286351
20	2	9	9.65	6.470621	5.030586	13.051658	8.184288
21	3	1	5.23	4.504601	9.135906	13.850217	9.163575
22	3	10	11.23	6.456360	6.083111	18.165172	10.234881
23	3	2	5.53	4.637955	8.967633	13.726397	9.110662
24	3	3	6.48	4.724024	8.148983	10.866801	7.913269
25	3	4	6.96	5.021238	7.660474	10.353324	7.678345
26	3	5	7.47	5.021238	7.660474	10.353324	7.678345
27	3	6	7.99	5.609985	6.674274	10.809030	7.697763
28	3	7	8.56	5.995751	5.990358	10.460304	7.482138
29	3	8	9.18	5.995751	5.990358	10.460304	7.482138
30	3	9	9.92	6.298887	4.879637	13.058693	8.079072

Figure 6.38 Return rate calculations of 30 portfolios.

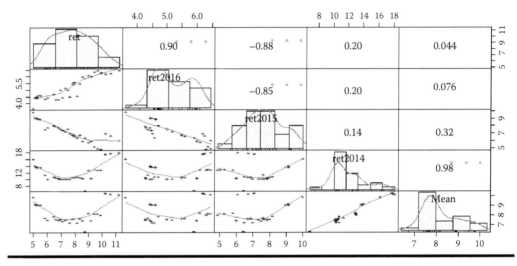

Figure 6.39 Correlation analysis of return rates.

For the expected returns and NAV curves provided by Machine Gene Investment, we cannot calculate them by the unknown data. It may involve the background algorithm, of which we have no idea.

6.5.4 Conclusion

In the above, we have analyzed Machine Gene Investment from the data perspective. The portfolios built by Machine Gene Investment are linear combinations, which fit the CAPM based on risk return. However, the portfolio number is limited, i.e. the target funds are limited; the algorithm is not transparent enough, so we cannot represent the result with the given data; the application lacks the continuous tracking of customers. Therefore, we have to understand Machine Gene Investment as a water testing based on the professional architecture of finance and targeted at speedy online launch. It can help individual investors experience the simple and high-efficient fortune management and help the financial managers release part of their pressure. But for professional investment managers, this is just a toy needing a lot of improvement.

Machine Gene Investment application is analyzed from data perspective here because I am personally interested in it. This does not represent the position of any company or any third party. Due to the limited data and my limited knowledge, if there is any one-sidedness, please help me correct it.

Epilogue

The body part of this book ends here. Thank you for reading. I hope this book can help you improve your skills and help you master programming in R language. Then you can discover rules and value in the real financial market by combining the skills and the financial knowledge you've learned and gain profits.

All the knowledge and cases introduced in this book are the conclusions of my practical experience. The financial knowledge here is merely a preliminary exploration of finance. The financial knowledge, similar to the IT knowledge, is a very huge system. We need to keep learning, so as to stay caught up and avoid being weeded out by the market. I will stick to my own way. I hope I can learn more about the financial logic and China's market in the next few years.

I have been exploring the road of transformation, and hope to make more friends in the industry. We, together, can create opportunities, create our future and prepare China to be the power of finance.

I will publish subsequent articles on my personal blog (http://fens.me). I may write another book or take a break. After all, writing a book in China is too painstaking. It took me 2 and a half years to finish this book and to release this pressure. I expect that programmers can successfully transform the financial market with IT technology and convert technology into real value. Please feel free to contact me and discuss this with me.

Appendix:
Docker Installation in Ubuntu

A.1 What Is Docker?

If someone works in the Internet circle and he/she does not know about Docker yet, he/she is out of date. Docker technology rose suddenly in the Internet technologies in 2014 and many companies have started a great deal of researches and applications of it in 2015 and 2016.

So what is Docker? Docker is an open source application container engine and it is a system-level lightweight virtualization technology, which provides solutions for the automatic deployment of applications.

You can quickly create a container and then develop and run your own applications in it. The configuration files can help with the automatic installation, deployment and upgrading of the application.

Docker is exalted in the industry, so it must have something outstanding and special.

■ Lightweight resources: The container is isolated on the process level. It uses the cores of the host machines, rather than virtualizes the whole operation system. Meanwhile, it does not need to call complicated operations for the virtualization and the system, so it saves a lot of expenses. It does not need the extra support of hypervisor, the virtual hardware or the extra whole set of system.
■ Portability: All the applications needed are in the container, which can be run in any host machine of Docker.
■ Predictable: The host machine and the container do not interfere with what is running in each other. They can call the programs by standardized interfaces.

Install Docker, or you are really out.

A.2 Docker Installation in Linux Ubuntu

There are only three steps to install Docker: download Docker, install it and check if it is successfully installed.

At present, Docker supports the environments of these three mainstream operation systems: Linux, Mac and Windows. The system environment of Linux used here is Linux Ubuntu 14.04.4 LTS 64bit. The download and installation of Docker in Ubuntu can simply be finished by apt-get.

Since Docker versions after 1.7.1 has specified its own source, we need to first configure the source of Docker in APT (advanced package tool).

Update the source of APT and install the libraries of https and ca licenses. Now these two libraries have been installed by default.

```
~ sudo apt-get update
~ sudo apt-get install apt-transport-https ca-certificates
```

Add the key of GPG to the APT configuration.

```
~ sudo apt-key adv --keyserver hkp://p80.pool.sks-keyservers.net:80
--recv-keys 58118E89F3A912897C070ADBF76221572C52609D
```

Add the source of Docker to the file /etc/apt/souces.list. My version is 14.04, corresponding to ubuntu-trusty.

```
~ sudo vi /etc/apt/sources.list

# Add to the last line
deb https://apt.dockerproject.org/repo ubuntu-trusty main
```

Next, we can install Docker by apt-get.

```
~ sudo apt-get update
~ sudo apt-get install docker-engine
```

After completing the installation, the system will initiate Docker by default.

```
# Check the service of Docker
~ service docker status
docker start/running, process 10013

# Check the progress of Docker
~ ps -aux|grep docker
root      10013  0.0  1.0 424948 40584 ?        Ssl  22:29   0:00 /usr/
bin/dockerd --raw-logs
root      10022  0.0  0.2 199680 10280 ?        Ssl  22:29   0:00 docker-
containerd -l unix:///var/run/docker/libcontainerd/docker-containerd.sock
--shimdocker-containerd-shim --metrics-interval=0 --start-timeout 2m
--state-dir /var/run/docker/libcontainerd/containerd --runtime
docker-runc

# Check the version of Docker
~ sudo docker version
Client:
Version: 1.12.1
 API version: 1.24
 Go version: go1.6.3
 Git commit: 23cf638
 Built: Thu Aug 18 05:22:43 2016
 OS/Arch: linux/amd64

Server:
Version: 1.12.1
 API version: 1.24
 Go version: go1.6.3
 Git commit: 23cf638
 Built: Thu Aug 18 05:22:43 2016
 OS/Arch: linux/amd64
```

Check if Docker has been successfully installed and run hello-world. If the information below appears, it means that the engine of Docker has been successfully installed.

```
~ sudo docker run hello-world
Unable to find image 'hello-world:latest' locally
latest: Pulling from library/hello-world
c04b14da8d14: Pull complete
Digest: sha256:0256e8a36e2070f7bf2d0b0763dbabdd67798512411de4cdcf9431a1fe
b60fd9
Status: Downloaded newer image for hello-world:latest

Hello from Docker!
This message shows that your installation appears to be working
correctly.

To generate this message, Docker took the following steps:
1. The Docker client contacted the Docker daemon.
2. The Docker daemon pulled the "hello-world" image from the Docker Hub.
3. The Docker daemon created a new container from that image which runs
the executable that produces the output you are currently reading.
4. The Docker daemon streamed that output to the Docker client, which
sent it to your terminal.
```

```
To try something more ambitious, you can run an Ubuntu container with:
$ docker run -it ubuntu bash
```

```
Share images, automate workflows, and more with a free Docker Hub
account:
https://hub.docker.com
```

```
For more examples and ideas, visit:
https://docs.docker.com/engine/userguide/
```

Note: When we execute the commands above, we usually encounter this error: Cannot connect to the Docker daemon. Is the docker daemon running on this host?

For example, just enter the command: docker run hello-world.

```
~ docker run hello-world
docker: Cannot connect to the Docker daemon. Is the docker daemon running
on this host?.
See 'docker run --help'.
```

This is caused by the authority problem. The Docker commands are tied with root authority by default. If sudo is not added, we will not have the authority.

A.3 Image Repository of Docker

For the command of docker run hello-world executed above, what does it mean?

Separate these three words. Docker represents the Docker program, run command and hello-world the image, so it means to use Docker to initiate the image of hello-world. For we do not have the image in the newly installed Docker, the run command will find an image named hello-world in the far-end repository of Docker, download it to the local machine and run it.

The official website for image repositories of Docker: https://hub.docker.com/.

In the official repositories of Docker, we can search for the systems, languages, technical frameworks, etc. that we are interested in, for many technologies have been dockerized. We can make convenient use of the containers already made by others and continue to work on the basis of the forefathers (Figure A.1).

Click on one item in the list and you will see a detailed introduction to this image. Take Ubuntu for example (Figure A.2).

If we want to download this image, we only need to follow the instruction and enter the command line: docker pull ubuntu.

```
~ sudo docker pull ubuntu
Using default tag: latest
latest: Pulling from library/ubuntu
2f0243478e1f: Pull complete
d8909ae88469: Pull complete
820f09abed29: Pull complete
01193a8f3d88: Pull complete
Digest: sha256:8e2324f2288c26e1393b63e680ee7844202391414dbd48497e9a4fd997
cd3cbf
Status: Downloaded newer image for ubuntu:latest
```

Figure A.1　Official repositories of Docker.

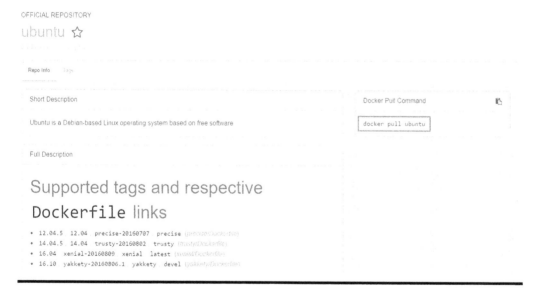

Figure A.2　Docker image of Ubuntu.

The downloaded image will be saved in the local repository. View the local image.

```
~ sudo docker images
REPOSITORY          TAG             IMAGE ID          CREATED
SIZE
```

```
ubuntu          latest          f8d79ba03c00        2 weeks ago        126.4 MB
hello-world     latest          c54a2cc56cbb        7 weeks ago        1.848 kB
```

Now there are two local images: hello-world and ubuntu.

A.4 Create Your Own Docker Image

We can create our own image and upload it to the official repositories, so that more people can use it. If you want to create your own Docker image, you only need to write a Dockerfile.

In the following, let's create a Docker allowing the Internet visitors. Crawl the latest eight articles on http://fens.me, put them into a list and then print the list to the console (Figure A.3).

Create a project directory.

```
~ mkdir fensme && cd fensme
```

To create a Dockerfile, we need the ubuntu image downloaded in the above and we need to install a curl repository for webpage crawling and for JSON (JavaScript Object Notation) data interpretation in jq repository.

```
~ vi Dockerfile

FROM ubuntu:latest
RUN apt-get update && apt-get install -y curl jq
CMD curl http://api.fens.me/blogs/ | jq .[]
```

Package it and create an image named fensme.

Figure A.3 Front page screenshot of http://fens.me.

```
# Package
~ sudo docker build -t fensme .

# View the image list
~ sudo docker images
REPOSITORY          TAG             IMAGE ID          CREATED
SIZE
fensme              latest          41b68972b35a      4 minutes ago
182.8 MB
ubuntu              latest          f8d79ba03c00      2 weeks ago
126.4 MB
hello-world         latest          c54a2cc56cbb      7 weeks ago
1.848 kB
```

Run the fensme image and the crawling of website data has been implemented.

```
~ sudo docker run fensme
  % Total      % Received % Xferd  Average Speed   Time      Time       Time
Current
                                 Dload  Upload   Total    Spent    Left  Speed
100  1421  100  1421      0      0     715      0  0:00:01  0:00:01
--:--:--     715
{
  "title": "R Interpretation of Autoregression Model",
  "date": 20160819,
  "link": "http://blog.fens.me/r-ar/",
  "img": "http://blog.fens.me/wp-content/uploads/2016/08/r-ar.png"
}
{
  "title": "A Summary of Common R Packages for Quantitative investment",
  "date": 20160810,
  "link": "http://blog.fens.me/r-quant-packages/",
  "img": "http://blog.fens.me/wp-content/uploads/2016/08/quant-packages.
png"
}
{
  "title": "R Calls C++ Across Boundary",
  "date": 20160801,
  "link": "http://blog.fens.me/r-cpp-rcpp",
  "img": "http://blog.fens.me/wp-content/uploads/2016/08/rcpp.png"
}
{
  "title": "R Interpretation of Multiple Linear Regression Model",
  "date": 20160727,
  "link": "http://blog.fens.me/r-multi-linear-regression/",
  "img": "http://blog.fens.me/wp-content/uploads/2016/07/reg-multi-liner.
png"
}
```

In this case, we have encapsulated a very simple crawler in Docker. Initiate it when you need it and it will write the result in the database. When the execution is completed, the system resources will be released and you do not need to take them into consideration again.

It is simple to dockerize a technology or a function and to build an individualized Docker.

A.5 Upload Docker Image to Public Repositories

The last step is to upload the Docker image we've created to the official repositories, so that others can use it.

First, we need to register an account on docker hub and then log in it (Figure A.4).

Create a repository of our own on docker hub (Figure A.5).

```
~ sudo docker login --username=bsspirit --email=bsspirit@163.com
Flag --email has been deprecated, will be removed in 1.13.
Password:
Login Succeeded
```

Figure A.4   Registration on docker hub.

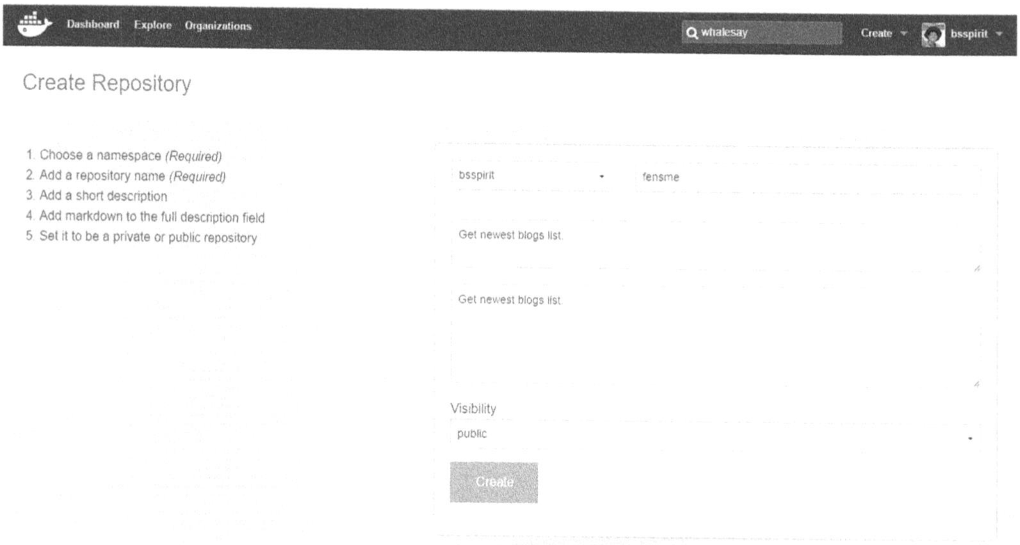

Figure A.5   Bind the account of docker hub in the local operation system.

Next, add a namespace to the fensme image you created just now and correspond it to the image named bsspirit/fensme on docker hub.

```
# Add a namespace to fensme
~ sudo docker tag 8496b10e857a bsspirit/fensme:latest

~ sudo docker images
REPOSITORY          TAG            IMAGE ID           CREATED
SIZE
bsspirit/fensme     latest         8496b10e857a       About a
minute ago    182.8 MB
fensme              latest         8496b10e857a       15 minutes
ago        182.8 MB
ubuntu              latest         f8d79ba03c00       2 weeks ago
126.4 MB
hello-world         latest         c54a2cc56cbb       7 weeks ago
1.848 kB
```

Upload the bsspirit/fensme image and then you can see your own image on docker hub.

```
~ sudo docker push bsspirit/fensme
The push refers to a repository [docker.io/bsspirit/fensme]
d9c50c22842b: Pushed
4699cbd1a947: Pushed
2bed5b3ec49f: Pushed
3834bde7e567: Pushed
d8d865b23727: Pushed
latest: digest: sha256:bfea736a92b6e602d6bbca867715b0e985f2e9bc3ea4a75b54
5d7e009e22ac2b size: 1362
```

Open the website of docker hub and refresh the page (Figure A.6).

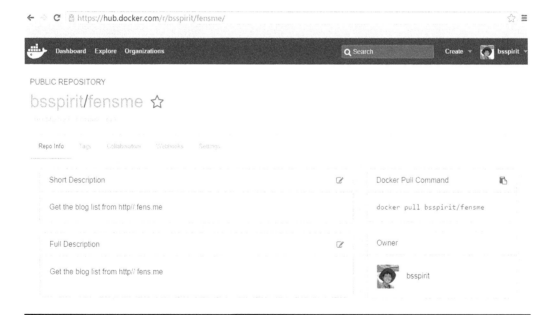

Figure A.6 fensme image.

In the end, if someone wants to use this docker image, he/she can just download it and run it as we introduced at the beginning.

```
~ sudo docker run bsspirit/fensme
```

After the above operations, we have completed the installation of Docker in Linux Ubuntu.

Index

Page numbers followed by *f* indicate figures

Milton Keynes UK
Ingram Content Group UK Ltd.
UKHW051931141024
449569UK00027B/1440